Applied Complex Variables for Scientists and Engineers

Second Edition

Yue Kuen Kwok

Applied Complex Variables for Scientists and Engineers

Second Edition

Yue Kuen Kwok

Hong Kong University of Science and Technology

CAMBRIDGE
UNIVERSITY PRESS

CAMBRIDGE
UNIVERSITY PRESS

University Printing House, Cambridge CB2 8BS, United Kingdom

One Liberty Plaza, 20th Floor, New York, NY 10006, USA

477 Williamstown Road, Port Melbourne, VIC 3207, Australia

314-321, 3rd Floor, Plot 3, Splendor Forum, Jasola District Centre, New Delhi - 110025, India

79 Anson Road, #06-04/06, Singapore 079906

Cambridge University Press is part of the University of Cambridge.

It furthers the University's mission by disseminating knowledge in the pursuit of education, learning and research at the highest international levels of excellence.

www.cambridge.org
Information on this title: www.cambridge.org/9780521701389

First published 2010

A catalogue record for this publication is available from the British Library

Library of Congress Cataloging in Publication data
Kwok, Y. K. (Yue-Kuen), 1957–
Applied complex variables for scientists and engineers / Yue Kuen Kwok. – 2nd ed.
p. cm.
Includes index.
ISBN 978-0-521-70138-9 (pbk.)
1. Functions of complex variables – Textbooks.   I. Title.

QA331.7.K88 2010
515´.9–dc22   2010008597

ISBN 978-0-521-70138-9 Paperback

Contents

Contents

Preface

This textbook is intended to be an introduction to complex variables for mathematics, science and engineering undergraduate students. The prerequisites are some knowledge of calculus (up to line integrals and Green's Theorem), though basic familiarity with differential equations would also be useful.

Complex function theory is an elegant mathematical structure on its own. On the other hand, many of its theoretical results provide powerful and versatile tools for solving problems in physical sciences and other branches of mathematics. The book presents the important analytical concepts and techniques in deriving most of the standard theoretical results in introductory complex function theory. I have included the proofs of most of the important theorems, except for a few that are highly technical. This book distinguishes itself from other texts in complex variables by emphasizing how to use complex variable methods. Throughout the text, many of the important theoretical results in complex function theory are followed by relevant and vivid examples in physical sciences. These examples serve to illustrate the uses and implications of complex function theory. They are drawn from a wide range of physical and engineering applications, like potential theory, steady state temperature problems, hydrodynamics, seepage flows, electrostatics and gravitation. For example, after discussing the mathematical foundations of the Laplace transform and Fourier transform, I show how to use the transform methods to solve initial-boundary problems arising from heat conduction and wave propagation problems. The materials covered in the book equip students with the analytical concepts of complex function theory together with the technical skills to apply complex variable methods to physical problems.

Throughout the whole textbook, both algebraic and geometric tools are employed to provide the greatest understanding, with many diagrams illustrating the concepts introduced. The book contains some 340 stimulating exercises, with solutions given to most of them. They are intended to aid students to grasp the concepts covered in the text and foster the skills in applying complex variable techniques to solve physical problems. Students are strongly advised to work through as many exercises as possible since mathematical knowledge can only be gained through active participation in the thinking and learning process.

The book begins by carefully exploring the algebraic, geometric and topological structures of the complex number field. In order to visualize the complex infinity, the Riemann sphere and the corresponding stereographic projection are introduced. Applications of complex numbers in electrical circuits are included.

Analytic functions are introduced in Chapter 2. The highlights of the chapter are the Cauchy–Riemann relations and harmonicity. The uses of complex functions in describing fluid flows and steady state heat distributions are illustrated.

In Chapter 3, the complex exponential function is introduced as an entire function which is equal to its derivative. The description of steady state temperature distributions by complex logarithm functions is illustrated. The mapping properties of complex trigonometric functions are examined. The notion of Riemann surfaces is introduced to help visualize multi-valued complex functions.

Complex integration forms the cornerstone of complex variable theory. The key results in Chapter 4 are the Cauchy–Goursat theorem and the Cauchy integral formulas. Other interesting results include Gauss' mean value theorem, Liouville's theorem and the maximum modulus theorem. The link of analytic functions and complex integration with the study of conservative fields is considered. Complex variable methods are seen to be effective analytical tools to solve conservation field models in potential flows, gravitational potentials and electrostatics.

Complex power series are the main themes in Chapter 5. We introduce different types of convergence of series of complex functions. The various tests that examine the convergence of complex series are discussed. The Taylor series theorem and Laurent series theorem show that a convergent power series is an analytic function within its disk or annulus of convergence, respectively. The notion of analytic continuation of a complex function is discussed. As an application, the solution to the potential flow over a perturbed circle is obtained as a power series in a perturbation parameter.

In Chapter 6, we start with the discussion of the classification of isolated singularities by examining the Laurent series expansion in a deleted neighborhood of the singularity. We then examine the theory of residues and illustrate the applications of the calculus of residues in the evaluation of complex integrals. The concept of the Cauchy principal value of an improper integral is introduced. Fourier transforms and Fourier integrals are considered. The residue calculus method is applied to compute the hydrodynamic lift and moment of an immersed obstacle.

The solutions of boundary value problems and initial-boundary value problems are considered in Chapter 7. The Poisson integral formula and the Schwarz integral formula for Dirichlet problems are derived. The inversion of the Laplace transform via the Bromwich contour integral is discussed. The Laplace transform techniques are applied to obtain the solutions of initial-boundary value problems arising from heat conduction and wave propagation models.

In the last chapter, we explore the rich geometric structure of complex variable theory. The geometric properties associated with mappings represented by complex functions are examined. The link between analyticity and conformality is derived. Various types of transformations that perform the mappings of regions are introduced. The bilinear and Schwarz–Christoffel transformations are discussed in full context. A wide range of physical examples are included to illustrate how to use these transformations to transform conservative field problems with complicated configurations into those with simple geometries. We also show how to use the hodograph transformations to solve seepage flow problems.

I would like to thank Ms Odissa Wong for her careful typing and editing of the manuscript, and her patience in entertaining the seemingly endless changes in the process. Also, I would like to thank the staff of Cambridge University Press for their editorial assistance in the production of this book. Last but not least, special thanks go to my wife Oi Chun and our two daughters, Grace and Joyce, for their forbearance while this book was written. Their love and care have been my main source of support in everyday life and work.

1

Complex Numbers

In this chapter, we explore the algebraic, geometric and topological structures of the complex number field $\mathbb{C}$. Readers are assumed to have some basic knowledge of the real number field $\mathbb{R}$. The construction of $\mathbb{C}$ serves as a resolution of the failure to find a real number x that satisfies the simple quadratic equation $x^2 + 1 = 0$.

There are similarities and differences between $\mathbb{C}$ and $\mathbb{R}$. For example, $\mathbb{R}$ is an ordered field while $\mathbb{C}$ is not. Also, the notions of infinity in $\mathbb{R}$ and $\mathbb{C}$ are different. Indeed, in order to visualize the complex infinity, the Riemann sphere and the corresponding stereographic projection are introduced. In addition, we introduce various topological properties of point sets in the complex number field. The chapter finishes with a discussion of the applications of complex numbers in alternating current circuit analysis.

1.1 Complex numbers and their representations

A *complex number* z is defined as an ordered pair

$$z = (x, y), \qquad (1.1.1)$$

where x and y are both real numbers. We commonly write

$$z = x + iy,$$

where i is a symbol yet to be defined. The operations of addition and multiplication of complex numbers will be defined in a meaningful manner, which will be seen to observe $i^2 = -1$ [see eq. (1.2.2)]. The set of all complex numbers is denoted by $\mathbb{C}$. The real numbers x and y in eq. (1.1.1) are called the real part and imaginary part of z, respectively. Symbolically, we write

$$\operatorname{Re} z = x, \quad \operatorname{Im} z = y.$$

1

If $y = 0$ in eq. (1.1.1), then $(x, 0) = x$ reduces to a real number. This shows that the set of real numbers $\mathbb{R}$ is a proper subset of $\mathbb{C}$. When $x = 0$ in eq. (1.1.1), $(0, y) = iy$ is called a pure imaginary number. Putting $x = 0$ and $y = 1$, we obtain the special number $i = (0, 1)$.

Since complex numbers are defined as ordered pairs, two complex numbers (x_1, y_1) and (x_2, y_2) are equal if and only if both their real parts and imaginary parts are equal, that is,

$$(x_1, y_1) = (x_2, y_2) \quad \text{if and only if} \quad x_1 = x_2 \text{ and } y_1 = y_2.$$

A complex number $z = (x, y)$ is defined by the pair of real numbers x and y, so it is natural to assume a one-to-one correspondence between the complex number $z = x + iy$ and the point (x, y) in the x-y plane. We refer to that plane as the *complex plane* or the *z-plane*.

Sometimes, it may be convenient to use polar coordinates

$$x = r\cos\theta, \qquad y = r\sin\theta. \tag{1.1.2}$$

Accordingly, we define the *modulus* of z to be

$$|z| = r = \sqrt{x^2 + y^2}. \tag{1.1.3}$$

The modulus $|z|$ is the distance between the origin and the point (x, y), which represents $z = x + iy$ in the complex plane. Obviously, Re $z \leq |z|$ and Im $z \leq |z|$. The polar representation of the complex number z is written as

$$z = x + iy = r(\cos\theta + i\sin\theta). \tag{1.1.4}$$

We call θ, denoted by arg z, the *argument* of z. Geometrically, arg z represents the angle between the positive x-axis and the line segment joining the origin and the point (x, y) (see Figure 1.1). Because of the periodicity of trigonometric functions, there are infinitely many values for arg z, each differing from the others by multiples of 2π. The *principal value* of arg z, denoted by Arg z, is the particular value of arg z chosen within the principal interval $(-\pi, \pi]$. Summarizing, we have

$$\arg z = \text{Arg } z + 2k\pi, \quad k \text{ is any integer and Arg } z \in (-\pi, \pi]. \tag{1.1.5}$$

For example: Arg $1 = 0$, Arg$(-1) = \pi$, arg$(-1 + i) = \frac{3\pi}{4} + 2k\pi$, where k is any integer. Note that arg 0 is not defined.

From eq. (1.1.2), we deduce a simple computational formula for finding Arg z:

$$\tan(\text{Arg } z) = \frac{y}{x}. \tag{1.1.6}$$

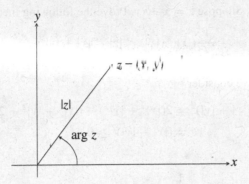

Figure 1.1. Vectorial representation of a complex number in the complex plane.

As a note of caution, $\tan^{-1}\left(\frac{y}{x}\right)$ returns a value in the interval $\left(-\frac{\pi}{2}, \frac{\pi}{2}\right]$. Therefore, we adjust the value for Arg z by adding π to $\tan^{-1}\left(\frac{y}{x}\right)$ if (x, y) lies in the second quadrant or subtracting π from $\tan^{-1}\left(\frac{y}{x}\right)$ if (x, y) lies in the third quadrant. For example: $\text{Arg}(1 - i) = -\frac{\pi}{4}$ while $\text{Arg}(-1 + i) = \frac{3\pi}{4}$, though $\tan^{-1}\left(\frac{-1}{1}\right) = \tan\left(\frac{1}{-1}\right) = -\frac{\pi}{4}$.

In the complex plane, any point that lies on the x-axis represents a real number. Therefore, the x-axis is termed the *real axis*. Similarly, any point on the y-axis represents an imaginary number, so the y-axis is called the *imaginary axis*.

The complex number $z = x + iy$ may also be regarded as the vector $x\mathbf{i} + y\mathbf{j}$ in the complex plane, where $\mathbf{i}$ and $\mathbf{j}$ are the respective unit vectors along the x-axis and y-axis. In this representation, $|z|$ is visualized as the length of the vector and arg z as the angle included between the vector and the positive real axis (see Figure 1.1).

The *complex conjugate* $\overline{z}$ of a given complex number $z = x + iy$ is defined by

$$\overline{z} = x - iy. \tag{1.1.7}$$

In the complex plane, the conjugate $\overline{z} = (x, -y)$ is the reflection of the point $z = (x, y)$ with respect to the real axis. From eqs. (1.1.3) and (1.1.6), we observe

$$|z| = |\overline{z}| \quad \text{and} \quad \text{Arg}\, z = -\text{Arg}\,\overline{z}.$$

The polar representation of $\overline{z}$ is seen to be

$$\overline{z} = r(\cos\theta - i\sin\theta).$$

Also, $|z|^2 = z\overline{z}$ and z is real if and only if $z = \overline{z}$.

Example 1.1.1   Suppose $z = x + iy$. Prove the following inequalities:

$$\frac{1}{\sqrt{2}}(|x| + |y|) \le |z| \le |x| + |y|.$$

Solution   First, we consider

$$2|z|^2 - (|x| + |y|)^2 = 2(|x|^2 + |y|^2) - (|x|^2 + |y|^2 + 2|x||y|)$$
$$= (|x| - |y|)^2 \ge 0,$$

so that

$$\frac{1}{2}(|x| + |y|)^2 \le |z|^2.$$

Since moduli are non-negative, we take the square root on both sides of the above inequality and obtain

$$\frac{1}{\sqrt{2}}(|x| + |y|) \le |z|.$$

Note that equality holds only if $|x| = |y|$. In an analogous manner, we consider

$$|z|^2 = |x|^2 + |y|^2 \le |x|^2 + |y|^2 + 2|x||y| = (|x| + |y|)^2,$$

so that upon taking the square root of both sides, we obtain

$$|z| \le |x| + |y|.$$

When either $x = 0$ or $y = 0$, equality holds in the above inequality. Combining the two inequalities, we obtain

$$\frac{1}{\sqrt{2}}(|x| + |y|) \le |z| \le |x| + |y|.$$

1.2 Algebraic properties of complex numbers

Given two complex numbers $z_1 = (x_1, y_1)$ and $z_2 = (x_2, y_2)$, we define the operations of addition $z_1 + z_2$ and multiplication $z_1 z_2$ to be

$$(x_1, y_1) + (x_2, y_2) = (x_1 + x_2, y_1 + y_2), \qquad (1.2.1a)$$
$$(x_1, y_1)(x_2, y_2) = (x_1 x_2 - y_1 y_2, x_1 y_2 + x_2 y_1). \qquad (1.2.1b)$$

When the complex numbers are restricted to real numbers, that is, $y_1 = 0$ and $y_2 = 0$, the above formulas reduce to

$$(x_1, 0) + (x_2, 0) = (x_1 + x_2, 0),$$
$$(x_1, 0)(x_2, 0) = (x_1 x_2, 0),$$

which agree with the usual operations of addition and multiplication in the real number field. These properties must be observed since the complex number field is an extension of the real number field.

Consider the square of the complex number $i = (0, 1)$. By performing multiplication according to the multiplication rule (1.2.1b), we have

$$i^2 = (0, 1)(0, 1) = (-1, 0) = -1. \tag{1.2.2}$$

If we write $(x_1, y_1) = x_1 + iy_1$ and $(x_2, y_2) = x_2 + iy_2$, the addition and multiplication formulas can be formally written as

$$(x_1 + iy_1) + (x_2 + iy_2) = (x_1 + x_2) + i(y_1 + y_2),$$
$$(x_1 + iy_1)(x_2 + iy_2) = (x_1 x_2 - y_1 y_2) + i(x_1 y_2 + x_2 y_1).$$

One observes that the formulas can be obtained by formally treating addition and multiplication of complex numbers as operations on real numbers, enforcing $i^2 = -1$, and collecting the real and imaginary parts separately.

The addition and multiplication of complex numbers obey the familiar commutative, associative and distributive rules like those for real numbers.

1. Addition is commutative:

$$z_1 + z_2 = z_2 + z_1. \tag{1.2.3a}$$

2. Addition is associative:

$$z_1 + (z_2 + z_3) = (z_1 + z_2) + z_3. \tag{1.2.3b}$$

3. Multiplication is commutative:

$$z_1 z_2 = z_2 z_1. \tag{1.2.3c}$$

4. Multiplication is associative:

$$z_1(z_2 z_3) = (z_1 z_2)z_3. \tag{1.2.3d}$$

5. Multiplication is distributive over addition:

$$z_1(z_2 + z_3) = z_1 z_2 + z_1 z_3. \tag{1.2.3e}$$

The proofs of these rules are straightforward and they are left as exercises.

According to rules (1.2.1a) and (1.2.1b), we have $0 + z = z$ and $1 \cdot z = z$ for any complex number z. Therefore, the numbers 0 and 1 retain their 'identity' properties in the complex number field. The additive and multiplicative inverses of z, denoted by $-z$ and $\frac{1}{z}$ respectively, are defined by

$$z + (-z) = 0 ,\tag{1.2.4a}$$

$$z\left(\frac{1}{z}\right) = 1 .\tag{1.2.4b}$$

Suppose $z = x + iy$. Then its additive inverse $-z = -x - iy$. For the multiplicative inverse $\frac{1}{z}$, it is seen that

$$\frac{1}{z} = \frac{\bar{z}}{z\bar{z}} = \frac{\bar{z}}{|z|^2} = \frac{x - iy}{x^2 + y^2} .$$

The subtraction of two complex numbers z_1 and z_2 can be defined via the additive inverse by

$$z_1 - z_2 = z_1 + (-z_2) = (x_1 - x_2) + i(y_1 - y_2) .$$

Similarly, the division operation $\dfrac{z_1}{z_2}$ (provided $z_2 \neq 0$) can be defined via the multiplicative inverse as

$$\frac{z_1}{z_2} = z_1\left(\frac{1}{z_2}\right) = \frac{(x_1 x_2 + y_1 y_2) + i(x_2 y_1 - x_1 y_2)}{x_2^2 + y_2^2} .$$

As a rule of thumb, while addition and subtraction are mostly performed using the cartesian forms, multiplication and division are more conveniently performed in polar representations. Suppose

$$z_1 = r_1(\cos\theta_1 + i \sin\theta_1) \text{ and } z_2 = r_2(\cos\theta_2 + i \sin\theta_2);$$

it can be shown by direct computation that

$$z_1 z_2 = r_1 r_2[\cos(\theta_1 + \theta_2) + i \sin(\theta_1 + \theta_2)],\tag{1.2.5a}$$

$$\frac{z_1}{z_2} = \frac{r_1}{r_2}[\cos(\theta_1 - \theta_2) + i \sin(\theta_1 - \theta_2)].\tag{1.2.5b}$$

One infers directly from the above equations that

$$\arg(z_1 z_2) = \arg z_1 + \arg z_2,\tag{1.2.6a}$$

$$\arg\left(\frac{z_1}{z_2}\right) = \arg z_1 - \arg z_2.\tag{1.2.6b}$$

Since $\arg z$ is multi-valued, eq. (1.2.6a) is interpreted as: $\arg(z_1 z_2)$ is equal to the sum of some value of $\arg z_1$ and some value of $\arg z_2$. It is relatively straightforward to verify the following properties of conjugates and moduli:

(i) $\overline{z_1 + z_2} = \overline{z_1} + \overline{z_2}$, (ii) $\overline{z_1 z_2} = \overline{z_1}\,\overline{z_2}$, (iii) $\overline{\dfrac{z_1}{z_2}} = \dfrac{\overline{z_1}}{\overline{z_2}}$,

(iv) $|z_1 z_2| = |z_1||z_2|$, (v) $\left|\dfrac{z_1}{z_2}\right| = \dfrac{|z_1|}{|z_2|}$.

In view of the satisfaction of the commutative, associative and distributive rules and the existence of additive and multiplicative inverses, the set of complex numbers $\mathbb{C}$ constitutes a field. Recall that the set of real numbers $\mathbb{R}$ is a proper subset of $\mathbb{C}$. However, the ordering property of $\mathbb{R}$ cannot be extended to the field $\mathbb{C}$. For any pair of real numbers a and b, the ordering property means $a < b$, $a = b$ or $a > b$. It is meaningless to write $z_1 < z_2$ when z_1 and z_2 are complex numbers unless both z_1 and z_2 happen to be real. To be an ordered field, the following property must be observed. Taking any non-zero complex number z in $\mathbb{C}$, we must have either z or $-z$ being positive, and $z^2 = (-z)^2$ being positive. However, we have $i^2 = (-i)^2 = -1$, which is negative. A contradiction is encountered, so $\mathbb{C}$ cannot be an ordered field.

1.2.1 De Moivre's theorem

Any complex number with unit modulus can be expressed as

$$\cos\theta + i\sin\theta.$$

Interestingly, this complex number is related to a complex exponential quantity by the following formula:

$$e^{i\theta} = \cos\theta + i\sin\theta. \tag{1.2.7}$$

This is the renowned *Euler formula*, the rigorous justification of which is presented when the formal definition of the exponential complex function is introduced in Section 3.1. From the Euler formula, we may deduce that

$$(\cos\theta + i\sin\theta)^n = (e^{i\theta})^n = e^{in\theta} = \cos n\theta + i\sin n\theta, \tag{1.2.8}$$

where n can be any integer. This result is the statement of *de Moivre's theorem*. To justify the theorem, we consider the following cases:

(i) The theorem is trivial when $n = 0$.
(ii) When n is a positive integer, the theorem can be proved easily by mathematical induction.

(iii) When n is a negative integer, let $m = -n$ so that m is a positive integer. We then have

$$(\cos\theta + i\sin\theta)^n = \frac{1}{(\cos\theta + i\sin\theta)^m} = \frac{1}{\cos m\theta + i\sin m\theta}$$

$$= \frac{\cos m\theta - i\sin m\theta}{(\cos m\theta + i\sin m\theta)(\cos m\theta - i\sin m\theta)}$$

$$= \cos m\theta - i\sin m\theta = \cos n\theta + i\sin n\theta.$$

How do we generalize the formula to $(\cos\theta + i\sin\theta)^s$, where s is a rational number? Let $s = p/q$, where p and q are irreducible integers. It is easily seen that the modulus of $(\cos\theta + i\sin\theta)^s$ remains unity. Let the polar representation of $(\cos\theta + i\sin\theta)^s$ take the form $\cos\phi + i\sin\phi$ for some ϕ. Now, we write

$$\cos\phi + i\sin\phi = (\cos\theta + i\sin\theta)^s = (\cos\theta + i\sin\theta)^{p/q}.$$

Taking the power of q of both sides of the above equation, we obtain

$$\cos q\phi + i\sin q\phi = \cos p\theta + i\sin p\theta,$$

which implies

$$q\phi = p\theta + 2k\pi \quad \text{or} \quad \phi = \frac{p\theta + 2k\pi}{q}, \quad k = 0, 1, \ldots, q - 1.$$

It suffices to limit the set of integers to be $\{0, 1, \cdots, q - 1\}$ since the value of ϕ corresponding to k beyond this set of integers equals one of those values defined in the above equation plus some multiple of 2π. Therefore, integers beyond the above set do not generate new distinct values of $\cos\phi + i\sin\phi$. As a result, there are q distinct roots of $(\cos\theta + i\sin\theta)^{p/q}$, namely,

$$\cos\left(\frac{p\theta + 2k\pi}{q}\right) + i\sin\left(\frac{p\theta + 2k\pi}{q}\right), \quad k = 0, 1, \ldots, q - 1.$$

Example 1.2.1 Find the three possible values of the cube root of the complex number

$$\frac{1 - i}{1 + i}.$$

Solution It is more convenient to use the polar representation. The polar forms of $1 - i$ and $1 + i$ are

$$1 - i = \sqrt{2}\left(\cos\frac{-\pi}{4} + i\sin\frac{-\pi}{4}\right) \quad \text{and} \quad 1 + i = \sqrt{2}\left(\cos\frac{\pi}{4} + i\sin\frac{\pi}{4}\right).$$

Using eq. (1.2.5b), we obtain

$$\frac{1-i}{1+i} = \frac{\sqrt{2}}{\sqrt{2}}\left[\cos\left(\frac{-\pi}{4}-\frac{\pi}{4}\right)+i\sin\left(\frac{-\pi}{4}-\frac{\pi}{4}\right)\right]=\cos\frac{-\pi}{2}+i\sin\frac{-\pi}{2}.$$

By de Moivre's theorem, the three possible values of the cube root of $\dfrac{1-i}{1+i}$ are

$$\cos\frac{-\pi}{6}+i\sin\frac{-\pi}{6},$$

$$\cos\left(\frac{-\pi}{6}+\frac{2\pi}{3}\right)+i\sin\left(\frac{-\pi}{6}+\frac{2\pi}{3}\right)=\cos\frac{\pi}{2}+i\sin\frac{\pi}{2}=i,$$

and

$$\cos\left(\frac{-\pi}{6}+\frac{4\pi}{3}\right)+i\sin\left(\frac{-\pi}{6}+\frac{4\pi}{3}\right)=\cos\frac{7\pi}{6}+i\sin\frac{7\pi}{6}.$$

Example 1.2.2   For any two complex numbers z_1 and z_2, show that

$$|z_1 - z_2|^2 + |z_1 + z_2|^2 = 2|z_1|^2 + 2|z_2|^2.$$

Hence, deduce that

$$|\alpha + \sqrt{\alpha^2 - \beta^2}| + |\alpha - \sqrt{\alpha^2 - \beta^2}| = |\alpha + \beta| + |\alpha - \beta|,$$

for any real numbers α and β.

Solution   For any two complex numbers z_1 and z_2,

$$|z_1 - z_2|^2 + |z_1 + z_2|^2 = (z_1 - z_2)(\overline{z_1} - \overline{z_2}) + (z_1 + z_2)(\overline{z_1} + \overline{z_2})$$

$$= (z_1\overline{z_1} - z_2\overline{z_1} - z_1\overline{z_2} + z_1\overline{z_2})$$

$$\quad + (z_1\overline{z_1} + z_2\overline{z_1} + z_1\overline{z_2} + z_2\overline{z_2})$$

$$= 2(z_1\overline{z_1} + z_2\overline{z_2}) = 2|z_1|^2 + 2|z_2|^2.$$

We let $z_1 = \alpha + \sqrt{\alpha^2 - \beta^2}$ and $z_2 = \alpha - \sqrt{\alpha^2 - \beta^2}$, where α and β are real, and apply the above relation to obtain

$$|\alpha + \sqrt{\alpha^2 - \beta^2}|^2 + |\alpha - \sqrt{\alpha^2 - \beta^2}|^2 = \frac{1}{2}\left(|2\sqrt{\alpha^2 - \beta^2}|^2 + |2\alpha|^2\right)$$

$$= 2|\alpha^2 - \beta^2| + 2\alpha^2.$$

Next, we consider

$$\left(|\alpha + \sqrt{\alpha^2 - \beta^2}| + |\alpha - \sqrt{\alpha^2 - \beta^2}|\right)^2 - (|\alpha + \beta| + |\alpha - \beta|)^2$$

$$= |\alpha + \sqrt{\alpha^2 - \beta^2}|^2 + |\alpha - \sqrt{\alpha^2 - \beta^2}|^2 + 2\beta^2 - (2\alpha^2 + 2\beta^2) - 2|\alpha^2 - \beta^2|$$

$$= |\alpha + \sqrt{\alpha^2 - \beta^2}|^2 + |\alpha - \sqrt{\alpha^2 - \beta^2}|^2 - 2|\alpha^2 - \beta^2| - 2\alpha^2 = 0,$$

so that either

$$|\alpha + \sqrt{\alpha^2 - \beta^2}| + |\alpha - \sqrt{\alpha^2 - \beta^2}| = |\alpha + \beta| + |\alpha - \beta|$$

or

$$|\alpha + \sqrt{\alpha^2 + \beta^2}| + |\alpha - \sqrt{\alpha^2 - \beta^2}| + |\alpha + \beta| + |\alpha - \beta| = 0.$$

Since modulus quantities are always non-negative, the latter identity holds only in the trivial case where $\alpha = \beta = 0$. Hence, we obtain

$$|\alpha + \sqrt{\alpha^2 - \beta^2}| + |\alpha - \sqrt{\alpha^2 - \beta^2}| = |\alpha + \beta| + |\alpha - \beta|,$$

for any real numbers α and β.

Example 1.2.3   Find the square roots of $a + ib$, where a and b are real constants.

Solution   Let $u + iv$ be a square root of $a + ib$, that is, $(u + iv)^2 = a + ib$. Equating the corresponding real and imaginary parts, we have

$$u^2 - v^2 = a \qquad \text{and} \qquad 2uv = b.$$

By eliminating v from the above equations, we obtain a quadratic equation for u^2:

$$4u^4 - 4au^2 - b^2 = 0.$$

The two real roots for u are found to be

$$u = \pm\sqrt{\frac{a + \sqrt{a^2 + b^2}}{2}}.$$

From the relation $v^2 = u^2 - a$, we obtain

$$v = \pm\sqrt{\frac{\sqrt{a^2 + b^2} - a}{2}}.$$

Since u and v each have two possible values, apparently there are four possible values for $u + iv$. However, there can be only two values of the square root of $a + ib$. By observing the relation $2uv = b$, one must choose u and v such that their product has the same sign as b. This leads to

$$u + iv = \pm\left(\sqrt{\frac{a + \sqrt{a^2 + b^2}}{2}} + i\frac{b}{|b|}\sqrt{\frac{\sqrt{a^2 + b^2} - a}{2}}\right),$$

provided that $b \neq 0$. The special case where $b = 0$ is trivial. As a numerical example, suppose $a = 3$ and $b = -4$. We then have

$$(3 - 4i)^{1/2} = \pm\left(\sqrt{\frac{9+5}{2}} - i\sqrt{\frac{5-3}{2}}\right) = \pm(2 - i).$$

Example 1.2.4 A complex number of the form $\alpha + i\beta$, where α and β are real integers, is called a *Gaussian integer*. A Gaussian integer a is said to be composite if and only if it can be factored into the form $a = bc$, where b and c are both Gaussian integers (excluding ± 1 and $\pm i$); otherwise it is prime. Show that, as Gaussian integers, 2 is composite but 3 is prime.

Solution The fact that 2 is composite is obvious since

$$2 = (1 + i)(1 - i).$$

It is easy to check that 3 cannot be expressed as the product of a real number and a complex number, or as the product of an imaginary number and a complex number. Assume 3 to be composite, then

$$3 = (\alpha_1 + i\beta_1)(\alpha_2 + i\beta_2) = (\alpha_1\alpha_2 - \beta_1\beta_2) + i(\alpha_1\beta_2 + \alpha_2\beta_1), \qquad \text{(i)}$$

where $\alpha_1, \alpha_2, \beta_1, \beta_2$ are non-zero integers. Without loss of generality, we assume $\alpha_1 \geq \alpha_2$. Since the imaginary part of 3 is zero, we have

$$\alpha_1\beta_2 + \alpha_2\beta_1 = 0,$$

or equivalently,

$$\frac{\alpha_1}{\alpha_2} = -\frac{\beta_1}{\beta_2} = k, \quad k \geq 1.$$

Putting the above relations into eq. (i), we obtain

$$k\left(\alpha_2^2 + \beta_2^2\right) = 3.$$

Since α_2 and β_2 are non-zero integers, and $k \geq 1$, the only possible solutions to the above equation are $\alpha_2 = \beta_2 = 1$ and $k = 3/2$. This gives $\alpha_1 = -\beta_1 = 3/2$, which is a non-integer. This is a contradiction, hence 3 cannot be composite.

Example 1.2.5 A complex number lies on or inside the unit circle in the complex plane if and only if its modulus is less than or equal to 1. A polynomial $P(z)$ is called a *simple von Neumann polynomial* if its roots all lie on or inside the unit circle and any root on the unit circle is simple; that is, has multiplicity 1.

Consider a quadratic polynomial of the form

$$P_2(z) = z^2 + a_1 z + a_0, \quad \text{where } a_0 \text{ and } a_1 \text{ are complex numbers.}$$

Show that

(a) when $|a_0| > 1$, $P_2(z)$ cannot be a simple von Neumann polynomial;

(b) when $|a_0| < 1$, $P_2(z)$ is a simple von Neumann polynomial if and only if

$$|a_1 - \overline{a_1} a_0| \le 1 - |a_0|^2;$$

(c) when $|a_0| = 1$, $P_2(z)$ is a simple von Neumann polynomial if and only if

$$a_1 = \overline{a_1} a_0 \quad \text{and} \quad |a_1| < 2.$$

Solution

(a) Let r_1 and r_2 denote the two roots of $P_2(z)$. Assume $P_2(z)$ to be a simple von Neumann polynomial. Then

$$|r_1 r_2| = |a_0| \le 1.$$

Hence, when $|a_0| > 1$, $P_2(z)$ cannot be a simple von Neumann polynomial.

(b) First, note that $a_0 = r_1 r_2$ and $-a_1 = r_1 + r_2$, so

$$a_1 - \overline{a_1} a_0 = r_1(|r_2|^2 - 1) + r_2(|r_1|^2 - 1). \tag{i}$$

Assume $P_2(z)$ to be a simple von Neumann polynomial. One then observes that

$$\begin{aligned}
0 &\le (1 - |r_1 r_2|)(1 - |r_1|)(1 - |r_2|) \\
&= 1 - |r_1|^2 |r_2|^2 - |r_1|(1 - |r_2|^2) - |r_2|(1 - |r_1|^2) \\
&\le 1 - |a_0|^2 - |a_1 - \overline{a_1} a_0|. \tag{ii}
\end{aligned}$$

When $|a_0| < 1$, the above inequality can be rearranged as

$$|a_1 - \overline{a_1} a_0| \le 1 - |a_0|^2. \tag{iii}$$

Conversely, suppose inequality (iii) holds and $|a_0| < 1$. One can establish the following inequality

$$(1 - |r_1|)(1 - |r_2|) \ge 0 \tag{iv}$$

by direct computation (see Problem 1.16). Since $|r_1| \, |r_2| = |a_0| < 1$, together with the result in inequality (iv), these lead to either

$$|r_1| < 1 \text{ and } |r_2| \le 1, \quad \text{or} \quad |r_1| \le 1 \text{ and } |r_2| < 1.$$

Therefore, the two roots must be on or inside the unit circle and any root on the unit circle must be simple. Hence, $P_2(z)$ is a simple von Neumann polynomial.

(c) When $P_2(z)$ is a simple von Neumann polynomial and $|a_0| = 1$, the two roots must satisfy $|r_1| = |r_2| = 1$ and $r_1 \neq r_2$. Equation (i) now becomes

$$a_1 = \overline{a_1} a_0.$$

In addition, since $|r_1| = |r_2| = 1$ and $r_1 \neq r_2$, we have

$$|a_1| = |r_1 + r_2| < |r_1| + |r_2| = 2.$$

Conversely, given the condition $|a_0| = 1$, there are only two possible cases of distribution of the roots. In one case, one root is inside and the other root is outside the unit circle. In the other case, both roots are on the unit circle.

The first case can be shown to be impossible, given that $a_1 - \overline{a_1} a_0 = 0$ and $|a_1| < 2$. To prove the claim, we note from eq. (i) that

$$0 = a_1 - \overline{a_1} a_0 = r_1(|r_2|^2 - 1) + r_2(|r_1|^2 - 1),$$

so $\text{Arg } r_1 = \text{Arg } r_2$ since $|r_1|^2 - 1$ and $|r_2|^2 - 1$ are real and of opposite signs. Now, consider

$$|a_1| = |r_1 + r_2| = |r_1| + |r_2| \quad (\text{since } \text{Arg } r_1 = \text{Arg } r_2)$$
$$= |r_1| + \frac{1}{|r_1|} \geq 2 \quad (\text{since } |r_1| \, |r_2| = 1),$$

contradicting the fact that $|a_1| < 2$.

Consider the second case, where both roots are on the unit circle. Given that $|a_1| < 2$, one can show that the two roots cannot be repeated. If otherwise, the condition of repeated roots requires $a_1^2 = 4a_0$. One then obtains $|a_1|^2 = 4|a_0| = 4$, giving $|a_1| = 2$, and this contradicts $|a_1| < 2$.

Combining the results, the two roots must lie on the unit circle and cannot be repeated, so $P_2(z)$ is a simple von Neumann polynomial.

1.3 Geometric properties of complex numbers

We have seen that a complex number $z = x + iy$ can be associated with a point (x, y) or a vector $x\mathbf{i} + y\mathbf{j}$ in the complex plane. In this section, we explore more properties of these geometric representations of complex numbers.

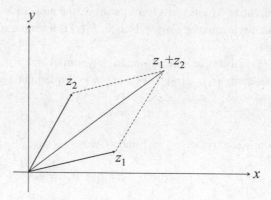

Figure 1.2. In the vectorial representation, the sum of two complex numbers can be constructed using the parallelogram law.

The distance between the two points representing z_1 and z_2 in the complex plane is given by

$$|z_1 - z_2| = |(x_1 - x_2) + i(y_1 - y_2)|$$
$$= \sqrt{(x_1 - x_2)^2 + (y_1 - y_2)^2}. \qquad (1.3.1)$$

For example, the locus of points z in the complex plane defined by the relation

$$|z - z_0| = r, \quad z_0 \text{ is complex and } r \text{ is real}, \qquad (1.3.2)$$

represents a circle centered at z_0 and with radius r.

What is the geometric interpretation of the addition of two complex numbers z_1 and z_2 in the complex plane? The sum $z_1 + z_2$ is represented by the point $(x_1 + x_2, y_1 + y_2)$ or the vector $(x_1 + x_2)\mathbf{i} + (y_1 + y_2)\mathbf{j}$. If we treat z_1 as the vector $x_1\mathbf{i} + y_1\mathbf{j}$ and z_2 as the vector $x_2\mathbf{i} + y_2\mathbf{j}$, then $z_1 + z_2$ can be found by adding the two vectors using the parallelogram law (see Figure 1.2). The difference $z_1 - z_2$ can be found similarly by treating $z_1 - z_2$ as $z_1 + (-z_2)$.

In Figure 1.2, the lengths of the sides of the two triangles are given by the moduli of the corresponding complex numbers representing the sides. Since the sum of any two sides of a triangle must be greater than or equal to the third side, we deduce immediately the renowned *triangle inequalities*. For any complex numbers z_1 and z_2, we have

$$|z_1 + z_2| \le |z_1| + |z_2|, \qquad (1.3.3a)$$

$$|z_1 - z_2| \ge \left||z_1| - |z_2|\right|. \qquad (1.3.3b)$$

The triangle inequalities can also be proved by an algebraic argument (see Example 1.3.1). By setting z_2 to be $-z_2$ in (1.3.3b), the above two inequalities can be combined into the form

$$\left| |z_1| - |z_2| \right| \leq |z_1 + z_2| \leq |z_1| + |z_2|.$$

Replacing z_2 by $z_2 + z_3$ in inequality (1.3.2a), we obtain

$$|z_1 + z_2 + z_3| \leq |z_1 + z_2| + |z_3| \leq |z_1| + |z_2| + |z_3|. \qquad (1.3.4)$$

The result can be extended to any finite number of complex numbers by induction. Accordingly, the generalized triangle inequality is given as

$$\left| \sum_{k=1}^{n} z_k \right| \leq \sum_{k=1}^{n} |z_k|, \qquad n = 2, 3, \ldots \qquad (1.3.5)$$

In Example 1.2.2, we establish the following relation

$$|z_1 - z_2|^2 + |z_1 + z_2|^2 = 2(|z_1|^2 + |z_2|^2).$$

What would be the geometric interpretation of the relation? Considering the vectorial representation of complex numbers, suppose we take the complex numbers z_1 and z_2 as the adjacent sides of a parallelogram. The modulus quantities $|z_1 + z_2|$ and $|z_1 - z_2|$ can be visualized as the lengths of the two diagonals of the parallelogram. The above relation shows that the sum of the squares of the lengths of the diagonals of a parallelogram is equal to the sum of the squares of the lengths of the four sides of the parallelogram.

In the complex plane, the various points $P_1, P_2, \ldots$ in a geometric figure can be represented by the complex numbers $z_1, z_2, \ldots$, respectively. A given geometric property of the figure, such as 'the triangle $P_1 P_2 P_3$ is equilateral' or 'the four points P_1, P_2, P_3, P_4 are concyclic', corresponds to some algebraic relation of the complex numbers. It is this type of correspondence between geometry and algebra that explains why complex numbers can be used as an effective tool in solving geometric problems.

For example, suppose we want to find the equation of the perpendicular bisector of the line segment joining the two points z_1 and z_2. Since any point z on the bisector is equidistant from z_1 and z_2, the equation of the bisector can be represented by

$$|z - z_1| = |z - z_2|. \qquad (1.3.6)$$

For a given equation $f(x, y) = 0$ of a geometric curve, if we set $x = (z + \bar{z})/2$ and $y = (z - \bar{z})/2i$, the equation can be expressed in terms of the pair of conjugate complex variables z and $\bar{z}$ as

$$f(x, y) = f\left(\frac{z + \bar{z}}{2}, \frac{z - \bar{z}}{2i} \right) = F(z, \bar{z}) = 0. \qquad (1.3.7)$$

For example, the unit circle centered at the origin as represented by the equation $x^2 + y^2 = 1$ can be expressed as $z\bar{z} = 1$.

The notion of the moduli of complex numbers is a useful tool for describing a region in the complex plane. Some examples are:

  (i) The set $\{z : |z - a| < r\}$, $a \in \mathbb{C}$, $r \in \mathbb{R}$, represents the set of points inside the circle centered at a and with radius r but excluding the boundary.
 (ii) The set $\{z : r_1 \leq |z - a| \leq r_2\}$, $a \in \mathbb{C}$, r_1 and $r_2 \in \mathbb{R}$, represents the annular region centered at a and bounded by circles of radii r_1 and r_2. Here, the boundary circles are included.
(iii) The set of points z such that $|z - \alpha| + |z - \beta| \leq 2d$, α and $\beta \in \mathbb{C}$ and $d \in \mathbb{R}$, is the set of all points on or inside the ellipse with foci α and β and with the length of the semi-major axis equal to d. What is the length of the semi-minor axis?

1.3.1 *n*th *roots of unity*

By definition, the nth roots of unity satisfy the equation

$$z^n = 1.$$

By de Moivre's theorem, the n distinct roots of unity are

$$z = e^{2k\pi i/n} = \cos\frac{2k\pi}{n} + i\sin\frac{2k\pi}{n}, \quad k = 0, 1, \ldots, n-1. \quad (1.3.8)$$

If we write $\omega_n = e^{2\pi i/n}$, then the nth roots are $1, \omega_n, \omega_n^2, \ldots, \omega_n^{n-1}$. Alternatively, if we pick any one of the roots and call it α, then the other roots are given by $\alpha\omega_n, \alpha\omega_n^2, \ldots, \alpha\omega_n^{n-1}$.

In the complex plane, the nth roots of unity correspond to the n vertices of a regular n-sided polygon inscribed inside the unit circle centered at the origin, with one vertex at the point $z = 1$. The vertices are equally spaced on the circumference of the circle. Figure 1.3 shows the regular octagon inscribed inside the unit circle when $n = 8$.

The above argument can be extended easily to the nth roots of an arbitrary complex number. Suppose the given complex number in polar form is represented by $r(\cos\phi + i\sin\phi)$. Its nth roots are given by

$$r^{1/n}\left(\cos\frac{\phi + 2k\pi}{n} + i\sin\frac{\phi + 2k\pi}{n}\right), \quad k = 0, 1, 2, \ldots, n-1,$$

where $r^{1/n}$ is the positive nth root of the positive real number r. The roots now lie on the circle $|z| = r^{1/n}$. These roots are equally spaced along the

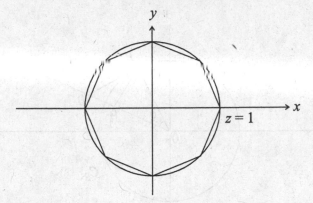

Figure 1.3. The vertices of the regular octagon inscribed inside the unit circle centered at the origin are the eighth roots of unity. One of the vertices of the octagon must be at $z = 1$.

circumference but one vertex must be at $r^{1/n}(\cos(\frac{\phi}{n}) + i \sin(\frac{\phi}{n}))$. For example, consider the cube roots of $1 + i$. In polar form, we have

$$1 + i = \sqrt{2}\left(\cos\left(\frac{\pi}{4}\right) + i \sin\left(\frac{\pi}{4}\right)\right)$$

so that the cube roots are $\sqrt[6]{2}(\cos(\frac{\pi}{12}) + i \sin(\frac{\pi}{12}))$, $\sqrt[6]{2}(\cos(\frac{3\pi}{4}) + i \sin(\frac{3\pi}{4}))$ and $\sqrt[6]{2}(\cos(\frac{17\pi}{12}) + i \sin(\frac{17\pi}{12}))$. They form the vertices of an equilateral triangle inscribed inside the circle $|z| = \sqrt[6]{2}$.

1.3.2 Symmetry with respect to a circle

Given a point α in the complex plane, we would like to construct the symmetry point of α with respect to the circle $C_R \colon |z| = R$. The symmetry point of α is defined to be $\beta = R^2/\overline{\alpha}$. Conversely, since we may write $\alpha = R^2/\overline{\beta}$, we can also consider α to be the symmetry point of β. The two points α and β are said to be symmetric with respect to the circle C_R (more details can be found in Subsection 8.2.2). We explore some of the geometric properties associated with a pair of symmetric points.

(i) We first assume that $|\alpha| < R$ so that the symmetry point β lies outside the circle C_R. By observing that

$$\text{Arg } \beta = \text{Arg } \frac{R^2}{\overline{\alpha}} = \text{Arg } \frac{1}{\overline{\alpha}} = -\text{Arg } \overline{\alpha} = \text{Arg } \alpha,$$

one concludes that α and β both lie on the same ray emanating from the origin. The symmetry point β can be constructed as follows: draw

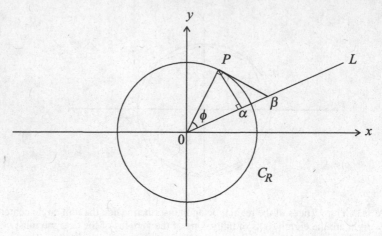

Figure 1.4. Construction of a pair of symmetric points with respect to the circle C_R.

the circle C_R and a ray L from the origin through α. We then draw a perpendicular to L through α which intersects the circle C_R at P. The point of intersection of the tangent line to the circle C_R at P and the ray L then gives β (see Figure 1.4).

The proof of the construction is simple. It is seen that α and β lie on the same ray through the origin so that $\operatorname{Arg}\alpha = \operatorname{Arg}\beta$. From the construction, it is observed that $|\beta|/R = R/|\alpha| = \sec\phi$, where ϕ is the angle between OP and the ray L, so $|\alpha|\,|\beta| = R^2$. Combining the results, α and β are related by $\beta = R^2/\overline{\alpha}$.

(ii) When $|\alpha| = R$, the symmetry point is just α itself.

(iii) Lastly, when $|\alpha| > R$, the symmetry point β will be inside the circle C_R. To reverse the method of construction in (i), we find a tangent to the circle which passes through α and call the point of tangency P. A ray L is then drawn from the origin through α and a perpendicular is dropped from P to L. The point of intersection of the perpendicular with the ray L then gives β.

Example 1.3.1   Show that for any two complex numbers z_1 and z_2,

$$|z_1 + z_2|^2 = |z_1|^2 + |z_2|^2 + 2\operatorname{Re}(z_1\overline{z_2}),$$

then deduce the combined triangle inequalities

$$\Big||z_1| - |z_2|\Big| \le |z_1 + z_2| \le |z_1| + |z_2|.$$

Solution Consider

$$|z_1 + z_2|^2 = (z_1 + z_2)(\overline{z_1} + \overline{z_2})$$
$$= z_1\overline{z_1} + z_2\overline{z_2} + z_1\overline{z_2} + z_2\overline{z_1}$$
$$= |z_1|^2 + |z_2|^2 + 2\,\mathrm{Re}(z_1\overline{z_2}).$$

By observing that $\mathrm{Re}(z_1\overline{z_2}) \le |z_1\overline{z_2}|$, we have

$$|z_1 + z_2|^2 \le |z_1|^2 + |z_2|^2 + 2|z_1\overline{z_2}|$$
$$= |z_1|^2 + |z_2|^2 + 2|z_1||z_2| = (|z_1| + |z_2|)^2.$$

Since moduli are non-negative, we take the square root of both sides and obtain

$$|z_1 + z_2| \le |z_1| + |z_2|. \tag{i}$$

To prove the other half of the triangle inequality, we write

$$|z_1| = |(z_1 + z_2) + (-z_2)| \le |z_1 + z_2| + |-z_2|,$$

giving

$$|z_1| - |z_2| \le |z_1 + z_2|. \tag{ii}$$

By interchanging z_1 and z_2 in the above inequality, we have

$$|z_2| - |z_1| \le |z_1 + z_2|. \tag{iii}$$

Combining the results in eqs. (i), (ii) and (iii), we obtain

$$\Big||z_1| - |z_2|\Big| \le |z_1 + z_2| \le |z_1| + |z_2|.$$

Example 1.3.2 Find the locus of the points z satisfying

$$|z - \alpha| + |z + \alpha| = 2r,$$

where α is complex and r is real. Discuss various possibilities and find the maximum and minimum values of $|z|$.

Solution The two moduli $|z - \alpha|$ and $|z + \alpha|$ represent the distances from z to α and $-\alpha$, respectively. By the triangle inequality, $|z - \alpha| + |z + \alpha| \ge 2|\alpha|$ and so $|\alpha| \le r$. We consider the following separate cases:

(a) $|\alpha| < r$. This represents an ellipse with foci at α and $-\alpha$. The length of the semi-major axis is r and the length of the semi-minor axis is $\sqrt{r^2 - |\alpha|^2}$. The angle of inclination of the major axis is $\mathrm{Arg}\,\alpha$. When $\alpha = 0$, the two foci coalesce and the ellipse becomes a circle centered at the origin and with radius r.

(b) $|\alpha| = r$. The ellipse collapses into a line segment joining α and $-\alpha$.

Further, from the geometric properties of an ellipse, we deduce that

  (i) max$\{|z|\}$ = length of the semi-major axis = r,
  (ii) min$\{|z|\}$ = length of the semi-minor axis = $\sqrt{r^2 - |\alpha|^2}$.

Example 1.3.3 Find the curve or region in the complex plane represented by each of the following equations or inequalities:

(a) Re $\dfrac{1}{z} = 2$,   (b) $|z + 1||z - 1| = 1$,   (c) $|z| + \text{Re}\, z \leq 1$,

(d) $0 < \text{Arg}\, \dfrac{z - i}{z + i} < \dfrac{\pi}{2}$,   (e) $\left|\dfrac{z - 1}{z + 1}\right| \leq 1$.

Solution

(a) Suppose we write $z = re^{i\theta}$, and so Re $\frac{1}{z} = \frac{\cos\theta}{r}$. The equation of the curve in polar form becomes

$$ r = \frac{\cos\theta}{2}. $$

Note that $r^2 = z\bar{z}$ and $r\cos\theta = \frac{z+\bar{z}}{2}$, so the polar equation can be rewritten as

$$ z\bar{z} = \frac{z + \bar{z}}{4} \quad \text{or} \quad \left(z - \frac{1}{4}\right)\left(\bar{z} - \frac{1}{4}\right) = \left(\frac{1}{4}\right)^2 \quad \text{or} \quad \left|z - \frac{1}{4}\right| = \frac{1}{4}. $$

The locus represents the circle centered at $(1/4, 0)$ and with radius $1/4$, but with the origin deleted since Re $\frac{1}{z}$ is not defined at $z = 0$.

(b) The curve is symmetrical with respect to both the x-axis and y-axis since the equation is invariant if we change z to $\bar{z}$ and $-\bar{z}$, respectively. It also passes through the origin. If we write the equation as $|z^2 - 1| = 1$ and let $z = re^{i\theta}$, we have

$$ |r^2 \cos 2\theta - 1 + ir^2 \sin 2\theta| = 1, $$

which can be simplified to the form

$$ r^2 = 2\cos 2\theta. $$

This is the standard form of a *lemniscate*. The polar representation reveals the symmetry of the curve with respect to both axes. Also, the curve is bounded within the circle $|z| = \sqrt{2}$ (see Figure 1.5).

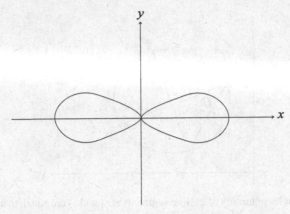

Figure 1.5. The shape of a lemniscate represented by the polar equation $r^2 = 2\cos 2\theta$.

(c) Suppose $z = x + iy$. The inequality can be written as

$$\sqrt{x^2 + y^2} \le 1 - x \Leftrightarrow y^2 \le 1 - 2x.$$

The above inequality represents the region on and inside the parabola $y^2 = 1 - 2x$.

(d) Let $z = x + iy$, then

$$\frac{z - i}{z + i} = \frac{(x^2 + y^2 - 1) - 2ix}{x^2 + (y + 1)^2}.$$

One observes that $\text{Arg} \frac{z-i}{z+i} \in \left(0, \frac{\pi}{2}\right)$ if and only if $\frac{z-i}{z+i}$ has both positive real and imaginary parts, that is, $x < 0$ and $x^2 + y^2 > 1$. Therefore, $0 < \text{Arg} \frac{z-i}{z+i} < \frac{\pi}{2}$ represents the region exterior to the unit circle $|z| = 1$ and on the left half-plane.

(e) Consider

$$\left|\frac{z - 1}{z + 1}\right| \le 1 \Leftrightarrow |z - 1|^2 \le |z + 1|^2$$

$$\Leftrightarrow (z - 1)(\bar{z} - 1) \le (z + 1)(\bar{z} + 1)$$

$$\Leftrightarrow z + \bar{z} \ge 0 \Leftrightarrow \text{Re } z \ge 0.$$

The region represented by the inequality is the right half-plane, including the y-axis.

Example 1.3.4 Suppose the four points z_1, z_2, z_3 and z_4 lie on a circle. Show that

$$\frac{(z_1 - z_3)(z_2 - z_4)}{(z_1 - z_4)(z_2 - z_3)}$$

is real.

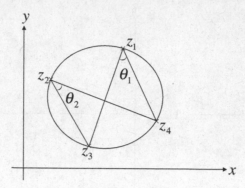

Figure 1.6. Angles of turning of the line segments at z_1 and z_2 are equal, that is, $\theta_1 = \theta_2$.

Solution Suppose the four points z_1, z_2, z_3 and z_4 are distributed on a circle as shown in Figure 1.6. Draw the two line segments, one joining z_1 with z_3 and the other joining z_1 with z_4. The angle of turning of these two line segments at z_1 (denoted by θ_1 in Figure 1.6) is given by Arg $\frac{z_1-z_4}{z_1-z_3}$. Similarly, we draw another pair of line segments joining z_2 with z_3 and z_2 with z_4. The angle of turning of the line segments at z_2 (denoted by θ_2 in Figure 1.6) is given by Arg $\frac{z_2-z_4}{z_2-z_3}$. Since the four points are concyclic, the two angles of turning are equal. Therefore, we have

$$\text{Arg} \, \frac{z_1 - z_4}{z_1 - z_3} = \theta_1 = \theta_2 = \text{Arg} \, \frac{z_2 - z_4}{z_2 - z_3},$$

and so

$$\frac{(z_1 - z_3)(z_2 - z_4)}{(z_1 - z_4)(z_2 - z_3)}$$

is real.

Remarks

(i) There are $3! = 6$ circular permutations of four distinct points on a circle. The above argument can be applied analogously to the other 5 circular permutations.

(ii) One can deduce the equation for the circle passing through three non-collinear points z_1, z_2, z_3 to be given by

$$\text{Im} \, \frac{(z_1 - z_3)(z_2 - z)}{(z_1 - z)(z_2 - z_3)} = 0.$$

1.4 Some topological definitions

In this section, some topological definitions commonly encountered in complex variable theory are introduced. These definitions are useful in our discussion of various topics related to analyticity, residue calculus, etc. in later chapters.

The set of points z such that $|z - z_0| < \epsilon$, where $z_0 \in \mathbb{C}, \epsilon \in \mathbb{R}$, contains points that are inside the circle centered at z_0 and with radius ϵ. We call it a *neighborhood* of z_0 and denote it by $N(z_0; \epsilon)$. A *deleted neighborhood* of z_0 is the point set $N(z_0; \epsilon) \setminus \{z_0\}$. We write it as $\widehat{N}(z_0; \epsilon)$.

A point z_0 is called a *limit point* or an *accumulation point* of a point set S if every neighborhood of z_0 contains a point of S other than z_0. That is, $\widehat{N}(z_0; \epsilon) \cap S \neq \phi$ for any $\epsilon > 0$. Since this is true for any neighborhood of z_0, S must contain infinitely many points. For example, consider the point set $\{z : |z| < 1\}$. The limit points are points on and inside the circle $|z| = 1$. Be aware that a limit point z_0 of S may or may not belong to the point set S.

Example 1.4.1 Find all the limit points of each of the following point sets and determine whether the point set contains all of its limit points.

(a) $E = \left\{z : z = (-1)^n \dfrac{n}{n+1}, \ n \text{ is an integer}\right\}$;

(b) $F = \left\{z : z = \dfrac{1}{m} + \dfrac{i}{n}, \ m \text{ and } n \text{ are integers}\right\}$.

Solution

(a) The point $z = 1$ is a limit point of E, the proof of which is as follows. Given any $\epsilon > 0$, we want to show that there exists a point in E other than $z = 1$ such that

$$\left|(-1)^n \frac{n}{n+1} - 1\right| < \epsilon.$$

If we choose n to be even and $n + 1 > \frac{1}{\epsilon}$, then the above inequality holds. Therefore, every neighborhood $N(1; \epsilon)$ contains at least one point in E other than $z = 1$. This completes the proof that $z = 1$ is a limit point. The proof that $z = -1$ is another limit point can be done similarly. The point set E has only two limit points. It is seen that E does not contain the two limit points.

(b) One may deduce judiciously that the set of limit points of F is the union of the two sets:

$$\left\{z : z = \frac{1}{m}, \ m \text{ is any integer}\right\} \cup \left\{z : z = \frac{i}{n}, \ n \text{ is any integer}\right\}.$$

It can be shown that every neighborhood of any limit point chosen from the above set contains a point in F other than the limit point itself. It is apparent that all limit points of F are contained in F.

Interior points, boundary points, and exterior points

A point z_0 is called an *interior point* of a point set S if there exists a neighborhood of z_0, all points of which belong to S. If every neighborhood of z_0 contains points of S and also points not belonging to S, then z_0 is called a *boundary point*. The set of all boundary points of the point set S is called the *boundary* of S. If a point is neither an interior nor a boundary point of S, then it is called an *exterior point* of S. Indeed, if z_0 is not a boundary point of S, then there exists a neighborhood of z_0 such that it is completely inside S or completely outside S. In the former case, it is an interior point; otherwise, it is an exterior point.

For example, consider the point set $S = \{z : |z| \leq 1\}$. Any point inside the unit circle is an interior point of S and any point on the circumference of the unit circle is a boundary point. The point $z = 1 + i$ is an exterior point. Actually, given any point that lies outside the circle $|z| = 1$, obviously it cannot be an interior point. Also, one can always find a neighborhood of this outside point that does not intersect the circle, so it cannot be a boundary point. These facts illustrate why any point lying outside the circle is an exterior point.

Open sets and closed sets

A point set which consists only of interior points is called an *open set*. Another way of looking at an open set is to observe that every point of this set has a neighborhood contained completely in the set. Intuitively, we may think of any two-dimensional set without boundary as an open set. A point set is said to be *closed* if it contains all its boundary points. For example, the set $\{z : |z| < 1\}$ is an open set while the set $\{z : |z| \leq 1\}$ is a closed set. The *closure* of a point set S is the closed set that contains all points in S together with the whole boundary of S. According to this definition, the closed set $\{z : |z| \leq 1\}$ is the closure of the open set $\{z : |z| < 1\}$.

Example 1.4.2 In this example, we examine whether a point set is open or closed or neither.

(a) Consider the point set

$$A = \{z : \text{Re}\, z > \text{Im}\, z\}.$$

Show that it is an open set. Find the closure of A. Is its complement A^c a closed set?

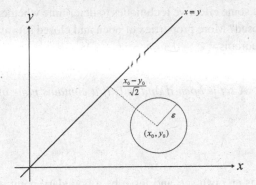

Figure 1.7. A neighborhood of (x_0, y_0) can always be found that lies completely inside set A, where $A = \{z : \operatorname{Re} z > \operatorname{Im} z\}$.

(b) Consider another point set

$$B = \{z : \operatorname{Re} z > \operatorname{Im} z \text{ if } \operatorname{Re} z \geq 0 \text{ and } \operatorname{Re} z \geq \operatorname{Im} z \text{ if } \operatorname{Re} z < 0\}.$$

Is this set open or closed or neither?

Solution

(a) Consider an arbitrary point $z_0 = (x_0, y_0)$ that lies in A. We would like to show that there exists a neighborhood of z_0 which lies completely inside A (see Figure 1.7). Note that the shortest distance from the point (x_0, y_0) to the line $x = y$ is $\frac{x_0 - y_0}{\sqrt{2}}$. We choose $\epsilon < \frac{x_0 - y_0}{\sqrt{2}}$ so that the neighborhood $N(z_0; \epsilon)$ lies completely inside A. Since A consists of interior points only, it is an open set.

The boundary points of A are points that lie along the line $\operatorname{Re} z = \operatorname{Im} z$, so the closure of A is given by

$$\overline{A} = \{z : \operatorname{Re} z \geq \operatorname{Im} z\}.$$

The complement of A is

$$A^c = \{z : \operatorname{Re} z \leq \operatorname{Im} z\},$$

which is seen to contain all its boundary points. Hence, A^c is closed.

(b) The set B is not open since a point (x_0, y_0) that satisfies $x_0 = y_0$ and $x_0 < 0$ is not an interior point. This is because any neighborhood of such a point contains points inside B and points outside B. Actually, it is a boundary point. On the other hand, B is not closed since it does not include all its boundary points that satisfy $x_0 = y_0$ and $x_0 \geq 0$.

Can we devise some effective techniques to determine whether a given point set is open or closed? More properties of open and closed sets are presented in the following theorems.

Theorem 1.4.1   *A set is open if and only if it contains none of its boundary points.*

Proof

"if" part

Suppose that D is an open set, and let p be a boundary point of D. Suppose p is in D. Then by virtue of the property of an open set, there is an open disc centered at p that lies completely inside D. This contradicts that p is a boundary point.

"only if" part

Suppose D is a set that contains none of its boundary points. For any $z_0 \in D$, z_0 cannot be a boundary point of D. Hence, there is some disc centered at z_0 that is either a subset of D or a subset of the complement of D. The latter is impossible since z_0 itself is in D. Hence, each point of D is an interior point so D is open.

Corollary   A set C is closed if and only if its complement $D = \{z : z \notin C\}$ is open.

To show the claim, by virtue of the definition of the boundary point, we observe that the boundary of a point set coincides exactly with the boundary of the complement of that set. Recall that a closed set contains all its boundary points. Its complement shares the same boundary, but these boundary points are not contained in the complement. By Theorem 1.4.1, the complement is open.

Recall that a closed set contains all its boundary points. But what is the relation between a closed set and the set of its limit points?

Theorem 1.4.2   *A set S is closed if and only if S contains all its limit points.*

Proof

We write $\widehat{N}(z; \epsilon)$ as the deleted ϵ-neighborhood of z, and S^c as the complement of S. Note that for $z \notin S$, $N(z; \epsilon) \cap S = \widehat{N}(z; \epsilon) \cap S$. We then have

S is closed $\Leftrightarrow$ S^c is open

$\qquad \Leftrightarrow$ given $z \notin S$, there exists $\epsilon > 0$ such that $N(z; r) \cap S$

$\qquad \Leftrightarrow$ given $z \notin S$, there exists $\epsilon > 0$ such that $\widehat{N}(z; \epsilon) \cap S = \phi$

$\qquad \Leftrightarrow$ no point of S^c is a limit point of S.

Consider the two point sets in Example 1.4.1. By virtue of Theorem 1.4.2, we can deduce immediately that set E is not closed while set F is closed. Is E an open set? Check whether the point $z = \frac{2}{3}$ lying in E is an interior point. If the answer is "no", then E cannot be open.

Compact sets

A *bounded set* is one that can be contained in a large enough circle centered at the origin, that is, $|z| < M$ for all points z in S where M is some sufficiently large constant. An *unbounded set* is one that is not bounded. A set which is both closed and bounded is called *compact*. For example, the set $\{z : \operatorname{Re} z \geq 1\}$ is closed but not bounded while the set $\{z : |z + 1| + |z - 1| \leq 3\}$ is compact since it is both closed and bounded.

Connectedness and domain

A set S is said to be *connected* if any two points of S can be joined by a continuous curve lying entirely inside S. For example, a neighborhood $N(z_0; \epsilon)$ is connected. An open connected set is called an *open region* or *domain*. For example, the set $\{z : \operatorname{Re} z \geq z\}$ is not a domain since it is not open. The set $\{z : 0 < \operatorname{Re} z < 1 \text{ or } 2 < \operatorname{Re} z < 3\}$ is also not a domain since it is open but not connected. To the point set S, suppose we add all of its boundary points. The new set $\overline{S}$ is the closure of S and the closure is a closed set. The closure of an open region is called a *closed region*. However, to an open region we may add none, some, or all of its boundary points. We simply call the resulting point set a *region*.

To illustrate the importance of connectedness in calculus, let us consider an example from real calculus. It is well known that if $f'(x) = 0$ for all x in an open interval (a, b), then f is constant throughout that interval. However, suppose the domain of definition of the function is not connected [say, $f(x)$ is defined over $(-1, 1) \cup (2, 3)$]. Then one can easily construct a function that is not constant in the domain of definition but where $f'(x) = 0$ throughout the domain of definition.

Jordan arc

Let $x(t)$ and $y(t)$ be real continuous functions of the real parameter t, $\alpha \leq t \leq \beta$. The set of points $z(t) = x(t) + iy(t)$ defines a *continuous arc* in the complex plane beginning at $z(\alpha)$ and ending at $z(\beta)$. A point z_0 is called a *multiple point* of the arc if the equation $z_0 = x(t) + iy(t)$ is satisfied by more than one value of t in $\alpha \leq t \leq \beta$. A continuous arc without multiple points is called a *Jordan arc*. If the arc has only one double point, corresponding to the initial and terminal values α and β of t, that is, $z(\alpha) = z(\beta)$, then it is called a *simple closed Jordan arc*.

Simply connected domains

A domain S is said to be *simply connected* if every simple closed Jordan arc in S can be shrunk to a point in S without passing through points not belonging to S. That is, S is a simply connected domain if for any closed Jordan arc lying in the domain S, the points inside the closed curve also belong to S. A domain that is not simply connected is said to be *multiply connected*. According to the definition, it is then always possible to construct some closed curve inside a multiply connected domain in such a manner that one or more points inside the curve do not belong to the domain. Intuitively, there are holes contained inside some Jordan arc lying completely in the domain. We have one hole in a doubly connected domain and two holes in a triply connected domain (see Figure 1.8). For example, the domain $\{z : 1 < |z| < 2\}$ is doubly connected.

Example 1.4.3 Let S be the set of points defined by $z = \frac{p}{m} + \frac{q}{n}i$, where p, q, m, n are positive integers, and $p < m, q < n$.

    (a) What are the limit points of S, if any?

    (b) Is S a closed set, or an open set, or neither?

    (c) Is S a compact set?

    (d) What is the closure of S?

    (e) Is the closure of S compact?

    (f) Is S a domain?

Solution

    (a) Every point of the form $\alpha + \beta i$, where α and β are real numbers inside the interval $[0, 1]$, is a limit point.

    (b) The set S does not contain the points $z = 1, z = i$, and $z = 1 + i$, and these points are limit points of S. Since it does not include

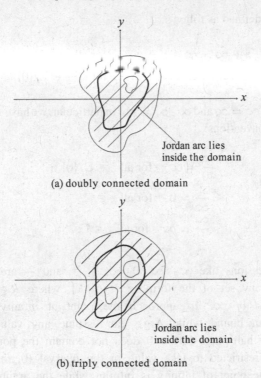

(a) doubly connected domain

(b) triply connected domain

Figure 1.8. The holes inside the Jordan arc correspond to points that do not belong to the domain. A doubly connected domain contains one hole while a triply connected domain contains two holes.

    all its limit points, S is not closed. None of the points in S is an interior point since the points in S are discrete, so S is not open.

(c) Since S is not closed, S cannot be compact.

(d) We obtain the closure of S by adding all its boundary points to S. It is seen that the closure of S is simply the set of its limit points.

(e) The closure of S is closed, and it is bounded by the circle $|z| = \sqrt{2}$. Hence, the closure is compact.

(f) Since S is not open, it is not a domain.

1.5 Complex infinity and the Riemann sphere

It is convenient to augment the complex plane with the point at infinity, denoted by ∞. The set $\mathbb{C} \cup \{\infty\}$ is called the *extended complex plane*. The algebra involving ∞ is defined as follows:

$$z + \infty = \infty + z = \infty, \quad \text{for all } z \in \mathbb{C},$$
$$z \cdot \infty = \infty \cdot z = \infty, \quad \text{for all } z \in \mathbb{C}/\{0\}.$$

We allow $\infty + \infty = \infty$ and $\infty \cdot \infty = \infty$. In particular, we have $-1 \cdot \infty = \infty$. We adopt the conventions:

$$\frac{z}{0} = \infty \quad \text{for all } z \in \mathbb{C}/\{0\},$$

$$\frac{z}{\infty} = 0 \quad \text{for all } z \in \mathbb{C}.$$

$$\frac{\infty}{z} = \infty \quad \text{for all } z \in \mathbb{C}.$$

However, expressions like $\infty - \infty$, $0 \cdot \infty$, $\frac{0}{0}$, $\frac{\infty}{0}$, and $\frac{\infty}{\infty}$ are not defined. Topologically, any set of the form $\{z : |z| > R\}$, where $R \geq 0$, is called a *neighborhood of* ∞. To approach the point at infinity, we let $|z|$ increase without bound while $\text{Arg}\, z$ can assume any value. Note that the open upper half-plane $\text{Im}\, z > 0$ does not contain the point at infinity since $\text{Arg}\, z$ is restricted to take value in the interval $(0, \pi)$. The modulus value of the point, of infinity is infinite while the argument value is indeterminate.

1.5.1 The Riemann sphere and stereographic projection

In order to visualize the point at infinity, we consider the *Riemann sphere* that has radius $\frac{1}{2}$ and is tangent to the complex plane at the origin (see Figure 1.9). We call the point of contact the south pole (denoted by S) and the point diametrically opposite S the north pole (denoted by N). Let z be an arbitrary complex number in the complex plane, represented by the point P. We draw the straight line PN that intersects the Riemann sphere at a unique point P', distinct from N. Conversely, to each point P' on the sphere, other than the north pole N, we draw the line $P'N$ which cuts uniquely one point P in the complex plane. Clearly there exists a one-to-one correspondence between points on the Riemann sphere, except N, and all the finite points in the complex plane. We assign the north pole N as the *point at infinity*. With such an assignment, we then establish a one-to-one correspondence between all the points on the Riemann sphere and all the points in the extended complex plane. This correspondence is known as the *stereographic projection*.

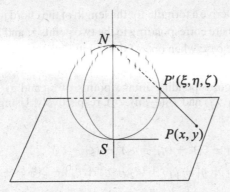

Figure 1.9. The Riemann sphere sitting on the complex plane.

Under the stereographic projection, meridians of the Riemann sphere are projected onto straight lines through the origin in the complex plane and latitudes are mapped onto concentric circles around the origin. The equator on the Riemann sphere is mapped onto the unit circle in the complex plane. Therefore, the southern hemisphere goes to the interior of the unit circle while the northern hemisphere goes to the exterior of the unit circle in the extended complex plane.

Consider an arbitrary point $P(x, y)$, $z = x + iy$, in the z-plane and let the corresponding point on the Riemann sphere be $P'(\xi, \eta, \zeta)$ (see Figure 1.9). From the geometry of the figure, the coordinates are related by

$$x = \frac{\xi}{1 - \zeta} \quad \text{and} \quad y = \frac{\eta}{1 - \zeta}. \tag{1.5.1}$$

Further, the equation of the Riemann sphere is given by

$$\xi^2 + \eta^2 + \left(\zeta - \frac{1}{2}\right)^2 = \left(\frac{1}{2}\right)^2 \quad \text{or} \quad \xi^2 + \eta^2 + \zeta^2 = \zeta. \tag{1.5.2}$$

The above three equations are now used to solve for ξ, η and ζ in terms of x and y. First, we substitute $\xi = x(1 - \zeta)$ and $\eta = y(1 - \zeta)$ into eq. (1.5.2) to obtain

$$\zeta = \frac{x^2 + y^2}{x^2 + y^2 + 1} = \frac{z\bar{z}}{|z|^2 + 1}. \tag{1.5.3a}$$

Once ζ is available, we use eq. (1.5.1) to obtain

$$\xi = \frac{x}{x^2 + y^2 + 1} = \frac{1}{2}\frac{z + \bar{z}}{|z|^2 + 1}, \tag{1.5.3b}$$

$$\eta = \frac{y}{x^2 + y^2 + 1} = \frac{1}{2i}\frac{z - \bar{z}}{|z|^2 + 1}. \tag{1.5.3c}$$

Example 1.5.1   Derive a formula for the length of the chord joining the images on the Riemann sphere corresponding to the two points z_1 and z_2 in the complex plane. Examine the case when one of the points is the point at infinity.

Solution   Let the corresponding image points of z_1 and z_2 on the Riemann sphere be $P_1(\xi_1, \eta_1, \zeta_1)$ and $P_2(\xi_2, \eta_2, \zeta_2)$, respectively. Using eqs. (1.5.2) and (1.5.3a,b,c), we have

$$
\begin{aligned}
d(z_1, z_2)^2 &= (\xi_1 - \xi_2)^2 + (\eta_1 - \eta_2)^2 + (\zeta_1 - \zeta_2)^2 \\
&= (\xi_1^2 + \eta_1^2 + \zeta_1^2) + (\xi_2^2 + \eta_2^2 + \zeta_2^2) - 2(\xi_1\xi_2 + \eta_1\eta_2 + \zeta_1\zeta_2) \\
&= (\zeta_1 + \zeta_2) - 2(\xi_1\xi_2 + \eta_1\eta_2 + \zeta_1\zeta_2) \\
&= \frac{|z_1|^2}{1 + |z_1|^2} + \frac{|z_2|^2}{1 + |z_2|^2} - 2\frac{x_1x_2 + y_1y_2 + |z_1|^2|z_2|^2}{(1 + |z_1|^2)(1 + |z_2|^2)} \\
&= \frac{|z_1 - z_2|^2}{(1 + |z_1|^2)(1 + |z_2|^2)}.
\end{aligned}
$$

Taking the positive square root of the expressions on both sides, we have

$$
d(z_1, z_2) = \frac{|z_1 - z_2|}{\sqrt{1 + |z_1|^2}\,\sqrt{1 + |z_2|^2}}.
$$

Suppose one of the points, say z_2, goes to infinity. We obtain

$$
d(z_1, \infty) = \frac{1}{\sqrt{1 + |z_1|^2}}.
$$

Example 1.5.2   Show that any small circle whose circumference contains the north pole N on the Riemann sphere corresponds to a straight line not passing through the origin in the complex plane.

Solution   A small circle whose circumference contains the north pole N on the Riemann sphere is determined by the intersection of the Riemann sphere $\xi^2 + \eta^2 + \zeta^2 - \zeta = 0$ and a plane through N. In terms of ξ, η and ζ, a plane passing through the north pole $N(0, 0, 1)$ has the general form

$$
A\xi + B\eta + C(\zeta - 1) = 0, \qquad C \neq 0.
$$

To find the image curve of this circle in the complex plane, we use the following transformations [see eqs. (1.5.3a,b,c)]:

$$
\xi = \frac{x}{1 + |z|^2}, \quad \eta = \frac{y}{1 + |z|^2} \quad \text{and} \quad \zeta = \frac{|z|^2}{1 + |z|^2}.
$$

Substituting the above relations into the equation of the plane through N, the equation of the image curve is found to be

$$A\lambda + D\mu \quad C = 0, \qquad C \neq 0$$

Thus, the image curve in the complex plane is shown to be a straight line not passing through the origin. The converse statement can be proved similarly by reversing the above argument.

1.6 Applications to electrical circuits

In this section, we discuss the application of complex numbers to alternating current circuit analysis. The effectiveness of the formulation using complex numbers lies in the close link between the algebraic and geometric properties of complex numbers.

An alternating current I with magnitude $\hat{I}$ and angular frequency ω is represented by

$$I = \hat{I}\cos\omega t = \text{Re}(\hat{I}e^{i\omega t}). \tag{1.6.1}$$

As an alternative representation, the current may be conveniently represented by the *phasor*

$$I = \hat{I}e^{i\omega t} \tag{1.6.2}$$

in the complex plane. The phasor is a complex number which revolves around the circle of radius $\hat{I}$ with angular frequency ω. Its projection onto the real axis gives the magnitude of the current (see Figure 1.10).

Suppose we connect in series two alternating currents of the same angular frequency ω but with different phases as represented by

$$I_1 = \hat{I}_1\cos(\omega t + \phi_1) \quad \text{and} \quad I_2 = \hat{I}_2\cos(\omega t + \phi_2).$$

Their resultant is then given by

$$I = \hat{I}_1\cos(\omega t + \phi_1) + \hat{I}_2\cos(\omega t + \phi_2).$$

Suppose we set I formally as $\hat{I}\cos(\omega t + \phi)$. What would be the relation between $\hat{I}, \phi$, and $\hat{I}_1, \hat{I}_2, \phi_1, \phi_2$?

Using the phasor notation, the above equation can be rewritten as

$$I = \text{Re}(\hat{I}e^{i(\omega t + \phi)}) = \text{Re}(\hat{I}_1e^{i(\omega t + \phi_1)} + \hat{I}_2e^{i(\omega t + \phi_2)}).$$

If we treat the phasors $\hat{I}_1e^{i(\omega t + \phi_1)}$ and $\hat{I}_2e^{i(\omega t + \phi_2)}$ as vectors, then $\hat{I}e^{i(\omega t + \phi)}$ is the vector sum of the two phasors. The magnitude and phase of the resultant

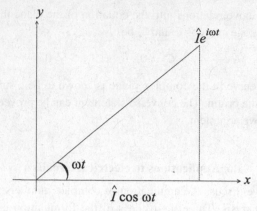

Figure 1.10. A phasor $\hat{I}e^{i\omega t}$ revolves around the circle of radius $\hat{I}$ with angular frequency ω.

phasor are given by

$$\hat{I}^2 = (\hat{I}_1 \cos\phi_1 + \hat{I}_2 \cos\phi_2)^2 + (\hat{I}_1 \sin\phi_1 + \hat{I}_2 \sin\phi_2)^2$$
$$= \hat{I}_1^{\,2} + \hat{I}_2^{\,2} + 2\hat{I}_1\hat{I}_2 \cos(\phi_1 - \phi_2)$$

and

$$\tan\phi = \frac{\hat{I}_1 \sin\phi_1 + \hat{I}_2 \sin\phi_2}{\hat{I}_1 \cos\phi_1 + \hat{I}_2 \cos\phi_2}.$$

The resultant phasor can be found graphically by the parallelogram law of vector addition. Since all the three phasors revolve with the same angular frequency ω, the parallelogram remains in the same shape at all times.

The general formula for finding the magnitude and phase of the resultant phasor of the sum of three or more current phasors with the same angular frequency can be derived analogously.

Next, we consider a closed alternating current circuit with an applied voltage V and the three basic circuit elements: resistor, inductor, and capacitor. This is usually called an *R-L-C* circuit. The voltage V (excitation) and the current I (response) are related by the formula

$$V = IZ, \tag{1.6.3}$$

where Z is called the *impedance* of the circuit. If V and I are represented by phasors, then Z becomes a complex quantity.

Suppose the alternating current $I = \hat{I} \cos\omega t$ is flowing through a pure resistor R. The voltage drop V_r across the resistor is given by

$$V_r = R\hat{I} \cos\omega t.$$

In this case, the voltage V_r is in phase with the current. When the current is flowing through a pure inductor L, the voltage drop V_i across the inductor is given by

$$V_i = L\frac{dI}{dt} = -\omega L\hat{I}\sin\omega t = \omega L\hat{I}\cos\left(\omega t + \frac{\pi}{2}\right).$$

That is, the voltage V_i leads the current by a phase angle of $\frac{\pi}{2}$. Suppose the same current is flowing through a pure capacitor C. The corresponding voltage drop V_c across the capacitor is given by

$$V_c = \frac{1}{C}\int I\,dt = \frac{\hat{I}}{\omega C}\sin\omega t = \frac{\hat{I}}{\omega C}\cos\left(\omega t - \frac{\pi}{2}\right).$$

In this case, the voltage V_c lags the current by a phase angle of $\frac{\pi}{2}$.

Suppose we connect the three elements in series and assume that steady state responses have been attained. The total voltage drop V is the sum of drops across the three circuit elements. We then have

$$V = \left[R\cos\omega t + \omega L\cos\left(\omega t + \frac{\pi}{2}\right) + \frac{1}{\omega C}\cos\left(\omega t - \frac{\pi}{2}\right)\right]\hat{I},$$

and in phasor notation, it takes the form

$$V = \hat{V}\,e^{i(\omega t + \phi)} = \left[R + i\left(\omega L - \frac{1}{\omega C}\right)\right]\hat{I}\,e^{i\omega t}.$$

According to eq. (1.6.3), the impedance of the R-L-C circuit is found to be

$$Z = R + i\left(\omega L - \frac{1}{\omega C}\right).$$

Example 1.6.1 In elementary circuit theory, the time-dependent response in an R-L-C series circuit is modeled by the second-order ordinary differential equation

$$L\frac{d^2 I}{dt^2} + R\frac{dI}{dt} + \frac{1}{C}I = 0,$$

where I is the instantaneous alternating current flowing through the circuit. Suppose the solution is assumed to be of the form

$$I = \hat{I}e^{\alpha t}.$$

Find an expression for α in terms of R, L, and C. Show that, as a complex quantity, α is never in the right half complex plane if R, L and C are non-negative.

Solution Suppose the governing equation admits a solution of the form

$$I = \hat{I}e^{\alpha t}.$$

Its first and second derivatives are

$$\frac{dI}{dt} = \alpha \hat{I}e^{\alpha t} \quad \text{and} \quad \frac{d^2 I}{dt^2} = \alpha^2 \hat{I}e^{\alpha t}.$$

Substituting into the differential equation, we obtain the following algebraic equation for α:

$$L\alpha^2 + R\alpha + \frac{1}{C} = 0.$$

Solving the quadratic equation for α, we have

$$\alpha = \frac{-R \pm \sqrt{R^2 - \frac{4L}{C}}}{2L}.$$

Consider the following cases:

(i) When $R = 0$, $\alpha = \pm\sqrt{\frac{1}{LC}}i$. The two roots are on the imaginary axis.

(ii) When $0 < \frac{R}{2L} < \frac{1}{\sqrt{LC}}$, $\alpha = -\frac{R}{2L} \pm \frac{\sqrt{\frac{4L}{C} - R^2}}{2L}i$. The two roots are in the left half complex plane.

(iii) When $\frac{R}{2L} = \frac{1}{\sqrt{LC}}$, $\alpha = -\frac{R}{2L}$ (a double root). The root lies on the negative real axis.

(iv) When $\frac{R}{2L} > \frac{1}{\sqrt{LC}}$, the two distinct roots both lie on the negative real axis.

Physically, the magnitude of the alternating current decays to zero as time tends to infinity whenever α has a negative real part (corresponding to $R > 0$).

1.7 Problems

1.1. Express the following complex numbers in the form $x + iy$:

(a) $(1 + 2i)^3$; (b) i^{17};

(c) $(1 + i)^n + (1 - i)^n$, n is any positive integer;

(d) $\frac{5}{-3 + 4i}$; (e) $\frac{i}{1 + i} + \frac{1 + i}{i}$.

1.2. Find the modulus of each of the following complex numbers:

(a) $-i\,(2 + i)\,(1 + 2i)\,(1 + i)$; (b) $\frac{(3 + i)\,(2 - i)}{(3 - i)\,(2 + i)}$;

(c) $\frac{(3 + 4i)\,(1 + i)^6}{(i)^5\,(2 + 4i)^2}$.

1.3. Let z_1 and z_2 be two complex numbers. Supposing $z_1 + z_2$ and $z_1 z_2$ are both real, show that either both z_1 and z_2 are real, or that one is the complex conjugate of the other.

1.4. Prove the commutative, associative and distributive rules for the addition and multiplication of complex numbers.

1.5. Show that if the product of two complex numbers is zero, then at least one of them is zero.

1.6. Show that

$$(1 + \cos\theta + i \sin\theta)^n = 2^n \cos^n \frac{\theta}{2} \left(\cos \frac{n\theta}{2} + i \sin \frac{n\theta}{2} \right).$$

1.7. Use de Moivre's theorem to show that

$$\sin(2m+1)\theta = \sin^{2m+1}\theta \, P_m(\cot^2\theta), \quad \text{for } 0 < \theta < \frac{\pi}{2},$$

where

$$P_m(x) = \sum_{k=0}^{m} (-1)^k C_{2k+1}^{2m+1} x^{m-k}, \quad C_r^n = \frac{n!}{(n-r)!r!}.$$

Hence, deduce that

$$\sum_{k=1}^{m} \cot^2 \frac{k\pi}{2m+1} = \frac{m(2m-1)}{3}.$$

1.8. A complex number, represented by $x + iy$, may also be visualized as a 2×2 matrix

$$\begin{pmatrix} x & y \\ -y & x \end{pmatrix}.$$

Verify that addition and multiplication of complex numbers defined via matrix operations are consistent with the usual addition and multiplication rules. What is the matrix representation corresponding to $(x + iy)^{-1}$?

1.9. For any pair of complex numbers z_1 and z_2, show that

(a) $\text{Re}(z_1 z_2) \le |z_1| \, |z_2|$;

(b) $|1 - \bar{z_1} z_2|^2 - |z_1 - z_2|^2 = (1 - |z_1|^2)(1 - |z_2|^2)$;

(c) $|z_1 + z_2| \ge \frac{1}{2} (|z_1| + |z_2|) \left| \frac{z_1}{|z_1|} + \frac{z_2}{|z_2|} \right|.$

1.10. Let a, b, u, v be complex numbers; prove that

    (a) $|au + bv|^2 + |\bar{b}u - \bar{a}v|^2 = (|a|^2 + |b|^2)(|u|^2 + |v|^2)$;

    (b) $|au + bv|^2 - |\bar{b}u + \bar{a}v|^2 = (|a|^2 - |b|^2)(|u|^2 - |v|^2)$.

1.11. Let z_1 and z_2 be complex numbers, and α_1 and α_2 be real numbers satisfying $\alpha_1^2 + \alpha_2^2 \neq 0$. Prove the following inequalities:

$$|z_1|^2 + |z_2|^2 - |z_1^2 + z_2^2| \leq 2\frac{|\alpha_1 z_1 + \alpha_2 z_2|^2}{\alpha_1^2 + \alpha_2^2} \leq |z_1|^2 + |z_2|^2 + |z_1^2 + z_2^2|.$$

1.12. Consider two complex numbers z_1 and z_2 such that $|z_1| = 1$ and $|z_2| \neq 1$. Show that

$$\left|\frac{z_1 - z_2}{1 - \bar{z}_2 z_1}\right| = 1.$$

1.13. If we write $z = re^{i\theta}$ and $w = Re^{i\phi}$, where $0 \leq r < R$, show that

$$\mathrm{Re}\left(\frac{w + z}{w - z}\right) = \frac{R^2 - r^2}{R^2 - 2Rr\cos(\theta - \phi) + r^2}.$$

This is called the Poisson kernel (see Subsection 7.1.1).

1.14. Show that if $r_1 e^{i\theta_1} + r_2 e^{i\theta_2} = re^{i\theta}$, then

$$r^2 = r_1^2 + 2r_1 r_2 \cos(\theta_1 - \theta_2) + r_2^2$$

$$\theta = \tan^{-1}\left(\frac{r_1 \sin\theta_1 + r_2 \sin\theta_2}{r_1 \cos\theta_1 + r_2 \cos\theta_2}\right).$$

Generalize the result to the sum of n complex numbers.

1.15. One may find the square roots of a complex number using polar representation. First, we write formally

$$z = r(\cos\theta + i\sin\theta).$$

Recalling the identities

$$\sin^2\frac{\theta}{2} = \frac{1 - \cos\theta}{2} \quad \text{and} \quad \cos^2\frac{\theta}{2} = \frac{1 + \cos\theta}{2},$$

show that

$$z^{1/2} = \pm\sqrt{r}\left(\sqrt{\frac{1 + \cos\theta}{2}} + i\sqrt{\frac{1 - \cos\theta}{2}}\right), \quad 0 \leq \theta \leq \pi.$$

Explain why the above formula becomes invalid for $-\pi < \theta < 0$. Find the corresponding formula under this case. Use the derived formula to compute $(3 - 4i)^{1/2}$. Compare the results with those obtained using the formula derived in Example 1.2.3.

1.16. This problem is related to the proof of inequality (iv) in Example 1.2.5. Using the same set of notation as in Example 1.2.5, suppose we write

$$\alpha_1 = |r_2|^2 - 1 \quad \text{and} \quad \alpha_2 = |r_1|^2 - 1$$

so that

$$a_1 - \bar{a}_1 a_0 = \alpha_1 r_1 + \alpha_2 r_2.$$

Show that

$$|a_1 - \bar{a}_1 a_0|^2 = \alpha_1^2(1 + \alpha_2) + \alpha_2^2(1 + \alpha_1) + \alpha_1\alpha_2(r_1\bar{r}_2 + r_2\bar{r}_1)$$
$$(1 - |a_0|^2)^2 = (\alpha_1 + \alpha_2 + \alpha_1\alpha_2)^2$$

so that

$$(1 - |a_0|^2)^2 - |a_1 - \bar{a}_1 a_0|^2$$
$$= (1 - |r_1|)(1 - |r_2|)(1 + |r_1|)(1 + |r_2|)(1 + r_1\bar{r}_2)(1 + r_2\bar{r}_1).$$

Given that $|a_1 - \bar{a}_1 a_0| \leq 1 - |a_0|^2$, use the above relation to deduce the following inequality

$$(1 - |r_1|)(1 - |r_2|) \geq 0.$$

1.17. Let z be a root of the following equation:

$$z^n + (z + 1)^n = 0,$$

where n is any positive integer. Show that

$$\text{Re}\, z = -\frac{1}{2}.$$

1.18. The nth order Chebyshev polynomial is defined by

$$T_n(x) = \cos(n \cos^{-1} x), \qquad n \text{ is a positive integer}; \ -1 \leq x \leq 1.$$

Using de Moivre's theorem, show that $T_n(x)$ has the formal polynomial representation

$$T_n(x) = \frac{1}{2}\left\{ \left[x + (x^2 - 1)^{1/2} \right]^n + \left[x - (x^2 - 1)^{1/2} \right]^n \right\}.$$

Hint: One needs not be concerned with the occurrence of fractional powers of $x^2 - 1$. Why?

1.19. Show that the distance of the point c from the line

$$\bar{a}z + a\bar{z} = b, \quad b \text{ is real,}$$

is given by $\dfrac{|\bar{a}c + a\bar{c} - 2b|}{2|a|}$.

Hint: Use the mapping $w = \bar{a}z$ so that the line

$$\bar{a}z + a\bar{z} = b$$

in the z-plane is mapped onto the line

$$\operatorname{Re} w = \frac{b}{2}$$

in the w-plane. The point c is mapped to $\bar{a}c$. Find the distance between the image point $\bar{a}c$ and the mapped line in the w-plane.

1.20. The triangle inequality can be generalized to n complex numbers. Given n complex numbers $z_1, z_2, \ldots, z_n$, show that

$$|z_1 + z_2 + \cdots + z_n| \le |z_1| + |z_2| + \cdots + |z_n|.$$

When does equality hold?

1.21. Let z_1, z_2, z_3, and z_4 be four arbitrary points in the complex plane. Show that

$$|z_1 - z_3|\,|z_2 - z_4| \le |z_1 - z_2|\,|z_3 - z_4| + |z_1 - z_4|\,|z_2 - z_3|.$$

1.22. Find the necessary and sufficient condition for any three points z_1, z_2 and z_3 to be collinear in the complex plane.

1.23. Suppose $z_1 + z_2 + z_3 = 0$ and $|z_1| = |z_2| = |z_3| = 1$. Show that $z_2 = \omega z_1$ and $z_3 = \omega^2 z_1$, where ω is a root of the quadratic equation $z^2 + z + 1 = 0$. Hence, show that z_1, z_2 and z_3 are the vertices of an equilateral triangle inscribed inside the unit circle $|z| = 1$.

1.24. For $z \ne 0$ and $-\pi < \operatorname{Arg} z \le \pi$, show that

$$|z - 1| \le \left| |z| - 1 \right| + |z||\operatorname{Arg} z|.$$

Hint: For $|z| > 1$, consider the following figure:

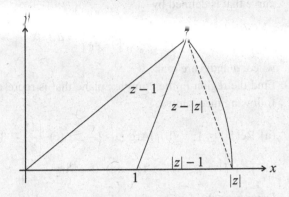

The circular arc is centered at $z = 0$ and passes through the two points z and $|z|$. Perform similar calculations for $|z| < 1$ and $|z| = 1$.

1.25. When $|\alpha| < 1$ and $|\beta| < 1$, show that

$$\left| \frac{\alpha - \beta}{1 - \bar{\alpha}\beta} \right| < 1.$$

1.26. Let $z^{1/2}$ denote the square root of z that has a positive imaginary part, where z is a non-real complex number. For any two non-real complex numbers z_1 and z_2 with the same modulus value, find a real number p such that

$$z_1 - z_2 = ipz_1^{1/2}z_2^{1/2}.$$

1.27. Show that the necessary and sufficient condition for the existence of z satisfying the following equation

$$|z - \alpha| + |z + \alpha| = 2|\beta|$$

is given by $|\alpha| \leq |\beta|$. Find the maximum and minimum values of $|z|$.

1.28. Show that the equation

$$|z| = \varepsilon(\operatorname{Re} z - k), \quad \varepsilon \text{ and } k \text{ are real},$$

represents a conic with its focus at the origin, eccentricity ε, and directrix along the line $\operatorname{Re} z = k$.

1.29. Determine the maximum and minimum distance from the origin to the curve that is defined by

$$\left| z + \frac{b}{z} \right| = a,$$

where a and b are real.

1.30. Find the region in the complex plane that is represented by each of the following inequalities:

(a) $\text{Re}(z^2) \leq 1$;   (b) $|\text{Arg } z| < \dfrac{\pi}{3}$;   (c) $\left| \dfrac{1}{z} \right| < 3$;

(d) $|z + 1| - |z - 1| < 1$;   (e) $0 < \text{Arg } \dfrac{z - 1}{z + 1} < \dfrac{\pi}{4}$.

1.31. Determine the family of curves represented by each of the following equations:

(a) $\left| \dfrac{z - z_1}{z - z_2} \right| = \lambda$;   (b) $\text{Arg } \dfrac{z - z_1}{z - z_2} = \alpha$, $-\pi < \alpha \leq \pi$.

1.32. Let z_1, z_2, z_3 and w_1, w_2, w_3 be the vertices of two triangles in the complex plane. Show that the necessary and sufficient condition for the two triangles to be similar is given by

$$\begin{vmatrix} 1 & 1 & 1 \\ z_1 & z_2 & z_3 \\ w_1 & w_2 & w_3 \end{vmatrix} = 0.$$

Hint: The given condition is equivalent to $\dfrac{z_2 - z_1}{z_3 - z_1} = \dfrac{w_2 - w_1}{w_3 - w_1}$.

1.33. If z_1, z_2 and z_3 represent the vertices of an equilateral triangle, show that

$$z_1^2 + z_2^2 + z_3^2 = z_1 z_2 + z_2 z_3 + z_3 z_1.$$

1.34. Show that the area of the triangle whose vertices are z_1, z_2, z_3 is given by the absolute value of

$$\frac{1}{2} |z_3 - z_2|^2 \text{ Im} \left(\frac{z_3 - z_1}{z_3 - z_2} \right).$$

1.35. Consider any triangle ABC whose sides are of length α, β, γ and for which the distances from the centroid to the vertices are λ, μ, ν. Show that

$$\frac{\alpha^2 + \beta^2 + \gamma^2}{\lambda^2 + \mu^2 + \nu^2} = 3.$$

Hint: Let the centroid of the triangle be at the origin and let z_1, z_2, and z_3 be the vertices; then $z_1 + z_2 + z_3 = 0$.

1.36. Let ω ($\omega \neq 1$) be any one of the nth roots of unity. Show that

$$1 + \omega + \omega^2 + \cdots + \omega^{n-1} = 0.$$

Hence, deduce the value of

$$1 + \omega^k + \omega^{2k} + \cdots + \omega^{(n-1)k}, \qquad k \text{ is any non-negative integer.}$$

Also, find the value of

$$1 + 2\omega + 3\omega^2 + 4\omega^3 + \cdots + n\omega^{n-1}.$$

1.37. Let $\omega = e^{\frac{i\pi}{n}}$ and z be any complex number. Show that

$$(z - \omega)(z - \omega^2) \cdots (z - \omega^{n-1}) = 1 + z + z^2 + \cdots + z^{n-1}.$$

1.38. Let ω be a cube root of unity and define

$$f(a, b, c) = a^2 + b^2 + c^2 - bc - ca - ab.$$

(a) Show that

$$f(a, b, c) = (a + \omega b + \omega^2 c)(a + \omega^2 b + \omega c).$$

(b) Prove that if n is not a multiple of 3, then

$$(b - c)^n + (c - a)^n + (a - b)^n$$

contains $f(a, b, c)$ as a factor.

(c) Suppose α, β and γ are complex numbers representing the vertices of an equilateral triangle in the complex plane. Explain why

$$\gamma - \alpha = \omega(\beta - \gamma) \text{ or } \gamma - \alpha = \omega^2(\beta - \gamma).$$

Hence, show that α, β and γ satisfy

$$f(\alpha, \beta, \gamma) = 0.$$

1.39. Suppose a, b, c and d are real numbers. Find the conditions under which the quadratic equation

$$x^2 + (a + ib)x + (c + id) = 0$$

has at least one real root.

1.40. Assume $|\alpha_k| < 1$ and $\lambda_k \geq 0$, $k = 1, 2, \ldots, n$, and

$$\lambda_1 + \lambda_2 + \cdots + \lambda_n = 1.$$

Here, $\alpha_1, \alpha_2, \ldots, \alpha_n$ are complex numbers. Show that

$$|\lambda_1\alpha_1 + \lambda_2\alpha_2 + \cdots + \lambda_n\alpha_n| < 1.$$

1.41. Let w_1 and w_2 be two complex numbers. Denote $\Delta w = w_2 - w_1$. Suppose $w_2 = rw_1$, where r is real and non-zero. Show that

$$\frac{1 + \cos(\operatorname{Arg} w_2 - \operatorname{Arg} \Delta w)}{1 + \cos(\operatorname{Arg} w_1 - \operatorname{Arg} \Delta w)} = \begin{cases} \frac{1}{r^2} & \text{when } 0 < r < 1 \\ 1 & \text{when } r > 1 \end{cases}.$$

When $r < 0$, show that it becomes undefined.

1.42. For each sequence $\{z_n\}$, find the corresponding limit points (if they exist):

(a) $z_n = (-1)^n$; (b) $z_n = (-1)^n \dfrac{1}{n}$; (c) $z_n = e^{\frac{n\pi i}{4}}$;

(d) $z_n = \dfrac{n}{n+1} e^{\frac{n\pi i}{4}}$.

1.43. Show that the whole complex plane and the empty set are both open. Are they both closed?

1.44. For each of the following point sets, find the interior points, exterior points, boundary points and limit points:

(a) $0 \leq \operatorname{Re}(iz) \leq 3$; (b) $0 \leq \operatorname{Arg} z < \frac{\pi}{4}$ and $|z| > 2$.

1.45. Classify the following sets according to the properties: open, closed, bounded, unbounded, compact.

(a) $S_1 = \{z : a \leq \operatorname{Re} z \leq b\}$; (b) $S_2 = \{z : 0 < \operatorname{Arg} z < \frac{\pi}{2}\}$;

(c) $S_3 = \{z : z = e^{\frac{2k\pi i}{5}}, \quad k = 0, 1, 2, 3, 4\}$.

1.46. Consider the set of points (x, y) which are solutions to

$$y = 0 \quad \text{and} \quad \sin\frac{\pi}{x} = 0$$

lying inside the punctured disk: $0 < |z| < 1, z = x + iy$. Does there exist any limit point of the point set? If so, find the limit point.

1.47. Find the image point of each of the following complex numbers on the Riemann sphere:

(a) 1; (b) i; (c) -1; (d) $-i$.

1.48. What is the relation between z_1 and z_2 if their images on the Riemann sphere are diametrically opposite each other?

1.49. Prove that the angle between two curves in the complex plane is equal to the angle between their image curves on the Riemann sphere.

1.50. Show that any circle or straight line in the complex plane corresponds to a circle on the Riemann sphere.

1.51. To find a particular solution $x_p(t)$ of the constant coefficient differential equation

$$\frac{d^2x}{dt^2} + c\frac{dx}{dt} + kx = F\cos\omega t,$$

one may consider the associated problem

$$\frac{d^2v}{dt^2} + c\frac{dv}{dt} + kv = Fe^{i\omega t},$$

and find its particular solution $v_p(t)$. Show that

$$x_p(t) = \mathrm{Re}\, v_p(t),$$

and use the result to find $x_p(t)$.

1.52. Show that the modulus and argument of the impedance of the R-L-C circuit are given by

$$|Z| = \sqrt{R^2 + \left(\omega L - \frac{1}{\omega C}\right)^2} \quad \text{and} \quad \mathrm{Arg}\, Z = \tan^{-1}\frac{\omega L - \frac{1}{\omega C}}{R},$$

respectively. Find the condition under which the voltage is in phase with the current.

2

Analytic Functions

In this chapter, we introduce functions of a complex variable and examine how some of their mathematical properties may differ from those of real-valued functions. The mapping properties associated with complex functions are illustrated. The theory of differentiation of complex functions and the concept of differentiability are developed. The highlights of the chapter are the Cauchy–Riemann relations and the definition of an analytic function. Analyticity plays a central role in complex variable theory. The relations between harmonic functions and analytic functions are established. We show how to solve the Poisson equation effectively using the formulation of complex conjugate variables. In addition, the application of complex differentiation in dynamics problems and the use of complex functions in describing fluid flows and steady state heat distribution are illustrated.

2.1 Functions of a complex variable

Let S be a set of complex numbers in the complex plane. For every point $z = x + iy \in S$, we specify the rule for assigning a corresponding complex number $w = u + iv$. This defines a function of the complex variable z, and the function is denoted by

$$w = f(z). \tag{2.1.1}$$

The set S is called the *domain of definition* of the function f and the collection of all values of w is called the *range* of f. Below are some examples of complex functions:

(1) $f_1(z) = z^2$; (2) $f_2(z) = \operatorname{Im} z$; (3) $f_3(z) = \operatorname{Arg} z$;

(4) $f_4(z) = \dfrac{z+3}{z^2+1}$.

46

Note that $f_3(z) = \operatorname{Arg} z$ is defined everywhere except at $z = 0$, and this function can assume all possible real values in the interval $(-\pi, \pi]$. The domain of definition of $f_1(z)$ ~~is seen to be~~ $\mathbb{C}\setminus\{i\ -i\}$, What is the range of this function?

A complex function of the complex variable z may be visualized as a pair of real functions of the two real variables x and y, where $z = x + iy$. Let $u(x, y)$ and $v(x, y)$ be the real and imaginary parts of $f(z)$, respectively. We may write

$$f(z) = u(x, y) + iv(x, y), \quad z = x + iy. \tag{2.1.2}$$

For example, consider the function

$$f_1(z) = z^2 = (x + iy)^2 = x^2 - y^2 + 2ixy;$$

its real and imaginary parts are the real functions

$$u(x, y) = x^2 - y^2 \quad \text{and} \quad v(x, y) = 2xy,$$

respectively.

In a single-valued complex function, only one value of w is assigned to each value of z. However, functions like $f(z) = z^{\frac{1}{2}}$ and $f(z) = \arg z$ are multi-valued. In complex variable theory, we may treat a multi-valued function as a collection of single-valued functions. Each member is called a *branch of the function*. In the above examples, $f(z) = z^{\frac{1}{2}}$ has two branches and $f(z) = \arg z$ has infinitely many branches. We usually choose one of the branches as the *principal branch of the multi-valued function*. For example, $\operatorname{Arg} z$ is chosen as the principal branch of $f(z) = \arg z$. A more detailed discussion on the characterization of the branches of multi-valued complex functions will be presented in Section 3.6.

Example 2.1.1   For each of the following functions, determine whether it is a many-to-one, one-to-one or one-to-many function:

(a) $w(z) = z + \dfrac{1}{z}$;   (b) $w(z) = \dfrac{iz + 4}{2z + 3i}$;   (c) $w(z) = z^{\frac{1}{2}}$.

Solution

(a) Consider $w = z + \frac{1}{z}$. It is obvious that

$$w\left(\frac{1}{z}\right) = \frac{1}{z} + \frac{1}{\frac{1}{z}} = w(z);$$

that is, both z and $\frac{1}{z}$ are mapped to the same point. In fact, the function may be expressed as a quadratic polynomial in z, where

$$z^2 - zw + 1 = 0.$$

For a given value of w, there are in general two values of z which satisfy the above relation. Therefore, the function $w(z) = z + \frac{1}{z}$ is many-to-one.

(b) For any two complex numbers z_1 and z_2, we have

$$
\begin{aligned}
&w(z_1) = w(z_2) \\
\Longleftrightarrow\ &\frac{iz_1 + 4}{2z_1 + 3i} = \frac{iz_2 + 4}{2z_2 + 3i} \\
\Longleftrightarrow\ &2iz_1z_2 + 8z_2 - 3z_1 + 12i = 2iz_1z_2 + 8z_1 - 3z_2 + 12i \\
\Longleftrightarrow\ &z_1 = z_2.
\end{aligned}
$$

Therefore, the function $w(z) = \frac{iz+4}{2z+3i}$ is one-to-one.

(c) Consider the following relation:

$$z = re^{i\theta} = re^{i(2\pi+\theta)}.$$

We have two possible values, namely,

$$w(z) = r^{\frac{1}{2}}e^{i\frac{\theta}{2}} \qquad \text{or} \qquad w(z) = r^{\frac{1}{2}}e^{i\frac{2\pi+\theta}{2}} = -r^{\frac{1}{2}}e^{i\frac{\theta}{2}}.$$

Therefore, the function $w(z) = z^{1/2}$ is one-to-many.

2.1.1 *Velocity of fluid flow emanating from a source*

The following example illustrates how a complex function arises from the description of a physical phenomenon. Consider the steady state flow of fluid in the z-plane emanating from a fluid source placed at the origin. We would like to find a complex function $v(z)$ which gives the velocity of the flow at any point z. Physically, the direction of the flow is radially outward from the fluid source and the speed is inversely proportional to the distance from the source (see Figure 2.1).

These physical properties can be described mathematically as

$$\text{Arg}\, v = \text{Arg}\, z \quad \text{and} \quad |v| = \frac{k}{|z|}, \tag{2.1.3}$$

where k is some positive real constant. If we write $z = re^{i\theta}$, the velocity function of the fluid source is given by

$$v(z) = |v|e^{i\text{Arg}\, v} = \frac{k}{|z|}\, e^{i\theta} = \frac{k}{\bar{z}}. \tag{2.1.4}$$

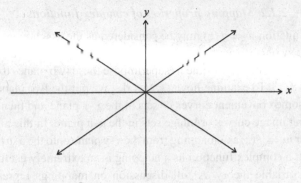

Figure 2.1. Fluid emanates from the source at the origin. The flow direction is radially outward from the source and the speed is inversely proportional to the distance from the source.

The constant k may be called the strength of the source. Physically, it is related to the amount of fluid flowing out from the source per unit time.

Example 2.1.2 Suppose we have a fluid source (i.e. fluid flowing out) at $z = \alpha$ and a fluid sink (i.e. fluid flowing in) at $z = \beta$ in the complex plane, both of the same strength k. Find the resulting velocity at an arbitrary point z. What happens if the source and the sink are approaching each other such that $(\alpha - \beta) \to 0$ and $k \to \infty$ but $\mu = k(\alpha - \beta)$ is kept finite?

Solution From eq. (2.1.4), the velocities at z due to the source and the sink are given by

$$\frac{k}{\overline{z} - \overline{\alpha}} \qquad \text{and} \qquad -\frac{k}{\overline{z} - \overline{\beta}},$$

respectively. Assuming that the superposition principle of velocities is applicable, the combined velocity at z is given by the sum of the two velocity functions, so

$$v(z) = k\left(\frac{1}{\overline{z} - \overline{\alpha}} - \frac{1}{\overline{z} - \overline{\beta}}\right) = \frac{k(\overline{\alpha} - \overline{\beta})}{(\overline{z} - \overline{\alpha})(\overline{z} - \overline{\beta})}.$$

Consider the limits $(\alpha - \beta) \to 0$ and $k \to \infty$, while $\mu = k(\alpha - \beta)$ is kept finite; such a configuration is called a *doublet*. The velocity of the flow fluid at z due to the doublet is found to be

$$v(z) = \frac{\overline{\mu}}{(\overline{z} - \overline{\alpha})^2}.$$

2.1.2 Mapping properties of complex functions

A complex function $w = f(z)$ may be considered as the assignment of a point (x, y) in the x-y plane to another point (u, v) in the u-v plane, where $z = x + iy$ and $w = u + iv$. It is not possible to superimpose these two planes to visualize the graph of $f(z)$. To examine how $f(z)$ works, we put the two planes side by side, select some convenient curves or sets in the x-y plane and then plot their corresponding image curves or image sets in the u-v plane. In this manner, we may consider $w = f(z)$ as a mapping from the x-y plane onto the u-v plane. The realization of a complex function as a mapping is an extremely useful concept in complex variable theory. A full discussion on mappings represented by complex functions will be presented in Chapter 8. The two examples presented below illustrate some interesting mapping properties of complex functions.

Example 2.1.3   The complex numbers $z = x + iy$ and $w = u + iv$ are represented by points P and Q in the z-plane and w-plane, respectively. Suppose P traverses along the vertical line $x = -\frac{1}{2}$, and z and w are related by

$$z = \frac{w}{1 - w}.$$

Find the trajectory traced out by Q.

Solution It is convenient to express the mapping relation as

$$w = \frac{z}{1 + z}.$$

Since P moves along $x = -\frac{1}{2}$, we may represent P by $z = -\frac{1}{2} + iy$, where y can take any real value. The modulus of the image point Q in the w-plane satisfies

$$\left| \frac{-\frac{1}{2} + iy}{\frac{1}{2} + iy} \right| = \sqrt{\frac{\left(-\frac{1}{2}\right)^2 + y^2}{\left(\frac{1}{2}\right)^2 + y^2}} = 1,$$

indicating that the locus of Q is the unit circle centered at the origin in the w-plane (see Figure 2.2). Furthermore, we observe that

$$\mathrm{Re}\left(\frac{-\frac{1}{2} + iy}{\frac{1}{2} + iy} \right) = \frac{y^2 - \frac{1}{4}}{y^2 + \frac{1}{4}} \quad \text{and} \quad \mathrm{Im}\left(\frac{-\frac{1}{2} + iy}{\frac{1}{2} + iy} \right) = \frac{y}{y^2 + \frac{1}{4}}.$$

Therefore, $z = -\frac{1}{2}$ (corresponding to $y = 0$) is mapped onto $w = -1$, and $z = \infty$ (corresponding to $y \to \pm\infty$) is mapped onto $w = 1$. Since $\mathrm{Im}\, w$ and y have the same sign, the portion of the line segment $y < 0$ is mapped onto the lower unit circle and the portion $y > 0$ is mapped onto the upper unit circle.

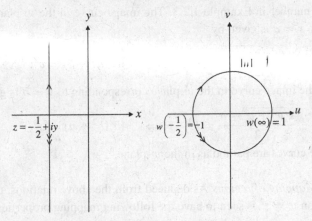

Figure 2.2. The mapping behavior of $z = w/(1-w)$.

Correspondingly, the region on the right hand side of the vertical line $x = -\frac{1}{2}$ is mapped onto the interior of the unit circle $|w| = 1$. The relation between the two loci traced by P and Q is shown in Figure 2.2.

Example 2.1.4 Consider the function $f(z) = z^2$ and write it as

$$f(z) = u(x, y) + iv(x, y), \quad \text{where } z = x + iy.$$

Find the curves in the x-y plane such that $u(x, y) = \alpha$ and $v(x, y) = \beta$. Also, find the curves in the u-v plane whose preimages in the x-y plane are $x = a$ and $y = b$. What is the image curve in the u-v plane of the closed curve $r = 2(1 + \cos\theta)$ in the x-y plane, where $z = re^{i\theta}$?

Solution For $z = x + iy$,

$$f(z) = (x + iy)^2 = x^2 - y^2 + 2ixy,$$

so that

$$u(x, y) = x^2 - y^2 \quad \text{and} \quad v(x, y) = 2xy.$$

For all points on the hyperbola $x^2 - y^2 = \alpha$ in the x-y plane, the corresponding image points in the w-plane are on the coordinate curve $u = \alpha$. Similarly, the points on the hyperbola $2xy = \beta$ are mapped onto the coordinate curve $v = \beta$.

To find the curves in the u-v plane whose preimage curves are the coordinate curves $x = a$ and $y = b$, we use the result obtained for the square roots of a complex number in Example 1.2.3. The image curve in the w-plane corresponding to $x = a$ is given by

$$\frac{u + \sqrt{u^2 + v^2}}{2} = a^2 \quad \Longleftrightarrow \quad 4a^2(a^2 - u) = v^2.$$

Similarly, the image curve in the w-plane corresponding to $y = b$ is given by

$$\frac{\sqrt{u^2 + v^2} - u}{2} = b^2 \quad \Longleftrightarrow \quad 4b^2(b^2 + u) = v^2.$$

Both image curves are parabolas in the w-plane.

Remark   *Mapping of regions* As deduced from the above relations, the complex function $w = z^2$ is seen to have the following mapping properties:

 (i) the line $x = 0$, $y > 0$ is mapped onto the line $u < 0$, $v = 0$;
 (ii) the line $0 \le x \le a$, $y = 0$ is mapped onto the line $0 \le u \le a^2$, $v = 0$; and
 (iii) the line $x = a$, $y > 0$ is mapped onto the upper portion of the parabola $4a^2(a^2 - u) = v^2$, $v > 0$.

Hence, the semi-infinite strip

$$\{(x, y) : 0 \le x \le a, \ y \ge 0\}$$

in the z-plane is mapped onto the semi-infinite parabolic wedge

$$\{(u, v) : 4a^2(a^2 - u) \ge v^2, \ v \ge 0\}$$

in the w-plane (see Figure 2.3).

The preimage curve $r = 2(1 + \cos\theta)$ is expressed in the polar form, so it is convenient to seek the polar representation of the image curve. Let $Re^{i\phi}$ be the polar representation of w. In polar form, the complex function $w = z^2$ becomes $Re^{i\phi} = r^2 e^{2i\theta}$. Consider a point on the curve

$$r = 2(1 + \cos\theta) = 4\cos^2\frac{\theta}{2};$$

we see that

$$r^2 = 16\cos^4\frac{\theta}{2}.$$

The corresponding image point in the w-plane can be represented as $16\cos^4\frac{\theta}{2}\, e^{2i\theta}$. By comparing like terms in $Re^{i\phi}$ and $16\cos^4\frac{\theta}{2}\, e^{2i\theta}$, we deduce that

$$\phi = 2\theta \quad \text{and} \quad R = 16\cos^4\frac{\theta}{2} = 16\cos^4\frac{\phi}{4}.$$

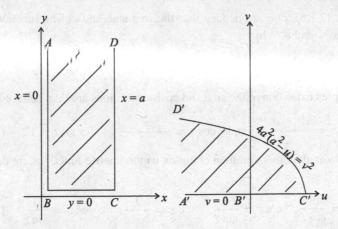

Figure 2.3. The complex function $w = z^2$ maps a semi-infinite strip in the z-plane onto a semi-infinite parabolic wedge in the w-plane.

Therefore, the polar form of the image curve in the w-plane is found to be

$$R = 16 \cos^4 \frac{\phi}{4}.$$

2.1.3 Definitions of the exponential and trigonometric functions

We illustrate how the complex exponential and trigonometric functions can be defined as natural extensions of their real counterparts. These functions are introduced earlier so that they can be used to serve as examples to illustrate the concepts of limit, continuity and differentiability of complex functions in later sections. In the next chapter, we show how to define the complex exponential and trigonometric functions using the first principles.

Let $z = x + iy$. The complex exponential function is defined by

$$e^z = e^x(\cos y + i \sin y), \quad \text{for all } z \text{ in } \mathbb{C}. \tag{2.1.5}$$

How would we justify the above definition? First, we require that the complex exponential function reduces to its real counterpart when z is real. In fact, if $y = 0$, e^z reduces to e^x, which is the real exponential function. By setting $x = 0$, eq. (2.1.5) becomes

$$e^{iy} = \cos y + i \sin y, \tag{2.1.6}$$

which is consistent with the Euler formula (see Subsection 1.2.1). The real and imaginary parts of the complex exponential functions are, respectively,

$$u(x, y) = e^x \cos y \quad \text{and} \quad v(x, y) = e^x \sin y.$$

From eq. (2.1.6), one can deduce that the real sine and cosine functions are related to e^{iy} and e^{-iy} by

$$\sin y = \frac{e^{iy} - e^{-iy}}{2i} \quad \text{and} \quad \cos y = \frac{e^{iy} + e^{-iy}}{2}.$$

The complex extension of the sine and cosine functions are then deduced to be

$$\sin z = \frac{e^{iz} - e^{-iz}}{2i} \quad \text{and} \quad \cos z = \frac{e^{iz} + e^{-iz}}{2}, \quad \text{for all } z \text{ in } \mathbb{C}. \quad (2.1.7)$$

Correspondingly, other common complex trigonometric functions are defined by

$$\tan z = \frac{\sin z}{\cos z}, \quad \cot z = \frac{1}{\tan z}, \quad \sec z = \frac{1}{\cos z}, \quad \operatorname{cosec} z = \frac{1}{\sin z}. \quad (2.1.8)$$

2.2 Limit and continuity of complex functions

Similar to the calculus of real variables, the differentiability of a complex function is defined using the concept of limit. Here, we start with the precise definition of the *limit of a complex function*.

2.2.1 Limit of a complex function

Let $w = f(z)$ be defined in the point set S and z_0 be a limit point of S. The mathematical statement

$$\lim_{z \to z_0} f(z) = L, \qquad z \in S, \qquad (2.2.1)$$

means that the value $w = f(z)$ can be made arbitrarily close to L if we choose z to be close enough, but not equal, to z_0. The formal definition of the limit of a function is stated as:

For any $\epsilon > 0$, there exists $\delta > 0$ (usually dependent on ϵ) such that

$$|f(z) - L| < \epsilon \quad \text{if} \quad 0 < |z - z_0| < \delta.$$

The limit L, if exists, must be unique. The value of L is independent of the direction along which $z \to z_0$ (see Figure 2.4).

The function $f(z)$ needs not be defined at z_0 in order for the function to have a limit at z_0. However, we do require z_0 to be a limit point of S so that it would never occur that $f(z)$ is not defined in some deleted neighborhood of z_0. For example, let us consider $\lim_{z \to 0} \frac{\sin z}{z}$. The domain of definition of $\frac{\sin z}{z}$ is $\mathbb{C} \setminus \{0\}$. Though $\frac{\sin z}{z}$ is not defined at $z = 0$, this is a limit point of the domain of definition. Hence, the above limit is well defined.

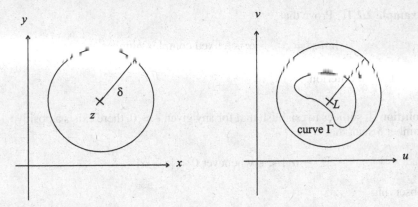

Figure 2.4. The region $0 < |z - z_0| < \delta$ in the z-plane is mapped onto the region enclosed by the curve Γ in the w-plane. The curve Γ lies completely inside the annulus $0 < |w - L| < \epsilon$.

If $L = \alpha + i\beta$, $f(z) = u(x, y) + iv(x, y)$, $z = x + iy$ and $z_0 = x_0 + iy_0$, then

$$|u(x, y) - \alpha| \leq |f(z) - L| \leq |u(x, y) - \alpha| + |v(x, y) - \beta|,$$
$$|v(x, y) - \beta| \leq |f(z) - L| \leq |u(x, y) - \alpha| + |v(x, y) - \beta|.$$

From the above inequalities, it is obvious that eq. (2.2.1) is equivalent to the following pair of limits

$$\lim_{(x,y) \to (x_0, y_0)} u(x, y) = \alpha \qquad (2.2.2a)$$

$$\lim_{(x,y) \to (x_0, y_0)} v(x, y) = \beta. \qquad (2.2.2b)$$

Therefore, the study of the limiting behavior of $f(z)$ is equivalent to that of a pair of real functions $u(x, y)$ and $v(x, y)$. Consequently, theorems concerning the limit of the sum, difference, product and quotient of complex functions hold as to those for real functions. Suppose that

$$\lim_{z \to z_0} f_1(z) = L_1 \quad \text{and} \quad \lim_{z \to z_0} f_2(z) = L_2$$

Then

$$\lim_{z \to z_0} (f_1(z) \pm f_2(z)) = L_1 \pm L_2, \qquad (2.2.3a)$$

$$\lim_{z \to z_0} f_1(z) f_2(z) = L_1 L_2, \qquad (2.2.3b)$$

$$\lim_{z \to z_0} \frac{f_1(z)}{f_2(z)} = \frac{L_1}{L_2}, \quad L_2 \neq 0. \qquad (2.2.3c)$$

Example 2.2.1 Prove that

$$\lim_{z \to \alpha} z^2 = \alpha^2, \quad \alpha \text{ is a fixed complex number,}$$

using the ϵ-δ criterion.

Solution It suffices to establish that for any given $\epsilon > 0$, there exists a positive number δ such that

$$|z^2 - \alpha^2| < \epsilon \text{ whenever } 0 < |z - \alpha| < \delta.$$

Observing

$$z^2 - \alpha^2 = (z - \alpha)(z + \alpha) = (z - \alpha)(z - \alpha + 2\alpha)$$

and applying the triangle inequality, we obtain

$$|z^2 - \alpha^2| = |z - \alpha||z - \alpha + 2\alpha| \le |z - \alpha|(|z - \alpha| + 2|\alpha|)$$

provided that z lies inside the deleted δ-neighborhood of α. Here, δ is chosen to be less than $\min\left(1, \frac{\epsilon}{1+2|\alpha|}\right)$, so

$$|z^2 - \alpha^2| \le |z - \alpha|(|z - \alpha| + 2|\alpha|) < \frac{\epsilon}{1 + 2|\alpha|}(1 + 2|\alpha|) = \epsilon.$$

Note that the choice of δ depends on ϵ and α.

Limit at infinity

The definition of limit holds even when z_0 or L is the point at infinity. We can simply replace the corresponding neighborhood of z_0 or L by the neighborhood of infinity. The mathematical statement

$$\lim_{z \to \infty} f(z) = L \tag{2.2.4}$$

can be understood as:

> For any $\epsilon > 0$, there exists $\delta(\epsilon) > 0$ such that
> $|f(z) - L| < \epsilon$ whenever $|z| > \frac{1}{\delta}$.

Here, z refers to a point in the finite complex plane and $|z| > \frac{1}{\delta}$ is visualized as a deleted neighborhood of ∞. Also, we must be cautious that the results in eqs. (2.2.3a,b,c) hold for z_0, L_1 and L_2 in the finite complex plane only.

Suppose we define $w = \frac{1}{z}$. Then $z \to \infty$ is equivalent to $w \to 0$. It is then not surprising to have the following properties on limit at infinity.

Theorem 2.2.1 *If z_0 and w_0 are points in the z-plane and the w-plane respectively, then*

(a) $\lim_{z \to z_0} f(z) = \infty$ *if and only if* $\lim_{z \to z_0} \frac{1}{f(z)} = 0$;

(b) $\lim_{z \to \infty} f(z) = w_0$ *if and only if* $\lim_{z \to 0} f\left(\frac{1}{z}\right) = w_0$.

Proof

(a) $\lim_{z \to z_0} f(z) = \infty$ implies that for any $\epsilon > 0$, there exists a positive number δ such that

$$|f(z)| > \frac{1}{\epsilon} \text{ whenever } 0 < |z - z_0| < \delta.$$

The above result may be rewritten as

$$\left|\frac{1}{f(z)} - 0\right| < \epsilon \text{ whenever } 0 < |z - z_0| < \delta,$$

so we obtain

$$\lim_{z \to z_0} \frac{1}{f(z)} = 0.$$

(b) $\lim_{z \to \infty} f(z) = w_0$ implies that for any $\epsilon > 0$, there exists $\delta > 0$ such that

$$|f(z) - w_0| < \epsilon \text{ whenever } |z| > \frac{1}{\delta}.$$

Replacing z by $\frac{1}{z}$, we obtain

$$\left|f\left(\frac{1}{z}\right) - w_0\right| < \epsilon \text{ whenever } 0 < |z - 0| < \delta,$$

so we obtain

$$\lim_{z \to 0} f\left(\frac{1}{z}\right) = w_0.$$

The above results provide the convenient tools to evaluate limits on infinity. For example,

$$\lim_{z \to 1} \frac{z - 1}{z^2 + 1} = 0 \quad \text{so} \quad \lim_{z \to 1} \frac{z^2 + 1}{z - 1} = \infty.$$

Also,

$$\lim_{z \to 0} \frac{1 + 4z^2}{5 + iz^2} = \lim_{z \to 0} \frac{\frac{1}{z^2} + 4}{\frac{5}{z^2} + i} = \frac{1}{5}.$$

would imply

$$\lim_{z \to \infty} \frac{z^2 + 4}{5z^2 + i} = \frac{1}{5}.$$

2.2.2 Continuity of a complex function

Continuity of a complex function is defined in the same manner as that of a real function. The complex function $f(z)$ is said to be *continuous* at z_0 if

$$\lim_{z \to z_0} f(z) = f(z_0). \tag{2.2.5}$$

The above statement implicitly implies the existence of both $\lim_{z \to z_0} f(z)$ and $f(z_0)$. Alternatively, the statement can be understood as:

For any $\epsilon > 0$, there exists $\delta > 0$ (usually dependent on ϵ) such that $|f(z) - f(z_0)| < \epsilon$ whenever $|z - z_0| < \delta$.

Since z_0 lies in the domain of definition of $f(z)$, the deleted δ-neighborhood of z_0 in the former definition of the limit of a function is now replaced by the δ-neighborhood of z_0 in this new definition.

A complex function is said to be continuous in a region R if it is continuous at every point in R.

Since continuity of a complex function is defined using the concept of limits, it can be shown similarly that eq. (2.2.5) is equivalent to

$$\lim_{(x,y) \to (x_0, y_0)} u(x, y) = u(x_0, y_0), \tag{2.2.6a}$$

$$\lim_{(x,y) \to (x_0, y_0)} v(x, y) = v(x_0, y_0). \tag{2.2.6b}$$

For example, consider $f(z) = e^z$; its real and imaginary parts are, respectively, $u(x, y) = e^x \cos y$ and $v(x, y) = e^x \sin y$. Since both $u(x, y)$ and $v(x, y)$ are continuous at any point (x_0, y_0) in the finite x-y plane, we conclude that e^z is continuous at any point $z_0 = x_0 + iy_0$ in $\mathbb{C}$.

Theorems on real continuous functions can be extended to complex continuous functions. If two complex functions are continuous at a point, then their sum, difference and product are also continuous at that point; and their quotient is continuous at any point where the denominator is non-zero. For example, since $g(z) = z^2$ is continuous everywhere, we conclude by the above remark that both $z^2 \pm e^z$ and z^2/e^z are continuous in $\mathbb{C}$. Examples of complex continuous functions in $\mathbb{C}$ are polynomials, exponential functions and trigonometric functions.

Another useful result is that a composition of continuous functions is continuous. If $f(z)$ is continuous at z_0 and $g(z)$ is continuous at ξ, and if $\xi = f(z_0)$, then the composite function $g(f(z))$ is continuous at $z = z_0$. Thus, functions like $\sin(z^2)$ and $\cos(z^2)$ are continuous functions in $\mathbb{C}$.

Since continuity of $f(z)$ implies continuity of its real and imaginary parts, the real function

$$|f(z)| = \sqrt{u(x, y)^2 + v(x, y)^2}, \quad f = u + iv \text{ and } z = x + iy,$$

is also continuous. By applying the well-known result on boundedness of a continuous real function in a closed and bounded region, we can deduce a related property on boundedness of the modulus of a continuous complex function. We state without proof the following theorem.

Theorem 2.2.2 *If $f(z)$ is continuous in a closed and bounded region R, then $|f(z)|$ is bounded in the region, that is,*

$$|f(z)| < M, \quad \text{for all } z \in R, \quad (2.2.7)$$

for some constant M. Also, $|f(z)|$ attains its maximum value at some point z_0 in R.

Example 2.2.2 Discuss the continuity of the following complex functions at $z = 0$:

(a) $f(z) = \begin{cases} 0 & z = 0 \\ \dfrac{\text{Re } z}{|z|} & z \neq 0 \end{cases}$;

(b) $f(z) = \dfrac{\text{Im } z}{1 + |z|}$.

Solution

(a) Let $z = x + iy, z \neq 0$. Then

$$\frac{\text{Re } z}{|z|} = \frac{x}{\sqrt{x^2 + y^2}}.$$

Suppose z approaches 0 along the half straight line $y = mx$ $(x > 0)$. Then

$$\lim_{\substack{z \to 0 \\ y = mx, \, x > 0}} \frac{\text{Re } z}{|z|} = \lim_{x \to 0^+} \frac{x}{\sqrt{x^2 + m^2 x^2}} = \lim_{x \to 0^+} \frac{x}{x \sqrt{1 + m^2}} = \frac{1}{\sqrt{1 + m^2}}.$$

Since the limit depends on m, $\lim_{z \to 0} f(z)$ does not exist. Therefore, $f(z)$ cannot be continuous at $z = 0$.

(b) Let $z = x + iy$. Then

$$\frac{\text{Im}\, z}{1 + |z|} = \frac{y}{1 + \sqrt{x^2 + y^2}}.$$

Now, consider the limit

$$\lim_{z \to 0} f(z) = \lim_{(x,y) \to (0,0)} \frac{y}{1 + \sqrt{x^2 + y^2}} = 0 = f(0).$$

Therefore, $f(z)$ is continuous at $z = 0$.

Uniform continuity

Suppose $f(z)$ is continuous in a region R. Then by definition, at each point z_0 inside R and for any $\epsilon > 0$, we can find $\delta > 0$ such that $|f(z) - f(z_0)| < \epsilon$ whenever $|z - z_0| < \delta$. Usually δ depends on ϵ and z_0 together. However, if we can find a single value of δ for each ϵ, independent of z_0 chosen in R, we say that $f(z)$ is *uniformly continuous* in the region R.

Example 2.2.3   Show that

(a) $f_1(z) = z^2$ is uniformly continuous in the region $|z| < R$, where $0 < R < \infty$.
(b) $f_2(z) = \frac{1}{z}$ is not uniformly continuous in the region $0 < |z| < 1$.

Solution

(a) It suffices to show that given any $\epsilon > 0$, we can find $\delta > 0$ such that $|z^2 - z_0^2| < \epsilon$ when $|z - z_0| < \delta$, where δ depends on ϵ but not on the particular point z_0 of the region. If z and z_0 are any two points inside $|z| < R$, then

$$|z^2 - z_0^2| = |z + z_0|\, |z - z_0| \le \{|z| + |z_0|\}|z - z_0| < 2R|z - z_0|. \quad \text{(i)}$$

This relation between $|f_1(z) - f_1(z_0)|$ and $|z - z_0|$ dictates the choice of $\delta = \frac{\epsilon}{2R}$, where δ depends on ϵ but not on z_0. Now, given any $\epsilon > 0$, suppose $|z - z_0| < \delta$. Then by inequality (i), we have

$$|f_1(z) - f_1(z_0)| = |z^2 - z_0^2| < 2R|z - z_0| < 2R\delta = \epsilon.$$

Hence, $f_1(z) = z^2$ is uniformly continuous in $|z| < R$.

(b) For z and z_0 inside $0 < |z| < 1$, we observe that

$$|f_2(z) - f_2(z_0)| = \left| \frac{1}{z} - \frac{1}{z_0} \right| = \frac{|z - z_0|}{|z_0|\, |z|},$$

and $|f_2(z) - f_2(z_0)|$ can be made to be larger than any positive number when z_0 becomes sufficiently close to 0. It is not possible to find δ that depends on ϵ but not z_0 such that for any given ϵ, we have

$$|f_2(z) - f_2(z_0)| < \epsilon$$

for $|z - z_0| < \delta$. Hence, $f_2(z) = \frac{1}{z}$ is not uniformly continuous in $0 < |z| < 1$.

Most of the theorems related to the properties of continuity for real functions can be extended to complex functions. However, this is not quite so when we consider differentiation.

2.3 Differentiation of complex functions

Let the complex function $f(z)$ be single-valued in a neighborhood of a point z_0. The derivative of $f(z)$ at z_0 is defined by

$$\frac{df}{dz}(z_0) = \lim_{z \to z_0} \frac{f(z) - f(z_0)}{z - z_0}$$
$$= \lim_{\Delta z \to 0} \frac{f(z_0 + \Delta z) - f(z_0)}{\Delta z}, \qquad \Delta z = z - z_0, \qquad (2.3.1)$$

provided that the above limit exists. The value of the limit must be independent of the path of z approaching z_0.

Since the derivative of a complex function is defined in a similar manner to that of a real function, many formulas for the computation of derivatives of complex functions are the same as those for the real counterparts. For example, the derivative of a complex polynomial is the same as the real case:

$$\frac{d}{dz} z^n = n z^{n-1}.$$

Standard theorems on differentiation in real calculus also hold in the complex counterpart. If the derivatives of two functions f and g exist at a point z, then

$$\frac{d}{dz}[cf(z)] = cf'(z), \qquad \text{where } c \text{ is a constant,}$$

$$\frac{d}{dz}[f(z) \pm g(z)] = f'(z) \pm g'(z),$$

$$\frac{d}{dz}[f(z)g(z)] = f'(z)g(z) + f(z)g'(z),$$

$$\frac{d}{dz}\left[\frac{f(z)}{g(z)}\right] = \frac{f'(z)g(z) - g'(z)f(z)}{[g(z)]^2}.$$

The chain rule for differentiation of composite functions also holds. Suppose f has a derivative at z_0 and g has a derivative at $f(z_0)$. Then the derivative of $g(f(z))$ at z_0 is given by

$$\frac{d}{dz}g(f(z_0)) = g'(f(z_0))\, f'(z_0).$$

Since we cannot graph a complex function in the usual sense as a real function, it is meaningless to visualize $f'(z_0)$ as the 'slope' of some curve as we do in the real case.

Like real calculus, existence of the derivative of a complex function at a point implies continuity of the function at the same point. Supposing $f'(z_0)$ exists, we consider

$$\lim_{z \to z_0} [f(z) - f(z_0)] = \lim_{z \to z_0} \frac{f(z) - f(z_0)}{z - z_0} \lim_{z \to z_0} (z - z_0) = 0$$

so that

$$\lim_{z \to z_0} f(z) = f(z_0).$$

This shows that $f(z)$ is continuous at z_0. However, continuity of $f(z)$ may not imply the differentiability of $f(z)$ at the same point.

It may occur that a complex function can be differentiable at a given point but not so in any neighborhood of that point (see Example 2.3.1).

Example 2.3.1   Show that the functions $\bar{z}$ and Re z are nowhere differentiable, while $|z|^2$ is differentiable only at $z = 0$.

Solution   According to definition (2.3.1), the derivative of $\bar{z}$ is given by

$$\frac{d}{dz}\bar{z} = \lim_{\Delta z \to 0} \frac{\bar{z} + \overline{\Delta z} - \bar{z}}{\Delta z} = \lim_{\Delta z \to 0} \frac{\overline{\Delta z}}{\Delta z} = \lim_{\Delta z \to 0} e^{-2i\,\text{Arg}\,\Delta z}.$$

The value of the limit depends on the path approaching z. Therefore, $\bar{z}$ is nowhere differentiable. Similarly,

$$\begin{aligned}
\frac{d}{dz}\text{Re}\, z &= \frac{d}{dz} \frac{1}{2}(z + \bar{z}) \\
&= \frac{1}{2} \lim_{\Delta z \to 0} \frac{(z + \bar{z} + \Delta z + \overline{\Delta z}) - (z + \bar{z})}{\Delta z} \\
&= \frac{1}{2} \lim_{\Delta z \to 0} \frac{\Delta z + \overline{\Delta z}}{\Delta z} = \frac{1}{2} + \frac{1}{2} \lim_{\Delta z \to 0} \frac{\overline{\Delta z}}{\Delta z}.
\end{aligned}$$

Again, Re z is shown to be nowhere differentiable. Lastly, the derivative of $|z|^2$ is given by

$$\frac{d}{dz}|z|^2 = \lim_{\Delta z \to 0} \frac{|z + \Delta z|^2 - |z|^2}{\Delta z} = \lim_{\Delta z \to 0} \left[\bar{z} + z\,\frac{\overline{\Delta z}}{\Delta z} + \overline{\Delta z} \right]$$

The above limit exists only when $z = 0$, that is, $|z|^2$ is differentiable only at $z = 0$.

2.3.1 Complex velocity and acceleration

A complex number z can be visualized geometrically as a position vector in the complex plane. Suppose $z(t)$ is considered as a position vector with the running parameter t. The terminal point of the position vector traverses a curve C in the complex plane. Similar to the differentiation of a vector function, we define the derivative of $z(t)$ with respect to t to be

$$\frac{dz}{dt} = \lim_{\Delta t \to 0} \frac{z(t + \Delta t) - z(t)}{\Delta t}.$$

Suppose we separate $z(t)$ into its real and imaginary parts and write $z(t) = x(t) + iy(t)$. Then the derivative of $z(t)$ can be expressed as

$$\frac{dz}{dt} = \frac{dx}{dt} + i\,\frac{dy}{dt}. \tag{2.3.2}$$

The derivative gives the direction of the tangent vector to the curve at t. If the parameter t is considered as the time variable, then $\frac{dz}{dt}$ represents the velocity with which the terminal point moves along the curve. Also, the second-order derivative $\frac{d^2z}{dt^2}$ gives the acceleration of the motion along the curve.

Example 2.3.2 Suppose the motion of a particle is described using the polar coordinates (r, θ) and its position in the complex plane is represented by

$$z(t) = r(t)e^{i\theta(t)}.$$

By differentiating $z(t)$ with respect to the time variable t, find the velocity and acceleration of the particle, separating them into their radial and tangential components.

Solution Starting with $z = re^{i\theta}$, where z, r and θ are all functions of t, we differentiate z with respect to the time variable t and obtain

$$u = \dot{z} = \dot{r}e^{i\theta} + ire^{i\theta}\dot{\theta}.$$

Here, u is called the *complex velocity* and the dot over a variable denotes differentiation of the variable with respect to t. Also, $e^{i\theta}$ and $ie^{i\theta}$ represent the unit vector in the radial direction and tangential direction, respectively. The radial component of velocity u_r and the tangential component of velocity u_θ are then given by

$$u_r = \dot{r} \qquad \text{and} \qquad u_\theta = r\dot{\theta}.$$

The complex velocity may be written as

$$u = (u_r + iu_\theta)e^{i\theta}.$$

The *complex acceleration* can be found by differentiating u again with respect to t. We obtain

$$a = \frac{du}{dt} = (\ddot{r} - r\dot{\theta}^2)e^{i\theta} + (2\dot{r}\dot{\theta} + r\ddot{\theta})ie^{i\theta}.$$

The radial component of acceleration a_r and the tangential component of acceleration a_θ are then given by

$$a_r = \ddot{r} - r\dot{\theta}^2 \qquad \text{and} \qquad a_\theta = 2\dot{r}\dot{\theta} + r\ddot{\theta}.$$

2.4 Cauchy–Riemann relations

This section discusses the necessary and sufficient conditions for the existence of the derivative of a complex function $f(z) = u(x, y) + iv(x, y)$, $z = x + iy$. The necessary conditions are given by the Cauchy–Riemann relations. The sufficient conditions require, additionally, the continuity of all first-order partial derivatives of u and v.

Let $f(z)$ be defined in a neighborhood of the point $z_0 = x_0 + iy_0$, and suppose it is differentiable at z_0, that is, the limit

$$f'(z_0) = \lim_{\Delta z \to 0} \frac{f(z_0 + \Delta z) - f(z_0)}{\Delta z}$$

exists. This limit is independent of the direction along which Δz approaches 0.

(i) First, we take $\Delta z \to 0$ in the direction parallel to the x-axis, that is, $\Delta z = \Delta x$. We then have

$$f(z_0 + \Delta z) - f(z_0) = u(x_0 + \Delta x, y_0) + iv(x_0 + \Delta x, y_0)$$
$$- u(x_0, y_0) - iv(x_0, y_0),$$

so that

$$f'(z_0) = \lim_{\Delta x \to 0} \frac{u(x_0 + \Delta x, y_0) - u(x_0, y_0)}{\Delta x}$$
$$+ i \lim_{\Delta x \to 0} \frac{v(x_0 + \Delta x, y_0) - v(x_0, y_0)}{\Delta x}$$
$$= \frac{\partial u}{\partial x}(x_0, y_0) + i \frac{\partial v}{\partial x}(x_0, y_0).$$

(ii) Next, we let $\Delta z \to 0$ in the direction parallel to the y-axis, that is, $\Delta z = i \Delta y$. Now, we have

$$f(z_0 + \Delta z) - f(z_0) = u(x_0, y_0 + \Delta y) + i v(x_0, y_0 + \Delta y)$$
$$- u(x_0, y_0) - i v(x_0, y_0),$$

so that

$$f'(z_0) = \lim_{\Delta y \to 0} \frac{u(x_0, y_0 + \Delta y) - u(x_0, y_0)}{i \Delta y}$$
$$+ i \lim_{\Delta y \to 0} \frac{v(x_0, y_0 + \Delta y) - v(x_0, y_0)}{i \Delta y}$$
$$= \frac{1}{i} \frac{\partial u}{\partial y}(x_0, y_0) + \frac{\partial v}{\partial y}(x_0, y_0).$$

Combining the above two equations, we obtain

$$f' = \frac{\partial u}{\partial x} + i \frac{\partial v}{\partial x} = \frac{\partial v}{\partial y} - i \frac{\partial u}{\partial y}. \tag{2.4.1}$$

Equating the respective real and imaginary parts gives

$$\frac{\partial u}{\partial x} = \frac{\partial v}{\partial y} \quad \text{and} \quad \frac{\partial v}{\partial x} = -\frac{\partial u}{\partial y}. \tag{2.4.2}$$

The results in eq. (2.4.2) are called the *Cauchy–Riemann relations*. They give the necessary conditions for the existence of the derivative of a complex function. The above results are summarized in the following theorem.

Theorem 2.4.1 *Suppose* $f(z) = u(x, y) + i v(x, y)$ *is differentiable at a point* $z_0 = x_0 + i y_0$. *Then at that point*

$$f'(z_0) = u_x(x_0, y_0) + i v_x(x_0, y_0) = v_y(x_0, y_0) - i u_y(x_0, y_0).$$

Accordingly, we have

$$u_x(x_0, y_0) = v_y(x_0, y_0) \quad \text{and} \quad v_x(x_0, y_0) = -u_y(x_0, y_0).$$

Readers should be wary that the satisfaction of the Cauchy–Riemann relations may not guarantee the existence of f' at the point (see Example 2.4.2). However, a partial converse can be salvaged by adding some extra assumptions.

Theorem 2.4.2 *Given* $f(z) = u(x, y) + iv(x, y)$, $z = x + iy$, *and assume that*

   (i) *the Cauchy–Riemann relations hold at a point* $z_0 = x_0 + iy_0$,
  (ii) u_x, u_y, v_x, v_y *are all continuous at the point* (x_0, y_0).

The derivative $f'(z_0)$ *then exists and it is given by*

$$f'(z_0) = u_x(x_0, y_0) + iv_x(x_0, y_0) = v_y(x_0, y_0) - iu_y(x_0, y_0).$$

Proof Since $u(x, y)$ and $v(x, y)$ have continuous first-order partial derivatives at (x_0, y_0) and satisfy the Cauchy–Riemann relations at the same point, we have

$$u(x, y) - u(x_0, y_0)$$
$$= u_x(x_0, y_0)(x - x_0) + u_y(x_0, y_0)(y - y_0) + \epsilon_1(|\Delta z|)$$
$$= u_x(x_0, y_0)(x - x_0) - v_x(x_0, y_0)(y - y_0) + \epsilon_1(|\Delta z|), \qquad \text{(i)}$$

and

$$v(x, y) - v(x_0, y_0)$$
$$= v_x(x_0, y_0)(x - x_0) + v_y(x_0, y_0)(y - y_0) + \epsilon_2(|\Delta z|)$$
$$= v_x(x_0, y_0)(x - x_0) + u_x(x_0, y_0)(y - y_0) + \epsilon_2(|\Delta z|), \qquad \text{(ii)}$$

where ϵ_1 and ϵ_2 satisfy

$$\lim_{|\Delta z| \to 0} \frac{\epsilon_1(|\Delta z|)}{|\Delta z|} = \lim_{|\Delta z| \to 0} \frac{\epsilon_2(|\Delta z|)}{|\Delta z|} = 0, \quad |\Delta z| = \sqrt{(x - x_0)^2 + (y - y_0)^2}.$$
$$\text{(iii)}$$

Adding eq. (i) and i times eq. (ii) together, we obtain

$$f(z) - f(z_0) = [u_x(x_0, y_0) + iv_x(x_0, y_0)](z - z_0) + \epsilon_1(|\Delta z|) + i\epsilon_2(|\Delta z|),$$

and subsequently,

$$\frac{f(z) - f(z_0)}{z - z_0} - [u_x(x_0, y_0) + iv_x(x_0, y_0)] = \frac{\epsilon_1(|\Delta z|) + i\epsilon_2(|\Delta z|)}{z - z_0}.$$

Note that

$$\left| \frac{\epsilon_1(|\Delta z|) + i\epsilon_2(|\Delta z|)}{z - z_0} \right| \leq \frac{\epsilon_1(|\Delta z|)}{|\Delta z|} + \frac{\epsilon_2(|\Delta z|)}{|\Delta z|};$$

and as $\Delta z \to 0$, so does $|\Delta z| \to 0$. It then follows from the results in eq. (iii)
that

$$\lim_{\Delta z \to 0} \frac{\epsilon_1(|\Delta z|) + i\epsilon_2(|\Delta z|)}{z - z_0} = 0.$$

Hence, we obtain

$$f'(z_0) = \lim_{z \to z_0} \frac{f(z) - f(z_0)}{z - z_0}$$

$$= u_x(x_0, y_0) + iv_x(x_0, y_0) = v_y(x_0, y_0) - iu_y(x_0, y_0).$$

Example 2.4.1 Express the Cauchy–Riemann relations in polar coordinates.

Solution Consider the polar coordinates $r^2 = x^2 + y^2$ and $\theta = \tan^{-1} \frac{y}{x}$. Differentiating r and θ with respect to both x and y, we obtain

$$\frac{\partial r}{\partial x} = \frac{x}{r} = \cos\theta, \qquad \frac{\partial r}{\partial y} = \frac{y}{r} = \sin\theta,$$

and

$$\frac{\partial\theta}{\partial x} = -\frac{y}{x^2 + y^2} = -\frac{1}{r}\sin\theta, \qquad \frac{\partial\theta}{\partial y} = \frac{x}{x^2 + y^2} = \frac{1}{r}\cos\theta.$$

Using the chain rule, the first-order partial derivatives of u are given by

$$\frac{\partial u}{\partial x} = \frac{\partial u}{\partial r}\frac{\partial r}{\partial x} + \frac{\partial u}{\partial\theta}\frac{\partial\theta}{\partial x} = \frac{\partial u}{\partial r}\cos\theta - \frac{1}{r}\frac{\partial u}{\partial\theta}\sin\theta, \qquad \text{(i)}$$

$$\frac{\partial u}{\partial y} = \frac{\partial u}{\partial r}\frac{\partial r}{\partial y} + \frac{\partial u}{\partial\theta}\frac{\partial\theta}{\partial y} = \frac{\partial u}{\partial r}\sin\theta + \frac{1}{r}\frac{\partial u}{\partial\theta}\cos\theta. \qquad \text{(ii)}$$

Similarly, the first-order partial derivatives of v are given by

$$\frac{\partial v}{\partial x} = \frac{\partial v}{\partial r}\cos\theta - \frac{1}{r}\frac{\partial v}{\partial\theta}\sin\theta, \qquad \text{(iii)}$$

$$\frac{\partial v}{\partial y} = \frac{\partial v}{\partial r}\sin\theta + \frac{1}{r}\frac{\partial v}{\partial\theta}\cos\theta. \qquad \text{(iv)}$$

Using one of the Cauchy–Riemann relations, we combine eqs. (i) and (iv) to give

$$\frac{\partial u}{\partial x} - \frac{\partial v}{\partial y}$$

$$= \left(\frac{\partial u}{\partial r} - \frac{1}{r}\frac{\partial v}{\partial\theta}\right)\cos\theta - \left(\frac{\partial v}{\partial r} + \frac{1}{r}\frac{\partial u}{\partial\theta}\right)\sin\theta = 0. \qquad \text{(v)}$$

Similarly, using eqs. (ii) and (iii) and applying the other Cauchy–Riemann relation, we have

$$\frac{\partial u}{\partial y} + \frac{\partial v}{\partial x}$$

$$= \left(\frac{\partial u}{\partial r} - \frac{1}{r}\frac{\partial v}{\partial \theta}\right)\sin\theta + \left(\frac{\partial v}{\partial r} + \frac{1}{r}\frac{\partial u}{\partial \theta}\right)\cos\theta = 0. \qquad \text{(vi)}$$

In order that eqs. (v) and (vi) are satisfied for all θ, we must have

$$\frac{\partial u}{\partial r} = \frac{1}{r}\frac{\partial v}{\partial \theta} \qquad \text{and} \qquad \frac{\partial v}{\partial r} = -\frac{1}{r}\frac{\partial u}{\partial \theta}.$$

These are the Cauchy–Riemann relations expressed in polar coordinates.

Example 2.4.2 Discuss the differentiability of the function

$$f(z) = f(x + iy) = \sqrt{|xy|}$$

at $z = 0$.

Solution Write $f(x + iy) = u(x, y) + iv(x, y)$ so that

$$u(x, y) = \sqrt{|xy|} \quad \text{and} \quad v(x, y) = 0.$$

Since $u(x, 0)$ and $u(0, y)$ are identically equal to zero, we have

$$u_x(0, 0) = u_y(0, 0) = 0.$$

Also, since $v(x, y)$ is identically zero, it is obvious that

$$v_x(0, 0) = v_y(0, 0) = 0.$$

Hence, the Cauchy–Riemann relations are satisfied at the point $(0, 0)$.

However, suppose z approaches the origin along the ray $x = \alpha t$, $y = \beta t$, $t > 0$, assuming that α and β cannot be zero simultaneously. For $z = \alpha t + i\beta t$, we then have

$$\frac{f(z) - f(0)}{z - 0} = \frac{f(z)}{z} = \frac{\sqrt{|\alpha\beta|}}{\alpha + i\beta}.$$

The limit of the above quantity as $z \to 0$ depends on the values of α and β, so the limit is non-unique. Therefore, $f(z)$ is not differentiable at $z = 0$, though the Cauchy–Riemann relations are satisfied at $z = 0$.

Let us check the continuity of u_x at $(0, 0)$. Since

$$\frac{\partial u}{\partial x} = \sqrt{|y|}\frac{d}{dx}\sqrt{|x|},$$

u_r fails to be continuous at $(0, 0)$. By virtue of Theorem 2.4.2, it is not surprising that $f(x) = \sqrt{|x\,y|}$ can fail to be differentiable at $z = 0$ since the Cauchy–Riemann relations are necessary but not sufficient for differentiability.

2.4.1 Conjugate complex variables

Let $z = x + iy$; its complex conjugate is $\overline{z} = x - iy$. Formally, we may treat the pair of conjugate complex variables, z and $\overline{z}$, as two independent variables. They are related to the real variables x and y by

$$x = \frac{z + \overline{z}}{2}, \quad y = \frac{z - \overline{z}}{2i}. \tag{2.4.3}$$

Take the example: $f(x, y) = x + ixy$, the function can be expressed in terms of z and $\overline{z}$ as

$$f(x, y) \equiv f(z, \overline{z}) = \frac{z + \overline{z}}{2} + \frac{z^2 - \overline{z}^2}{4}.$$

Applying the transformation rules in calculus, we have

$$\frac{\partial}{\partial x} = \frac{\partial}{\partial z} + \frac{\partial}{\partial \overline{z}} \quad \text{and} \quad \frac{\partial}{\partial y} = i \left(\frac{\partial}{\partial z} - \frac{\partial}{\partial \overline{z}} \right). \tag{2.4.4}$$

Here, $\frac{\partial}{\partial z}$ and $\frac{\partial}{\partial \overline{z}}$ are visualized as the symbolic derivatives with respect to z and $\overline{z}$, respectively. Suppose we write

$$f(x, y) = u(x, y) + iv(x, y);$$

the Cauchy–Riemann relations can be expressed as

$$\frac{\partial f}{\partial x} = \frac{\partial u}{\partial x} + i\frac{\partial v}{\partial x} = \frac{\partial v}{\partial y} - i\frac{\partial u}{\partial y} = -i\frac{\partial f}{\partial y}.$$

Using the relations between the differential operators given in eq. (2.4.4), the above equation can be reduced to

$$\frac{\partial f}{\partial \overline{z}} = 0. \tag{2.4.5}$$

This is an elegant representation of the Cauchy–Riemann relations.

As an application of the result, one can conclude by a simple calculation that the complex function

$$f(z) = \text{Re } z = \frac{z + \overline{z}}{2}$$

is nowhere differentiable since

$$\frac{\partial}{\partial \overline{z}} \text{Re } z = \frac{1}{2} \neq 0.$$

As another example, recall that the complex function

$$f(z) = |z|^2 = z\overline{z}$$

is differentiable only at $z = 0$. This is because it satisfies eq. (2.4.5) only at $z = 0$ as

$$\frac{\partial}{\partial \overline{z}} z\overline{z} = z$$

equals zero only at $z = 0$.

2.5 Analyticity

Analyticity is one of the central concepts in the calculus of complex variables. It may be visualized as some form of an extended notion of differentiability. Analyticity is a property defined over an open set while differentiability is confined to one single point. Let us start with the definition of a function that is analytic at a point.

Definition 2.5.1 A function $f(z)$ is said to be analytic at some point z_0 if it is differentiable at every point inside a certain neighborhood of z_0. In other words, $f(z)$ is analytic at z_0 if and only if there exists a neighborhood $N(z_0; \epsilon)$, $\epsilon > 0$, such that $f'(z)$ exists for all $z \in N(z_0; \epsilon)$.

Remark   Some authors use the terms *holomorphic* or *regular* as synonyms for *analytic*.

Since $z_0 \in N(z_0; \epsilon)$, analyticity at z_0 implies differentiability at z_0. The converse statement is not true, that is, differentiability of $f(z)$ at z_0 does not guarantee the analyticity of $f(z)$ at z_0. For example, the function $f(z) = |z|^2$ is nowhere differentiable except at the origin, hence $f(z) = |z|^2$ is not analytic at $z = 0$. A point at which f is not analytic is called a *singular point* (or singularity) of f.

Analyticity in a domain

We say that a function is analytic in a region R if it is analytic at every point in R. We would like to show that the region of analyticity must be open. Recall that an open region is called a domain, so analyticity of a function is defined in a domain. To prove the claim, consider a function that is analytic at some point z_0 inside R. According to the requirement of analyticity at a point, there exists a neighborhood around z_0 that lies completely inside the region. This would imply implicitly that z_0 must be an interior point of the region. The region of analyticity contains only interior points, so analytic functions are defined in domains only.

Suppose a function f is said to be analytic in a closed region, say $|z| \le r$. Then it is implicit that the function is analytic in some domain $\mathcal{D}$ containing the closed region. If a complex function is analytic in the entire complex plane, then the function is called an *entire function*. Examples of entire functions are polynomials, exponential functions and trigonometric functions.

It is relatively straightforward to show that the existence of f' for all points inside a domain implies analyticity of f in the domain. Since any point in the domain of analyticity of a complex function is an interior point, there exists a neighborhood of that point that is contained completely inside the domain of analyticity. This implies implicitly that f' exists for all points inside the neighborhood, so f is analytic at that point. In other words, in order to show that $f(z)$ is analytic in a domain $\mathcal{D}$, it suffices to show either

(i) $f'(z)$ exists for all z in $\mathcal{D}$, or
(ii) the real and imaginary parts of $f(z)$ have continuous first-order partial derivatives, and these derivatives satisfy the Cauchy–Riemann relations at every point inside $\mathcal{D}$.

Suppose two complex functions are analytic in some domain $\mathcal{D}$. Then their sum, difference and product are all analytic in $\mathcal{D}$. Also, their quotient is analytic in $\mathcal{D}$ given that the denominator function is non-zero at all points in $\mathcal{D}$. The composition of two analytic functions is also analytic. Suppose $f_1(z)$ is analytic at z_0 while $f_2(z)$ is non-analytic at z_0. Is the sum $f_1(z) + f_2(z)$ analytic at z_0? Let us assume

$$S(z) = f_1(z) + f_2(z)$$

to be analytic at z_0. Then $S(z) - f_1(z)$ is analytic at z_0 since the difference of two functions analytic at z_0 remains analytic at that point. This leads to a contradiction since

$$S(z) - f_1(z) = f_2(z)$$

is non-analytic at z_0. Hence, $f_1(z) + f_2(z)$ fails to be analytic at z_0. However, given two functions that are non-analytic at z_0, it may be possible that their sum becomes analytic at z_0. For example, $f_1(z) = \sin z + \frac{1}{z}$ and $f_2(z) = e^z - \frac{1}{z}$ are both non-analytic at $z = 0$. However, their sum

$$f_1(z) + f_2(z) = \sin z + e^z$$

is an entire function.

Can we find a non-constant analytic function $f(z)$ such that both $f(z)$ and $\overline{f(z)}$ are analytic in a domain $\mathcal{D}$? This is not possible since we cannot find a non-constant $f(z)$ such that the Cauchy–Riemann relations are satisfied for both $f(z)$ and $\overline{f(z)}$ at every point in $\mathcal{D}$. To prove the claim, we write

$$f(z) = u(x, y) + i v(x, y), \quad z = x + iy,$$
$$\overline{f(z)} = u(x, y) - i v(x, y).$$

Their respective pairs of Cauchy-Riemann relations are given by

$$u_x = v_y, \quad u_y = -v_x;$$

and

$$u_x = -v_y, \quad u_y = v_x.$$

Combining the above relations, we obtain

$$u_x = u_y = v_x = v_y = 0$$

so that

$$f'(z) = u_x(x, y) + i v_x(x, y) = 0$$

in $\mathcal{D}$. Observing the connectedness property of the domain $\mathcal{D}$, we deduce that $f(z)$ is constant throughout $\mathcal{D}$.

Example 2.5.1   Find the domains in which the function

$$f(z) = |x^2 - y^2| + 2i|xy|, \quad z = x + iy,$$

is analytic.

Solution   The functional values of $f(z)$ depend on the signs of $x^2 - y^2$ and xy. When $x^2 - y^2 > 0$ and $xy > 0$, $f(z) = z^2$; also when $x^2 - y^2 < 0$ and $xy < 0$, $f(z) = -z^2$. Both functions are known to be analytic. However, when $x^2 - y^2 > 0$ and $xy < 0$, $f(z) = x^2 - y^2 - 2ixy$, the Cauchy–Riemann relations are not satisfied, and so $f(z)$ fails to be analytic. Also, for $x^2 - y^2 < 0$ and $xy > 0$, the function becomes $f(z) = -(x^2 - y^2) + 2ixy$, which is non-analytic.

Thus, the domains of analyticity of $f(z)$ occupy alternative sectors, each subtending an angle of $\frac{\pi}{4}$ in the complex plane. Also, any neighborhood drawn around any point on the boundary rays: $x = \pm y$, $x = 0$ and $y = 0$ must overlap with some region where $f(z)$ is not differentiable. Hence, $f(z)$ is not analytic along these rays.

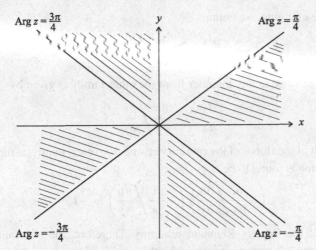

Figure 2.5. Domains of analyticity (shown in shaded areas) of $f(z) = |x^2 - y^2| + 2i|xy|$.

Summary

The function is analytic within the following domains (see Figure 2.5):

$$0 < \text{Arg } z < \tfrac{\pi}{4}, \qquad \tfrac{\pi}{2} < \text{Arg } z < \tfrac{3\pi}{4},$$

$$-\pi < \text{Arg } z < -\tfrac{3\pi}{4} \quad \text{and} \quad -\tfrac{\pi}{2} < \text{Arg } z < -\tfrac{\pi}{4}.$$

Example 2.5.2   Given an analytic function

$$w = f(z) = u(x, y) + iv(x, y), \quad z = x + iy,$$

the equations

$$u(x, y) = \alpha \quad \text{and} \quad v(x, y) = \beta, \quad \alpha \text{ and } \beta \text{ are constants,}$$

define two families of curves in the complex plane. Show that the two families are mutually orthogonal to each other.

Solution   Consider a particular member from the first family

$$u(x, y) = \alpha_1, \quad \text{for some constant } \alpha_1.$$

The slope of the curve, given by $\frac{dy}{dx}$, can be found by differentiating the above equation with respect to x. This gives

$$\frac{\partial u}{\partial x} + \frac{\partial u}{\partial y} \frac{dy}{dx} = 0,$$

and upon rearranging, we obtain

$$\frac{dy}{dx} = -\frac{\partial u}{\partial x} \bigg/ \frac{\partial u}{\partial y}.$$

Similarly, the slope of any member from the other family is given by

$$\frac{dy}{dx} = -\frac{\partial v}{\partial x} \bigg/ \frac{\partial v}{\partial y}.$$

The product of the slopes of the two curves, one from each family, at their point of intersection is found to be

$$\left(-\frac{\partial u}{\partial x} \bigg/ \frac{\partial u}{\partial y}\right)\left(-\frac{\partial v}{\partial x} \bigg/ \frac{\partial v}{\partial y}\right) = -1,$$

by virtue of the Cauchy–Riemann relations. Therefore, the two families of curves are mutually orthogonal to each other.

Remark   The image curves of these two families of curves are mutually orthogonal to each other in the w-plane since they are simply the horizontal lines and vertical lines in the w-plane. Orthogonal families of curves are mapped onto orthogonal families of image curves. This preservation of the orthogonality property under mapping stems from analyticity of the complex function $f(z)$. This interesting geometric property is closely related to the *conformal property* of an analytic function (see Section 8.1).

2.6 Harmonic functions

A real-valued function $\phi(x, y)$ of two real variables x and y is said to be *harmonic* in a given domain $\mathcal{D}$ in the x-y plane if ϕ has continuous partial derivatives up to the second order in $\mathcal{D}$ and satisfies the *Laplace equation*

$$\phi_{xx}(x, y) + \phi_{yy}(x, y) = 0. \tag{2.6.1}$$

Interestingly, analytic functions are closely related to harmonic functions. Suppose

$$f(z) = u(x, y) + iv(x, y), \quad z = x + iy,$$

is analytic in $\mathcal{D}$; we will show that both the component functions $u(x, y)$ and $v(x, y)$ are harmonic in $\mathcal{D}$. We need one result to prove the above claim. It will be shown in Section 4.3 that if a complex function is analytic at a point, then its real and imaginary parts have continuous partial derivatives of all orders at that point (see Remark (ii) in Theorem 4.3.2).

Suppose f is analytic in $\mathcal{D}$. Then the Cauchy–Riemann relations hold throughout $\mathcal{D}$; that is

$$u_x = v_y \quad \text{and} \quad u_y = -v_x,$$

The analyticity of $f(z)$ in $\mathcal{D}$ dictates the continuity of partial derivatives of $u(x, y)$ and $v(x, y)$ of all orders in $\mathcal{D}$. Differentiating both sides of the Cauchy–Riemann relations with respect to x, we obtain

$$u_{xx} = v_{yx} \quad \text{and} \quad v_{xx} = -u_{yx}.$$

Similarly, differentiation of the Cauchy–Riemann relations with respect to y gives

$$u_{xy} = v_{yy} \quad \text{and} \quad v_{xy} = -u_{yy}.$$

Since the above partial derivatives are all continuous, we have

$$u_{xy} = u_{yx} \quad \text{and} \quad v_{xy} = v_{yx}.$$

Combining the above relations, we obtain

$$v_{yy} = u_{xy} = u_{yx} = -v_{xx} \quad \text{so} \quad v_{xx} + v_{yy} = 0; \qquad (2.6.2a)$$

and similarly,

$$-u_{yy} = v_{xy} = v_{yx} = u_{xx} \quad \text{so} \quad u_{xx} + u_{yy} = 0. \qquad (2.6.2b)$$

Therefore, both $u(x, y)$ and $v(x, y)$ are harmonic in $\mathcal{D}$.

2.6.1 Harmonic conjugate

Given two harmonic functions $\phi(x, y)$ and $\psi(x, y)$ that satisfy the Cauchy–Riemann relations throughout a domain $\mathcal{D}$, with

$$\phi_x = \psi_y \quad \text{and} \quad \phi_y = -\psi_x, \qquad (2.6.3)$$

we call ψ a *harmonic conjugate* of ϕ in $\mathcal{D}$.

There is a close link between analyticity and harmonic conjugacy. Suppose $f = u + iv$ is analytic in $\mathcal{D}$. We have shown that u and v are harmonic in $\mathcal{D}$. Also, from Theorem 2.4.1, u and v satisfy the Cauchy–Riemann relations. Therefore, v is a harmonic conjugate of u in $\mathcal{D}$. How about the validity of the converse statement? If v is a harmonic conjugate of u in $\mathcal{D}$, is the complex function $f = u + iv$ analytic in $\mathcal{D}$? The proof of the converse statement follows directly from Theorem 2.4.2. These results can be succinctly summarized by the following theorem.

Theorem 2.6.1 *A complex function* $f(z) = u(x, y) + i v(x, y)$, $z = x + iy$, *is analytic in a domain* $\mathcal{D}$ *if and only if v is a harmonic conjugate of u in* $\mathcal{D}$.

Note that harmonic conjugacy is not a symmetric relation because of the minus sign in the second Cauchy–Riemann relation. In fact, while ψ is a harmonic conjugate of ϕ, $-\phi$ is a harmonic conjugate of ψ.

Given that $\phi(x, y)$ is harmonic in a *simply connected domain* $\mathcal{D}$, it can be shown that it is always possible to obtain its harmonic conjugate $\psi(x, y)$ by integration. Starting from the differential form

$$d\psi = \psi_x \, dx + \psi_y \, dy,$$

and using the Cauchy–Riemann relations, we have

$$d\psi = -\phi_y \, dx + \phi_x \, dy.$$

The right-hand side of the above equation contains known terms ϕ_x and ϕ_y since $\phi(x, y)$ has been given. To obtain ψ, we integrate along some path Γ joining a fixed point (x_0, y_0) to (x, y), that is,

$$\psi(x, y) = \int_\Gamma -\phi_y \, dx + \phi_x \, dy. \tag{2.6.4}$$

The above line integral is an exact differential provided that

$$-(-\phi_y)_y + (\phi_x)_x = 0. \tag{2.6.5}$$

The required condition is satisfied since it is simply the harmonicity property of ϕ. Further, since $\mathcal{D}$ is a simply connected domain, the value of the above line integral is independent of the path Γ chosen. To simplify the computation, we choose the path that consists of horizontal and vertical line segments as shown in Figure 2.6. This gives

$$\psi(x, y) = \int_{x_0}^{x} -\phi_y(x, y_0) \, dx + \int_{y_0}^{y} \phi_x(x, y) \, dy. \tag{2.6.6}$$

The choice of a different starting point (x_0, y_0) of the integration path simply leads to a different additive constant in $\psi(x, y)$. Indeed, a harmonic conjugate is unique up to an additive constant since the governing equations for $\psi(x, y)$ involve derivatives of $\psi(x, y)$ only [see eq. (2.6.3)].

Example 2.6.1 Let $w = \phi + i\psi$ be an analytic function so that both ϕ and ψ are harmonic. Suppose

$$\nabla^2 \phi = \frac{\partial^2 \phi}{\partial x^2} + \frac{\partial^2 \phi}{\partial y^2} = 0 \text{ in } \mathcal{D}, \text{ and } \phi = g \text{ on } \partial\mathcal{D},$$

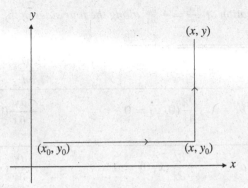

Figure 2.6. To ease the computation in performing the line integration, the path of integration is chosen to be composed of horizontal and vertical line segments.

where $\partial \mathcal{D}$ denotes the boundary of $\mathcal{D}$. Show that the boundary condition for ψ is given by

$$\frac{\partial \psi}{\partial n} = -\frac{\partial \phi}{\partial s} \quad \text{on } \partial \mathcal{D},$$

where $\frac{\partial}{\partial s}$ and $\frac{\partial}{\partial n}$ denote the operators that perform differentiation with respect to arc length s and outward normal n, respectively. Verify the result for the harmonic function

$$\phi(x, y) = \frac{\sin x}{\sin a} \frac{\sinh y}{\sinh b}$$

in the rectangle $0 \leq x \leq a$ and $0 \leq y \leq b$.

Solution From the Cauchy–Riemann relations, we have

$$\frac{\partial \phi}{\partial x} = \frac{\partial \psi}{\partial y} \quad \text{and} \quad \frac{\partial \phi}{\partial y} = -\frac{\partial \psi}{\partial x}.$$

On the boundary $\partial \mathcal{D}$, the tangential derivative of ϕ and the normal derivative of ψ are related by

$$-\frac{\partial \phi}{\partial s} = -\left(\ell \frac{\partial \phi}{\partial x} + m \frac{\partial \phi}{\partial y} \right) = m \frac{\partial \psi}{\partial x} - \ell \frac{\partial \psi}{\partial y} = \frac{\partial \psi}{\partial n},$$

where ℓ and m are the direction cosines of the unit tangent vector $\mathbf{t}$.

Table 2.1. *Verification of* $\frac{\partial \psi}{\partial n} = -\frac{\partial \phi}{\partial s}$ *along the four sides of the rectangular domain.*

$\phi = g$ on $\partial \mathcal{D}$	$-\dfrac{\partial g}{\partial s}$	$\dfrac{\partial \psi}{\partial n}$
$x = 0,\ 0 \le y \le b,$ $g(0, y) = 0$	$\dfrac{\partial g}{\partial y}(0, y) = 0$	$-\dfrac{\partial \psi}{\partial x}(0, y) = 0$
$y = 0,\ 0 \le x \le a,$ $g(x, 0) = 0$	$-\dfrac{\partial g}{\partial x}(x, 0) = 0$	$-\dfrac{\partial \psi}{\partial y}(x, 0) = 0$
$x = a,\ 0 \le y \le b,$ $g(a, y) = \dfrac{\sinh y}{\sinh b}$	$-\dfrac{\partial g}{\partial y}(a, y) = -\dfrac{\cosh y}{\sinh b}$	$\dfrac{\partial \psi}{\partial x}(a, y) = -\dfrac{\cosh y}{\sinh b}$
$y = b,\ 0 \le x \le a,$ $g(x, b) = \dfrac{\sin x}{\sin a}$	$\dfrac{\partial g}{\partial x}(x, b) = \dfrac{\cos x}{\sin a}$	$\dfrac{\partial \psi}{\partial y}(x, b) = \dfrac{\cos x}{\sin a}$

The harmonic conjugate to $\phi(x, y)$ is given by (unique up to an additive constant)

$$
\begin{aligned}
\psi(x, y) &= \int_{(0,0)}^{(x,y)} -\phi_y \, dx + \phi_x \, dy \\
&= \int_0^x -\phi_y(x, 0) \, dx + \int_0^y \phi_x(x, y) \, dy \\
&= \frac{1}{\sin a \sinh b} \left[\int_0^x -\sin x \, dx + \int_0^y \cos x \, \sinh y \, dy \right] \\
&= \frac{\cos x \, \cosh y - 1}{\sin a \, \sinh b}.
\end{aligned}
$$

The verification of $\frac{\partial \psi}{\partial n} = -\frac{\partial \phi}{\partial s}$ on the four sides of the rectangle is revealed in Table 2.1.

Example 2.6.2   Given the harmonic function

$$u(x, y) = e^{-x} \cos y + xy,$$

find the family of curves that is orthogonal to the family

$$u(x, y) = \alpha, \quad \alpha \text{ is constant.}$$

Solution First, it is observed that $u(x, y)$ is harmonic in the whole complex plane. Let $v(x, y)$ denote the harmonic conjugate of $u(x, y)$ (unique up to an additive constant). From the result in Example 2.5.2, the required orthogonal family of curves is given by

$$v(x, y) = \beta, \quad \beta \text{ is constant.}$$

Besides the line integration method given in eq. (2.6.6), we would like to illustrate three other methods of finding the harmonic conjugate.

Method One

From the first Cauchy–Riemann relation, we have

$$\frac{\partial v}{\partial y} = \frac{\partial u}{\partial x} = -e^{-x} \cos y + y.$$

Integrating with respect to y, we obtain

$$v(x, y) = -e^{-x} \sin y + \frac{y^2}{2} + \eta(x),$$

where $\eta(x)$ is an arbitrary function arising from integration. Using the second Cauchy–Riemann relation, we have

$$\frac{\partial v}{\partial x} = e^{-x} \sin y + \eta'(x) = -\frac{\partial u}{\partial y} = e^{-x} \sin y - x.$$

Comparing like terms, we obtain

$$\eta'(x) = -x,$$

and subsequently,

$$\eta(x) = -\frac{x^2}{2} + C, \text{ where } C \text{ is an arbitrary constant.}$$

Hence, a harmonic conjugate is found to be (taking C to be zero for convenience)

$$v(x, y) = -e^{-x} \sin y + \frac{y^2 - x^2}{2}.$$

The corresponding analytic function, $f = u + iv$, is seen to be

$$f(z) = e^{-z} - \frac{iz^2}{2}, \quad z = x + iy,$$

which is an entire function.

Method Two

It is readily seen that

$$e^{-x}\cos y = \operatorname{Re} e^{-z} \quad \text{and} \quad xy = \frac{1}{2}\operatorname{Im} z^2.$$

A harmonic conjugate of $\operatorname{Re} e^{-z}$ is $\operatorname{Im} e^{-z}$, while that of $\frac{1}{2}\operatorname{Im} z^2$ is $-\frac{1}{2}\operatorname{Re} z^2$. Therefore, a harmonic conjugate of $u(x, y)$ can be taken to be

$$v(x, y) = \operatorname{Im} e^{-z} - \frac{1}{2}\operatorname{Re} z^2 = -e^{-x}\sin y + \frac{y^2 - x^2}{2}.$$

Method Three

It is known that

$$f'(z) = \frac{\partial u}{\partial x}(x, y) - i\frac{\partial u}{\partial y}(x, y), \quad z = x + iy.$$

Putting $y = 0$, we obtain

$$f'(x) = \frac{\partial u}{\partial x}(x, 0) - i\frac{\partial u}{\partial y}(x, 0).$$

By replacing x by z formally, this gives

$$f'(z) = \frac{\partial u}{\partial x}(z, 0) - i\frac{\partial u}{\partial y}(z, 0).$$

In the present problem, we have

$$\frac{\partial u}{\partial x}(z, 0) = -e^{-z} \quad \text{and} \quad \frac{\partial u}{\partial y}(z, 0) = z$$

so that

$$f'(z) = -e^{-z} - iz.$$

Integrating with respect to z, we obtain

$$f(z) = e^{-z} - \frac{iz^2}{2}.$$

2.6.2 *Steady state temperature distribution*

One of the most interesting examples of applying the theory of harmonic functions to physical modeling is the study of two-dimensional steady state temperature fields. When steady state prevails, the temperature function inside a two-dimensional body $T(x, y)$ can be shown to be harmonic. To verify the claim, we need to examine the fundamental physical process in heat conduction.

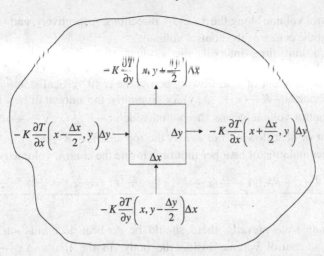

Figure 2.7. An infinitesimal control volume of widths Δx and Δy is contained inside a two-dimensional body. The heat fluxes across the four sides of the rectangular control volume are shown.

The heat flux Q across a surface inside a solid body at a point is defined as the amount of heat flowing normal to the surface per unit area per unit time at that point. The empirical law of heat conduction states that the flux across a surface is proportional to the normal temperature gradient at the point on the surface. We consider a two-dimensional solid body bounded by a simple closed curve Γ in the complex plane. Let $T(x, y)$ denote the steady state temperature at a point (x, y) inside Γ. The mathematical statement of the law of heat conduction can be written as

$$Q = -K \frac{\partial T}{\partial n}, \quad K > 0, \tag{2.6.7}$$

where the constant K is called the thermal conductivity of the material of the solid body. For a homogeneous solid, K may be assumed to be constant throughout the body. The negative sign indicates that the heat flux flows from the high temperature to the low temperature region.

Physical argument dictates that steady state temperature distribution prevails if there is no heat source or sink present inside the body and there is no net heat flux across the bounding surface. It is reasonable to assume that the temperature function $T(x, y)$ and all its partial derivatives up to second order are continuous at all points inside the body.

We consider an infinitesimal control volume of rectangular shape inside a two-dimensional body as shown in Figure 2.7. Let Δx and Δy be the widths

of the control volume along the x- and y-directions, respectively, and (x, y) be
the point at the center of the control volume.

Within a unit time interval, the amount of heat flowing across the
left vertical side *into* the rectangular control volume is $-K\frac{\partial T}{\partial x}(x - \frac{\Delta x}{2}, y)\Delta y$.
Likewise, the amount of heat flowing across the right vertical side *out of* the
control volume is $-K\frac{\partial T}{\partial x}(x + \frac{\Delta x}{2}, y)\Delta y$. Similarly, the amount of heat flowing
into the control volume across the bottom side is $-K\frac{\partial T}{\partial y}(x, y - \frac{\Delta y}{2})\Delta x$, and
the amount of heat flowing *out* across the top side is $-K\frac{\partial T}{\partial y}(x, y + \frac{\Delta y}{2})\Delta x$.
The net accumulation of heat per unit time inside the control volume is

$$K\Delta x\Delta y\left[\frac{\frac{\partial T}{\partial x}(x + \frac{\Delta x}{2}, y) - \frac{\partial T}{\partial x}(x - \frac{\Delta x}{2}, y)}{\Delta x} + \frac{\frac{\partial T}{\partial y}(x, y + \frac{\Delta y}{2}) - \frac{\partial T}{\partial y}(x, y - \frac{\Delta y}{2})}{\Delta y}\right].$$

When steady state prevails, there should be no heat accumulation in any
infinitesimal control volume inside the body. In the limits $\Delta x \to 0$ and
$\Delta y \to 0$, we then obtain

$$\frac{\partial^2 T}{\partial x^2} + \frac{\partial^2 T}{\partial y^2} = 0. \tag{2.6.8}$$

Hence, the steady state temperature function $T(x, y)$ is a harmonic function.

Example 2.6.3   Isothermal curves in a temperature field are curves that connect points with the same temperature value. Supposing the isothermal curves
of a steady state temperature field are given by the family of parabolas

$$y^2 = \alpha^2 + 2\alpha x, \quad \alpha \text{ is real positive,}$$

in the x-y plane, find the general solution of the temperature function $T(x, y)$.

Solution   First, we solve for the parameter α in the equation of the isothermal
curves. This gives

$$\alpha = -x + \sqrt{x^2 + y^2},$$

where the positive sign is chosen since $\alpha > 0$. A naive guess may suggest that
the temperature function $T(x, y)$ is given by

$$T(x, y) = -x + \sqrt{x^2 + y^2}.$$

However, since $T(x, y)$ has to be harmonic, the above function cannot be a
feasible solution.

The systematic approach to finding the temperature function is to set

$$T(x, y) = f(t),$$

where $t = \sqrt{x^2 + y^2} - x$, and f is some function to be determined such that $T(x, y)$ is harmonic. To solve for $f(t)$, we first compute

$$\frac{\partial^2 T}{\partial x^2} = f''(t)\left(\frac{x}{\sqrt{x^2+y^2}} - 1\right)^2 + f'(t)\frac{y^2}{(x^2+y^2)^{3/2}},$$

$$\frac{\partial^2 T}{\partial y^2} = f''(t)\frac{y^2}{x^2+y^2} + f'(t)\frac{x^2}{(x^2+y^2)^{3/2}}.$$

The requirement that $T(x, y)$ satisfies the Laplace equation leads to

$$2\left(1 - \frac{x}{\sqrt{x^2+y^2}}\right)f''(t) + \frac{1}{\sqrt{x^2+y^2}}f'(t) = 0 \quad \text{or} \quad \frac{f''(t)}{f'(t)} = -\frac{1}{2t}.$$

The above equation can be integrated twice to give

$$f(t) = C_1\sqrt{t} + C_2,$$

where C_1 and C_2 are arbitrary constants. The temperature function is then given by

$$T(x, y) = f(t) = C_1\sqrt{\sqrt{x^2 + y^2} - x} + C_2.$$

When expressed in polar coordinates, the temperature function becomes

$$T(r, \theta) = C_1\sqrt{r(1 - \cos\theta)} + C_2 = C_1\sqrt{2r}\sin\frac{\theta}{2} + C_2.$$

Remark   Since $T(r, \theta)$ can be expressed as $\sqrt{2}C_1\text{Im }z^{1/2} + C_2$, the harmonic conjugate of $T(r, \theta)$ is easily seen to be

$$F(r, \theta) = -C_1\sqrt{2r}\cos\frac{\theta}{2} + C_3 = -C_1\sqrt{x + \sqrt{x^2 + y^2}} + C_3,$$

where C_3 is another arbitrary constant. Using the result in Example 2.5.2, the family of curves defined by

$$x + \sqrt{x^2 + y^2} = \beta \quad \text{or} \quad y^2 = \beta^2 - 2\beta x, \quad \beta > 0,$$

are orthogonal to the isothermal curves $y^2 = \alpha^2 + 2\alpha x$, $\alpha > 0$.

Physically, the direction of heat flux is normal to the isothermal lines. Therefore, the family of curves orthogonal to the isothermal lines are called the *flux lines*. These flux lines indicate the flow directions of heat in the steady state temperature field.

2.6.3 Poisson's equation

A partial differential equation of the form

$$\nabla^2 \phi(x, y) = h(x, y) \tag{2.6.9}$$

is called the *Poisson equation*. The use of conjugate complex variables may facilitate the solution of this type of equation. Recall from eq. (2.4.4) that

$$\frac{\partial}{\partial x} = \frac{\partial}{\partial z} + \frac{\partial}{\partial \bar{z}} \quad \text{and} \quad \frac{\partial}{\partial y} = i\left(\frac{\partial}{\partial z} - \frac{\partial}{\partial \bar{z}}\right),$$

so that

$$\frac{\partial^2}{\partial x^2} = \frac{\partial^2}{\partial z^2} + 2\frac{\partial^2}{\partial z \partial \bar{z}} + \frac{\partial^2}{\partial \bar{z}^2}, \quad \frac{\partial^2}{\partial y^2} = -\left(\frac{\partial^2}{\partial z^2} - 2\frac{\partial^2}{\partial z \partial \bar{z}} + \frac{\partial^2}{\partial \bar{z}^2}\right).$$

In terms of the pair of conjugate complex variables, the Poisson equation can be expressed as

$$4\frac{\partial^2 \phi}{\partial z \partial \bar{z}} = h\left(\frac{z + \bar{z}}{2}, \frac{z - \bar{z}}{2i}\right). \tag{2.6.10}$$

The general solution to eq. (2.6.10) can be obtained by integrating the equation with respect to $\bar{z}$ and z successively.

Example 2.6.4   Find the general solution to the Poisson equation

$$\nabla^2 \phi(x, y) = 8(x^2 - y^2).$$

Solution   In terms of the pair of conjugate complex variables, the Poisson equation can be expressed as

$$4\frac{\partial^2 \phi}{\partial z \partial \bar{z}} = 4(z^2 + \bar{z}^2).$$

Integrating the above equation with respect to z and $\bar{z}$ successively, we obtain

$$\phi(z, \bar{z}) = \frac{z^3 \bar{z}}{3} + \frac{z \bar{z}^3}{3} + F(z) + G(\bar{z}),$$

where $F(z)$ and $G(\bar{z})$ are arbitrary functions in z and $\bar{z}$, respectively. In terms of x and y, the solution takes the form

$$\phi(x, y) = \frac{2}{3}(x^4 - y^4) + F(x + iy) + G(x - iy).$$

Note that $F(x + iy) + G(x - iy)$ is the general solution to the homogeneous equation $\nabla^2 \phi = 0$, while $\frac{2}{3}(x^4 - y^4)$ is a particular solution corresponding to the non-homogeneous term $8(x^2 - y^2)$.

2.7 Problems

2.1. Letting $z = x + iy$, find the real and imaginary parts of the following functions:

(a) $2z^3 - 3z$;   (b) $\dfrac{1}{z}$;   (c) $\dfrac{i+z}{i-z}$.

2.2. Represent each of the following functions in terms of z and $\bar{z}$:

(a) $w = x^2 - y^2 + 2ixy$;

(b) $w = x(x^2 - 3y^2) - y(3x^2 - y^2)i$;

(c) $w = \dfrac{2(x^2 + y^2) - (x + iy)}{4(x^2 + y^2) - 4x + 1}$.

2.3. For the function

$$w = f(z) = z + \frac{1}{z}, \quad z = x + iy \quad \text{and} \quad w = u + iv,$$

find the image curve corresponding to the circle $|z| = r_0$. Also, find the preimage curves in the x-y plane corresponding to the coordinates curves $u = \alpha$ and $v = \beta$ in the u-v plane.

2.4. An *isometry* is a function $f : \mathbb{C} \to \mathbb{C}$ such that

$$|f(z_1) - f(z_2)| = |z_1 - z_2|,$$

for all z_1 and z_2 in $\mathbb{C}$. Define

$$g(z) = \frac{f(z) - f(0)}{f(1) - f(0)}.$$

(a) Show that $g(z)$ is also an isometry if f is an isometry.

(b) By observing $g(1) = 1$ and $g(0) = 0$, show that the real parts of $g(z)$ and z are equal, for all z in $\mathbb{C}$; and $g(i) = i$ or $-i$.

(c) Show that

    i. if $g(i) = i$, then $g(z) = z$;

    ii. if $g(i) = -i$, then $g(z) = \bar{z}$.

(d) Prove that any isometry f must be of the form

$$f(z) = \alpha z + \beta \quad \text{or} \quad f(z) = \alpha \bar{z} + \beta,$$

where α and β are constants, and $|\alpha| = 1$.

2.5. This problem finds the velocity of fluid flow emanating from a *vortex*. Physically, the direction of the velocity of fluid flow due to a vortex is tangential to the concentric circles drawn with the vortex as the center, and the magnitude is inversely proportional to the distance from the vortex. Suppose the vortex is placed at the point α in the complex plane. Find the velocity function that describes the velocity of fluid flow at the point z due to the vortex. How does the strength of the vortex enter into the velocity function?

2.6. Suppose

$$\lim_{z \to z_0} f(z) = A \quad \text{and} \quad \lim_{z \to z_0} g(z) = B.$$

Show that

(i) $\displaystyle\lim_{z \to z_0} f(z)g(z) = AB$, (ii) $\displaystyle\lim_{z \to z_0} \frac{f(z)}{g(z)} = \frac{A}{B}$, $B \neq 0$.

Given that $\lim_{z \to z_0} f(z)$ exists, show

$$\overline{\lim_{z \to z_0} f(z)} = \lim_{z \to z_0} \overline{f(z)}$$

directly from the definition of the limit of a complex function.

2.7 Show that

$$\lim_{z \to \infty} f(z) = \infty$$

if and only if

$$\lim_{z \to 0} \frac{1}{f\left(\frac{1}{z}\right)} = 0.$$

2.8. For each of the following functions, examine whether the function is continuous at $z = 0$:

(a) $f(z) = \begin{cases} 0 & z = 0 \\ \dfrac{\operatorname{Re} z}{|z|} & z \neq 0 \end{cases}$; (b) $f(z) = \begin{cases} 0 & z = 0 \\ \dfrac{(\operatorname{Re} z)^2}{|z|} & z \neq 0 \end{cases}$.

2.9. Suppose a complex function $f(z)$ is continuous in a region R; show that its modulus $|f(z)|$ is also continuous within the same region.

2.10. Show that $f(z) = \frac{1}{z}$ is continuous in $0 < |z| < 1$ (see Example 2.2.3).

2.11. Suppose a function $f(z)$ is continuous and nonzero at a point z_0. Show that $f(z)$ is nonzero throughout some neighborhood of that point.

2.12. The motion of a particle in the complex plane is given by

$$z(t) = z_1 \cos^2 t + z_2 \sin^2 t,$$

where t is the time variable and z_1 and z_2 are some fixed complex numbers. Describe the path traversed by the particle.

2.13. If a particle moves with the instantaneous speed v along any plane curve C, show that the normal component of the acceleration at any point on

C is given by v^2/r, where r is the radius of curvature of the curve C at that point.

2.14. Suppose the trajectory of a particle in the complex plane is described by

$$z(t) = a \cos \omega t + ib \sin \omega t.$$

Show that the acceleration at any point is always directed toward the origin. Find the equation of the periodic trajectory.

2.15. For each of the following functions, determine the region in the complex plane where the corresponding Cauchy–Riemann relations are satisifed:

(a) $w = 3 - z + 2z^2$; (b) $w = \dfrac{1}{z}$; (c) $w = |z|^2 z$; (d) $w = z^5$.

2.16. Consider the function $f(z) = xy^2 + ix^2y$, $z = x + iy$. Find the region where

(a) the Cauchy–Riemann relations are satisfied;
(b) the function is differentiable;
(c) the function is analytic.

2.17. Show that the function $f(z) = z \operatorname{Re} z$ is nowhere differentiable except at the origin; hence find $f'(0)$.

2.18. Find a complex function that is continuous for $|z| < 1$ but differentiable only at the origin. Find another complex function that is continuous in a region but differentiable only along certain lines in the region of continuity.

2.19. Discuss the analyticity of each of the following functions. If the function is analytic, find its derivative.

(a) $f(z) = z|z|$;
(b) $f(z) = x^2 - y^2 - 3x + 2 + i(2x - 3)y$, $z = x + iy$;
(c) $f(z) = \dfrac{x+y}{x^2+y^2} + i\,\dfrac{x-y}{x^2+y^2}$, $z = x + iy$.

2.20. Let $f(z) = (x - y)^2 + 2i(x + y)$.

(a) Show that the Cauchy–Riemann relations are satisfied only along the curve $x - y = 1$.
(b) Deduce that $f(z)$ has a derivative along that curve and find the derivative value. Then explain why $f(z)$ is nowhere analytic.

2.21. (a) Show that the function

$$f(z) = \sqrt{|\operatorname{Im}(z^2)|}$$

satisfies the Cauchy–Riemann relations at $z = 0$, but is not differentiable at $z = 0$.

(b) Consider the function

$$f(z) = \begin{cases} e^{-\frac{1}{z^4}} & z \neq 0 \\ 0 & z = 0 \end{cases}.$$

Show that the Cauchy–Riemann relations are satisfied at all points in the complex plane, but $f(z)$ is not analytic (and not continuous) at $z = 0$.

2.22. Consider the following function

$$f(z) = \begin{cases} (1+i)\dfrac{\text{Im}(z^2)}{|z|^2} & z \neq 0 \\ 0 & z = 0 \end{cases}.$$

Show that the Cauchy–Riemann relations are satisfied at $z = 0$. Is $f(z)$ differentiable at $z = 0$?

2.23. Determine the set on which

$$f(z) = \begin{cases} z & \text{if } |z| \leq 1 \\ z^2 & \text{if } |z| > 1 \end{cases}$$

is analytic and compute its derivative.

2.24. Suppose the complex function

$$f(z) = u(x, y) + iv(x, y), \quad z = x + iy,$$

satisfies the following conditions at $z_0 = x_0 + iy_0$:

(a) $u(x, y)$ and $v(x, y)$ are differentiable at the point (x_0, y_0), and

(b) $\displaystyle\lim_{z \to z_0} \left| \frac{f(z) - f(z_0)}{z - z_0} \right|$ exists.

Show that either $f(z)$ or $\overline{f(z)}$ is differentiable at z_0.

2.25. Let $f(z) = u(x, y) + iv(x, y)$, $z = x + iy$. Determine the validity of the following statements.

(a) Suppose $u(x, y)$ and $v(x, y)$ are differentiable at (x_0, y_0); then $f(z)$ is also differentiable at z_0, $z_0 = x_0 + iy_0$.

(b) Suppose $f(z)$ is analytic in a domain $\mathcal{D}$, and u is a real constant; then $f(z)$ is constant throughout $\mathcal{D}$.

2.26. Suppose $f(z)$ is analytic inside the domain $\mathcal{D}$. Show that $f(z)$ is constant inside $\mathcal{D}$ if it satisfies any one of the following conditions:

(a) $|f(z)|$ is constant inside $\mathcal{D}$;

(b) $\text{Re } f(z) - [\text{Im } f(z)]^2$ inside $\mathcal{D}$.

2.27. Let $f(z)$ be analytic in a domain $\mathcal{D}$, and let α and β be two points inside $\mathcal{D}$. Assuming that all points on the line segment L joining α and β lie inside $\mathcal{D}$, show that there exist z_1 and z_2 on L such that

$$\frac{f(\beta) - f(\alpha)}{\beta - \alpha} = \text{Re } f'(z_1) + i \text{ Im } f'(z_2).$$

This result somewhat resembles the mean value theorem in real variable calculus.

Hint: Parametrize the line joining z_1 and z_2 by

$$z(t) = \alpha + (\beta - \alpha)t, \quad 0 \le t \le 1.$$

Write $f = u + iv$ and consider the function value of $f(z)$ along L as a function of t. Set

$$f(z(t)) = u(\alpha + (\beta - \alpha)t) + iv(\alpha + (\beta - \alpha)t), \quad 0 \le t \le 1.$$

Next, consider

$$\frac{f(\beta) - f(\alpha)}{\beta - \alpha} = \frac{u(\beta) - u(\alpha)}{\beta - \alpha} + i\frac{v(\beta) - v(\alpha)}{\beta - \alpha},$$

then apply the mean value theorem in real variable calculus to u and v by treating them as real variable functions of t.

2.28. Let Γ be a two-dimensional curve defined by the differentiable functions

$$x = x(t) \quad \text{and} \quad y = y(t), \quad t_1 \le t \le t_2,$$

in the z-plane, where $z = x + iy$. Let $w = f(z)$ be an analytic function in a domain containing Γ. Show that

$$\frac{dw}{dt} = f'(z)\frac{dz}{dt}.$$

2.29. Suppose $f(z)$ is differentiable at $z = z_0$. Let $z_0 + \Delta z$ approach z_0 along a line making an angle α with the x-axis so that $\Delta z = \Delta s\, e^{i\alpha}$, where s is the arc length variable. Show that

$$f'(z_0) = e^{-i\alpha}\left(\frac{du}{ds} + i\frac{dv}{ds}\right),$$

where

$$\frac{du}{ds} = \nabla_\alpha u \quad \text{and} \quad \frac{dv}{ds} = \nabla_\alpha v$$

are the directional derivatives of u and v in the direction chosen. Similarly, show that

$$f'(z_0) = e^{-i\alpha} \left(\nabla_{\alpha+\frac{\pi}{2}} v - i \nabla_{\alpha+\frac{\pi}{2}} u \right).$$

Use the above results to deduce the Cauchy–Riemann relations.

2.30. Find an analytic function $f(z)$ whose real part $u(x, y)$ is·

(a) $u(x, y) = y^3 - 3x^2 y$, $f(i) = 1 + i$;

(b) $u(x, y) = \dfrac{y}{x^2 + y^2}$, $f(1) = 0$;

(c) $u(x, y) = (x - y)(x^2 + 4xy + y^2)$.

2.31. Find the orthogonal trajectories of each of the following families of curves:

(a) $x^3 y - xy^3 = \alpha$;

(b) $2e^{-x} \sin y + x^2 - y^2 = \alpha$;

(c) $(r^2 + 1) \cos \theta = \alpha r$.

2.32. Determine the values of the parameters appearing in the following functions such that the functions become analytic:

(a) $f(z) = (x^2 + axy + by^2) + i(cx^2 + dxy + y^2)$, $z \in \mathbb{C}$;

(b) $f(z) = \dfrac{x + k - iy}{x^2 + y^2 + 2x + 1}$, $z \in \mathbb{C}\backslash\{-1\}$;

(c) $f(z) = e^x(\cos \ell y + i \sin \ell y)$, $z \in \mathbb{C}$.

2.33. The pair of real functions

$$u(x, y) = x^2 - y^2 \qquad \text{and} \qquad v(x, y) = \frac{y}{x^2 + y^2}$$

are harmonic; however, the function $f(z) = u(x, y) + iv(x, y)$, $z = x + iy$, is non-analytic. Explain why.

2.34. If $u(x, y)$ is a harmonic function, determine whether u^2 is harmonic.

2.35. If $f(z)$ is analytic and does not vanish in a domain $\mathcal{D}$, determine whether the modulus function $|f(z)|$ is harmonic inside $\mathcal{D}$.

2.36. If v is a harmonic conjugate of u in a domain $\mathcal{D}$, is uv harmonic in $\mathcal{D}$? Give an explanation to your answer.

2.37. If u and v are harmonic in a domain $\mathcal{D}$, prove that

$$\left(\frac{\partial u}{\partial y} - \frac{\partial v}{\partial x} \right) + i \left(\frac{\partial u}{\partial x} + \frac{\partial v}{\partial y} \right)$$

is analytic in $\mathcal{D}$.

2.38. Let $\theta = APB$, where A and B are the fixed points $(-a, 0)$ and $(a, 0)$, respectively and P is the variable point $z = x + iy$. Show that $\theta(x, y)$ is a harmonic function. Find the corresponding harmonic conjugate v such that $\theta + iv$ is an analytic function.

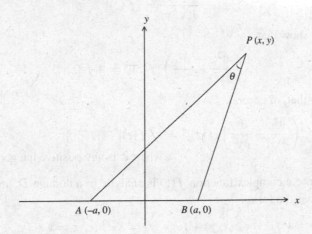

2.39. From the polar form of the Cauchy–Riemann relations (see Example 2.4.1), derive the Laplace equation in polar form.

2.40. Find a particular solution to each of the following Poisson equations:

(a) $\nabla^2 u(x, y) = xye^y$; (b) $\nabla^2 u(r, \theta) = r^2 \cos \theta$.

2.41. Consider

$$w(z) = f(x, y) = f\left(\frac{z + \bar{z}}{2}, \frac{z - \bar{z}}{2i}\right) = \phi(z, \bar{z}), \quad z = x + iy,$$

where the pair of conjugate complex variables z and $\bar{z}$ are considered to be independent variables. Show that

$$\frac{\partial}{\partial z} = \frac{1}{2}\left(\frac{\partial}{\partial x} - i\frac{\partial}{\partial y}\right) \quad \text{and} \quad \frac{\partial}{\partial \bar{z}} = \frac{1}{2}\left(\frac{\partial}{\partial x} + i\frac{\partial}{\partial y}\right).$$

Let $u(x, y)$ and $v(x, y)$ denote the real and imaginary parts of $w(z)$, respectively. Prove that

(a) $\dfrac{\partial \phi}{\partial z} = \dfrac{1}{2}\left[(u_x + v_y) + i(-u_y + v_x)\right];$

(b) $\dfrac{\partial \phi}{\partial \bar{z}} = \dfrac{1}{2}\left[(u_x - v_y) + i(u_y + v_x)\right];$

(c) $\dfrac{dw}{dz} = \dfrac{\partial \phi}{\partial z} + \dfrac{\partial \phi}{\partial \bar{z}}e^{-2i\alpha}, \quad \alpha = \text{Arg } dz.$

2.42. Suppose $f(z) = u(x, y) + iv(x, y)$ is analytic in a domain $\mathcal{D}$, where $z = x + iy$. Show that the Jacobian of the transformation from (x, y) to (u, v) is given by

$$\frac{\partial(u, v)}{\partial(x, y)} = |f'(z)|^2.$$

Also, show that

$$\left(\frac{\partial^2}{\partial x^2} + \frac{\partial^2}{\partial y^2}\right) |f(z)|^2 = 4|f'(z)|^2.$$

Show that, in general,

$$\left(\frac{\partial^2}{\partial x^2} + \frac{\partial^2}{\partial y^2}\right) |f(z)|^k = k^2 |f(z)|^{k-2} |f'(z)|^2,$$

where k is any positive integer.

2.43. Suppose a complex function $f(z)$ is analytic in a domain $\mathcal{D}$, and that it assumes the form

$$f(z) = u(x, y) + iv(x, y) = Xe^{iY}, \quad z = x + iy,$$

where X and Y are real, and Y is independent of x. Show that X must be independent of y.

2.44. Suppose the isothermal lines of a steady state temperature field are the family of curves

$$x^2 + y^2 = \alpha, \quad \alpha > 0.$$

Find the general solution of the temperature function, and the equation of the family of flux lines.

3

Exponential, Logarithmic and Trigonometric Functions

The full details of the properties of the complex exponential, logarithmic, trigonometric and hyperbolic functions are discussed in this chapter. The mapping properties of some of these complex functions are also examined. These complex functions are defined as a natural extension of their real counterparts (actually, they can be reduced to their corresponding real functions by setting $y = 0$ in the independent complex variable $z = x + iy$). However, some of the properties of these complex functions may not be shared by their real counterparts. For example, the complex exponential function is periodic but the real one is not; the equation $\sin z = 2$ has solutions only if z is complex.

The logarithmic function, the inverse trigonometric and hyperbolic functions, and the generalized power functions are multi-valued functions. The notion of the Riemann surface for a multi-valued function is introduced in the final section. The Riemann surface can be considered as an extension of the complex plane to a surface which has more than one 'sheet'. By virtue of the construction of the Riemann surface, the multi-valued function assumes only one value corresponding to each point on the Riemann surface.

3.1 Exponential functions

Recall that the complex exponential function is defined by (see Section 2.1.3)

$$e^z = e^x(\cos y + i \sin y), \quad z = x + iy. \tag{3.1.1}$$

Its real and imaginary parts are, respectively,

$$u(x, y) = e^x \cos y \quad \text{and} \quad v(x, y) = e^x \sin y. \tag{3.1.2}$$

93

The derivative of e^z is found to be

$$\frac{d}{dz}e^z = \frac{\partial}{\partial x}u(x, y) + i\frac{\partial}{\partial x}v(x, y)$$
$$= e^x \cos y + ie^x \sin y$$
$$= e^z, \qquad \text{for all } z \text{ in } \mathbb{C}, \qquad (3.1.3)$$

that is, the derivative of e^z is equal to itself. This is one of the fundamental properties of the exponential function. Since the derivative of e^z exists for all z in the whole z-plane, the exponential function is an entire function. Also, it can be verified that

$$e^{z_1+z_2} = e^{z_1}e^{z_2}, \qquad (3.1.4)$$

another basic property of the exponential function. The modulus of e^z is non-zero since

$$|e^z| = e^x \neq 0, \qquad \text{for all } z \text{ in } \mathbb{C}, \qquad (3.1.5)$$

and so $e^z \neq 0$ for all z in the complex z-plane. The range of the complex exponential function is the entire complex plane except the zero value. Further, it can be shown that

$$e^{z+2k\pi i} = e^z, \quad \text{for any } z \text{ and integer } k, \qquad (3.1.6)$$

that is, e^z is periodic with the fundamental period $2\pi i$. Interestingly, the complex exponential function is periodic while its real counterpart is not.

3.1.1 Definition from the first principles

From the first principles, it seems natural to define the complex exponential function as a complex function $f(z)$ that satisfies the following defining properties:

(1) $f(z)$ is entire,
(2) $f'(z) = f(z)$,
(3) $f(x) = e^x$, x is real.

We would like to show how the definition of the complex exponential functon can be derived from these properties. Let $f(z) = u(x, y) + iv(x, y)$, $z = x + iy$. From property (1), u and v are seen to satisfy the Cauchy–Riemann relations. Combining properties (1) and (2), we obtain the following relations:

$$u_x + iv_x = v_y - iu_y = u + iv.$$

First, we observe that

$$u_u = u \text{ and } v_x = v.$$

From these two relations, we deduce that

$$u = e^x g(y) \text{ and } v = e^x h(y),$$

where $g(y)$ and $h(y)$ are arbitrary functions of y. In addition, we also have the relations

$$v_y = u \quad \text{and} \quad u_y = -v,$$

from which we deduce that the arbitrary functions are related by

$$h'(y) = g(y) \quad \text{and} \quad -g'(y) = h(y).$$

By eliminating $g(y)$ in the above relations, we obtain

$$h''(y) = -h(y).$$

The general solution of the above equation is given by

$$h(y) = A \cos y + B \sin y,$$

where A and B are arbitrary constants. Furthermore, using $g(y) = h'(y)$, we have

$$g(y) = -A \sin y + B \cos y.$$

To determine the arbitrary constants A and B, we use property (3) where

$$e^x = u(x, 0) + i v(x, 0) = g(0)e^x + ih(0)e^x = Be^x + i Ae^x.$$

We then obtain $B = 1$ and $A = 0$. Putting all the results together, the complex exponential function is found to be

$$f(z) = e^z = e^x \cos y + i e^x \sin y,$$

which agrees with the earlier definition given in eq. (3.1.1).

Example 3.1.1 Find all roots of the equation

$$e^z = i.$$

Solution Equating the real and imaginary parts of e^z and i leads to the following pair of equations:

$$e^x \cos y = 0 \quad \text{and} \quad e^x \sin y = 1.$$

Since $e^x \neq 0$ for all x, we deduce from the first equation that

$$y = \frac{\pi}{2} + k\pi, \quad k \text{ is any integer.}$$

From the second equation, since $e^x > 0$ so that $\sin y > 0$, the value of y is restricted to

$$y = \frac{\pi}{2} + 2k\pi, \quad k \text{ is any integer.}$$

Substituting y into the second equation, we have $e^x = 1$ and so $x = 0$. The roots of the given equation are then found to be

$$z = \left(2k + \frac{1}{2}\right)\pi i, \quad k \text{ is any integer.}$$

Example 3.1.2   Consider the following function:

$$f(z) = e^{\frac{\alpha}{z}}, \quad \alpha \text{ is real.}$$

Show that $|f(z)|$ is constant on the circle $x^2 + y^2 - ax = 0$, where a is a real constant.

Solution   The equation of the given circle can be written in the standard form

$$\left(x - \frac{a}{2}\right)^2 + y^2 = \left(\frac{a}{2}\right)^2,$$

which reveals that the circle is centered at $(a/2, 0)$ and has radius $a/2$. A possible parametric representation of the circle is

$$x = \frac{a}{2}(1 + \cos\theta) \quad \text{and} \quad y = \frac{a}{2}\sin\theta, \quad -\pi < \theta \leq \pi.$$

Geometrically, the parameter θ is the angle between the positive x-axis and the line joining the center $(a/2, 0)$ to the point (x, y). Correspondingly, the complex representation of the circle can be expressed as

$$z = \frac{a}{2}(1 + e^{i\theta}), \quad -\pi < \theta \leq \pi.$$

The modulus of $f(z)$ when z lies on the circle is found to be

$$|f(z)| = \left| e^{\frac{2\alpha}{a(1+e^{i\theta})}} \right| = \left| e^{\frac{2\alpha}{a} \frac{1+\cos\theta - i\sin\theta}{2(1+\cos\theta)}} \right| = e^{\frac{\alpha}{a}}.$$

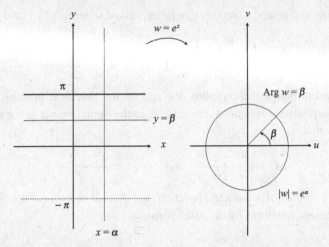

Figure 3.1. Mapping properties of $w = e^z$. Vertical lines are mapped onto concentric circles centered at the origin and horizontal lines are mapped onto rays through the origin.

The modulus value is equal to a constant with no dependence on θ, that is, independent of the choice of the point on the circle.

3.1.2 Mapping properties of the complex exponential function

Since the complex exponential function is periodic with fundamental period $2\pi i$, it is a many-to-one function. Suppose the domain of interest is chosen to be the infinite strip $-\pi < \text{Im } z \leq \pi$. Then the mapping $w = e^z$ becomes one-to-one. The vertical line $x = \alpha$ is mapped onto the circle $|w| = e^\alpha$, while the horizontal line $y = \beta$ is mapped onto the ray Arg $w = \beta$ (see Figure 3.1).

When the vertical line $x = \alpha$ moves further to the left, the mapped circle $|w| = e^\alpha$ shrinks to a smaller radius. When the horizontal line in the z-plane moves vertically from $y = -\pi$ to $y = \pi$, the image ray in the w-plane traverses in the anticlockwise sense from Arg $w = -\pi$ to Arg $w = \pi$. In particular, the whole x-axis is mapped onto the whole positive u-axis, and the portion of the y-axis, $-\pi < y \leq \pi$, is mapped onto the unit circle $|w| = 1$.

3.2 Trigonometric and hyperbolic functions

Using the Euler formula

$$e^{iy} = \cos y + i \sin y,$$

the real sine and cosine functions can be expressed in terms of e^{iy} and e^{-iy} as follows:

$$\sin y = \frac{e^{iy} - e^{-iy}}{2i} \quad \text{and} \quad \cos y = \frac{e^{iy} + e^{-iy}}{2}.$$

It is natural to define the complex sine and cosine functions in terms of the complex exponential functions e^{iz} and e^{-iz} in the same manner as for the real functions, that is,

$$\sin z = \frac{e^{iz} - e^{-iz}}{2i} \quad \text{and} \quad \cos z = \frac{e^{iz} + e^{-iz}}{2}. \tag{3.2.1}$$

The other complex trigonometric functions are defined in terms of the complex sine and cosine functions by the usual formulas:

$$\tan z = \frac{\sin z}{\cos z}, \quad \cot z = \frac{\cos z}{\sin z}, \quad \sec z = \frac{1}{\cos z}, \quad \operatorname{cosec} z = \frac{1}{\sin z}. \tag{3.2.2}$$

The complex sine and cosine functions are entire since they are formed by the linear combination of the entire functions e^{iz} and e^{-iz}. The functions $\tan z$ and $\sec z$ are analytic in any domain that does not include points where $\cos z = 0$. Similarly, the functions $\cot z$ and $\operatorname{cosec} z$ are analytic in any domain excluding those points z such that $\sin z = 0$.

Let $z = x + iy$. Then

$$e^{iz} = e^{-y}(\cos x + i \sin x) \quad \text{and} \quad e^{-iz} = e^{y}(\cos x - i \sin x).$$

The real and imaginary parts of the complex sine and cosine functions are seen to be

$$\sin z = \sin x \cosh y + i \cos x \sinh y,$$
$$\cos z = \cos x \cosh y - i \sin x \sinh y.$$

Moreover, their moduli are found to be

$$|\sin z| = \sqrt{\sin^2 x + \sinh^2 y}, \quad |\cos z| = \sqrt{\cos^2 x + \sinh^2 y}.$$

Since $\sinh y$ is unbounded at large values of y, the above modulus values can increase (as y does) without bound. While the real sine and cosine functions are always bounded between -1 and 1, their complex counterparts are unbounded.

Next, we define the complex hyperbolic functions in the same manner as their real counterparts. They are defined as

$$\sinh z = \frac{e^{z} - e^{-z}}{2}, \quad \cosh z = \frac{e^{z} + e^{-z}}{2}, \quad \tanh z = \frac{\sinh z}{\cosh z}. \tag{3.2.3}$$

The other hyperbolic functions cosech z, sech z and coth z are defined as the reciprocal of sinh z, cosh z and tanh z, respectively.

In fact, the hyperbolic functions are closely related to the trigonometric functions. Supposing z is replaced by iz in eq. (3.2.3), we obtain

$$\sinh iz = i \sin z.$$

Similarly, one can show that

$$\sin iz = i \sinh z, \quad \cosh iz = \cos z, \quad \cos iz = \cosh z.$$

The real and imaginary parts of sinh z and cosh z are found to be

$$\sinh z = \sinh x \cos y + i \cosh x \sin y,$$
$$\cosh z = \cosh x \cos y + i \sinh x \sin y,$$

and their moduli are given by

$$| \sinh z | = \sqrt{\sinh^2 x + \sin^2 y},$$
$$| \cosh z | = \sqrt{\cosh^2 x - \sin^2 y}.$$

The complex hyperbolic functions sinh z and cosh z are periodic with fundamental period $2\pi i$, and tanh z is periodic with fundamental period πi. Thus the complex hyperbolic functions are periodic, unlike their real counterparts.

A zero α of a function $f(z)$ satisfies $f(\alpha) = 0$. Like their real counterparts, the zeros of sin z are $k\pi$ and the zeros of cos z are $k\pi + \frac{\pi}{2}$, k is any integer. While the real cosh has no zero and the real sinh has only one zero at $z = 0$, the complex cosh and sinh have infinitely many zeros.

To find the zeros of sinh z, we observe that

$$\sinh z = 0 \Leftrightarrow | \sinh z | = 0 \Leftrightarrow \sinh^2 x + \sin^2 y = 0.$$

Hence, x and y must satisfy sinh $x = 0$ and sin $y = 0$, thus giving $x = 0$ and $y = k\pi$, k is any integer. The zeros of sinh z are $z = k\pi i$, k is any integer. Similarly, the zeros of cosh z are $z = (k + \frac{1}{2})\pi i$, k is any integer.

Knowing the derivative of e^z, the derivatives of the trigonometric and hyperbolic functions can be found easily. Indeed, the derivative formulas for the complex trigonometric and hyperbolic functions are exactly the same as those for their real counterparts. In addition, the compound angle formulas for real trigonometric and hyperbolic functions also hold for their complex counter-parts. For example,

$$\cos(z_1 \pm z_2) = \cos z_1 \cos z_2 \mp \sin z_1 \sin z_2,$$
$$\sinh(z_1 \pm z_2) = \sinh z_1 \cosh z_2 \pm \cosh z_1 \sinh z_2,$$
$$\tan(z_1 \pm z_2) = \frac{\tan z_1 \pm \tan z_2}{1 \mp \tan z_1 \tan z_2}.$$

Example 3.2.1 Show that $\cos \bar{z} = \overline{\cos z}$.

Solution Let $z = x + iy$, and consider

$$\overline{e^{iz}} = \overline{e^{i(x+iy)}} = \overline{e^{-y}e^{ix}} = e^{-y}e^{-ix} = e^{-i(x-iy)} = e^{-i\bar{z}}.$$

Similarly, we have

$$\overline{e^{-iz}} = e^{i\bar{z}}.$$

Now, consider

$$\overline{\cos z} = \overline{\frac{1}{2}(e^{iz} + e^{-iz})} = \frac{1}{2}\,(e^{-i\bar{z}} + e^{i\bar{z}}) = \cos \bar{z}.$$

Remark   The above result is a manifestation of the *reflection principle*, which states:

> Suppose a function f is analytic in some domain $\mathcal{D}$ which includes part of the real axis and $\mathcal{D}$ is symmetric about the real axis. Further, $f(x)$ is real whenever x is a point on that part of the real axis. We then have
>
> $$f(\bar{z}) = \overline{f(z)} \quad \text{for any } z \text{ in } \mathcal{D}.$$

The conditions required in the reflection principle are seen to be satisfied by $f(z) = \cos z$, since the domain of analyticity is the whole complex plane and the complex cosine function reduces to the real cosine function when the argument is real.

By virtue of the reflection principle, a similar property about con-jugation holds for all complex trigonometric and hyperbolic functions. More detailed discussion of the reflection principle can be found in Section 5.5.1.

Example 3.2.2 Let $z = x + iy$. Derive the following lower bounds:

(a) $|\sin z| \geq \dfrac{|e^y - e^{-y}|}{2}$; (b) $|\tan z| \geq \dfrac{|e^y - e^{-y}|}{e^y + e^{-y}}$.

Solution

(a) Consider

$$|\sin z| = \left| \frac{e^{iz} - e^{-iz}}{2i} \right| = \frac{|e^{iz} - e^{-iz}|}{2}.$$

Use the triangle inequality to give

$$|e^{iz} - e^{-iz}| \geq \left| |e^{iz}| - |e^{-iz}| \right| = |e^y - e^{-y}|,$$

and finally

$$|\sin z| \geq \frac{|e^y - e^{-y}|}{2}.$$

(b) From the relation

$$|\tan z| = \frac{|e^{iz} - e^{-iz}|}{|e^{iz} + e^{-iz}|},$$

and using the triangle inequality, we obtain

$$|e^{iz} + e^{-iz}| \leq |e^{iz}| + |e^{-iz}| = e^y + e^{-y}.$$

Combining with the result in (a), we then have

$$|\tan z| \geq \frac{|e^y - e^{-y}|}{e^y + e^{-y}}.$$

Example 3.2.3 Suppose z moves to infinity along a ray through the origin. Discuss the possible values for

$$\lim_{z \to \infty} \tan z.$$

Solution Let $z = re^{i\theta}$, where θ is fixed for a given ray through the origin. Consider the following cases:

(i) $0 < \theta < \pi$

We have

$$\tan z = -i \frac{e^{2iz} - 1}{e^{2iz} + 1}$$

and observe that

$$|e^{2iz}| = e^{-2r\sin\theta} \quad \text{and} \quad \sin\theta > 0.$$

As $r \to \infty$, $e^{-2r\sin\theta} \to 0$ and so $\displaystyle\lim_{\substack{r\to\infty \\ z=re^{i\theta}}} \tan z = i$.

(ii) $-\pi < \theta < 0$

Now, we write

$$\tan z = -i\,\frac{1 - e^{-2iz}}{1 + e^{-2iz}}$$

and observe that

$$|e^{-2iz}| = e^{2r\sin\theta} \quad \text{and} \quad \sin\theta < 0.$$

As $r \to \infty$, $e^{2r\sin\theta} \to 0$ and so $\displaystyle\lim_{\substack{r\to\infty \\ z=re^{i\theta}}} \tan z = -i$.

(iii) $\theta = 0$ or $\theta = \pi$.

In these cases,

$$\tan z = \begin{cases} \tan r & \text{when } \theta = 0 \\ -\tan r & \text{when } \theta = \pi \end{cases}.$$

As $r \to \infty$, the limit of $\tan r$ does not exist and so the same holds for $\tan z$.

3.2.1 Mapping properties of the complex sine function

Consider the complex sine function

$$w = \sin z = \sin x \cosh y + i \cos x \sinh y, \quad z = x + iy.$$

Suppose we write $w = u + iv$. Then

$$u = \sin x \cosh y \quad \text{and} \quad v = \cos x \sinh y.$$

To comprehend the mapping properties of $w = \sin z$, we find the images of the coordinates lines $x = \alpha$ and $y = \beta$. When $x = \alpha$, $u = \sin\alpha\cosh y$ and $v = \cos\alpha\sinh y$. By eliminating y in the above equations, we obtain

$$\frac{u^2}{\sin^2\alpha} - \frac{v^2}{\cos^2\alpha} = 1, \tag{3.2.4}$$

which represents a hyperbola in the w-plane (see Figure 3.2a). When $0 < \alpha < \pi/2$, $u = \sin\alpha\cosh y > 0$ for all values of y, so the line $x = \alpha$ is mapped onto the right-hand branch of the hyperbola. Likewise, the line

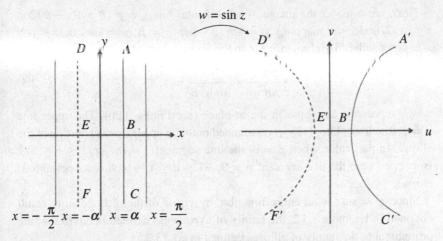

Figure 3.2.a Vertical lines are mapped onto hyperbolas under the mapping $w = \sin z$.

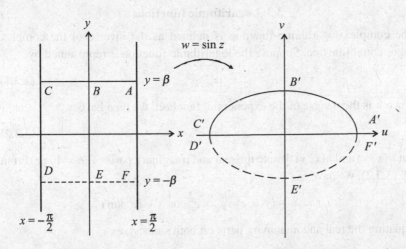

Figure 3.2.b Horizontal lines are mapped onto ellipses under the mapping $w = \sin z$.

$x = -\alpha$ is mapped onto the left-hand branch of the same hyperbola. In particular, when $\alpha = \pi/2$, the line $x = \pi/2$ is mapped onto the line segment $v = 0$, $u \geq 1$, which is a degenerated hyperbola (see Example 3.4.2). Also, the y-axis is mapped onto the v-axis. We conclude that the infinite strip $\{0 \leq \operatorname{Re} z \leq \pi/2\}$ is mapped to the right half-plane $\{u \geq 0\}$; and by symmetry, the other infinite strip $\{-\pi/2 \leq \operatorname{Re} z \leq 0\}$ is mapped to the left half-plane $\{u \leq 0\}$.

Next, we consider the image of a horizontal line $y = \beta$ ($\beta > 0$), $-\pi/2 \le x \le \pi/2$ under the mapping $w = \sin z$. When $y = \beta$, $u = \sin x \cosh \beta$ and $v = \cos x \sinh \beta$. By eliminating x in the above equations, we obtain

$$\frac{u^2}{\cosh^2 \beta} + \frac{v^2}{\sinh^2 \beta} = 1, \tag{3.2.5}$$

which represents an ellipse in the w-plane (see Figure 3.2b). The upper line $y = \beta$ (the lower line $y = -\beta$) is mapped onto the upper (lower) portion of the ellipse. In particular, when $\beta = 0$, the line segment $y = 0$, $-\pi/2 \le x \le \pi/2$ is mapped onto the line segment $v = 0$, $-1 \le u \le 1$, which is a degenerated ellipse.

Since $w = \sin z$ is an entire function, by virtue of the orthogonality result obtained in Example 2.5.2, the family of hyperbolas defined in eq. (3.2.4) are orthogonal to the family of ellipses defined in eq. (3.2.5).

3.3 Logarithmic functions

The complex logarithmic function is defined as the inverse of the complex exponential function. Suppose the logarithmic function is represented by

$$w = \log z. \tag{3.3.1}$$

Since it is the inverse of the exponential function, we then have

$$z = e^w. \tag{3.3.2}$$

Let $u(x, y)$ and $v(x, y)$ denote the real and imaginary parts of $w = \log z$. From eq. (3.3.2), we have

$$z = x + iy = e^{u+iv} = e^u \cos v + ie^u \sin v.$$

Equating the real and imaginary parts on both sides gives

$$x = e^u \cos v \quad \text{and} \quad y = e^u \sin v.$$

It then follows that

$$e^{2u} = x^2 + y^2 = |z|^2 = r^2 \quad \text{and} \quad v = \tan^{-1} \frac{y}{x}. \tag{3.3.3}$$

Using the polar form $z = re^{i\theta}$, we deduce from eq. (3.3.3) that

$$u = \ln r = \ln |z|, \quad r \ne 0 \quad \text{and} \quad v = \theta = \arg z.$$

Putting the results together, we have

$$w = \log z = \ln |z| + i \arg z, \quad z \ne 0. \tag{3.3.4}$$

Remark To avoid confusion, we follow the convention that 'ln' refers to real logarithm while 'log' refers to complex logarithm.

Recall that $\arg z$ is multi-valued; so then is $\log z$. For a fixed z, there are infinitely many possible values of $\log z$, each differing by a multiple of $2\pi i$. Among the possible values of $\arg z$, we then choose some value θ_0 and restrict $\arg z$ to $\theta_0 < \theta \le \theta_0 + 2\pi$. In this way, we obtain a branch of $\arg z$, and correspondingly a branch of $\log z$. Within a branch, the function $\arg z$ is single-valued. In Section 1.1, we chose the principal value of $\arg z$, denoted by $\mathrm{Arg}\, z$, as the branch where $-\pi < \mathrm{Arg}\, z \le \pi$. This particular branch of $\log z$ corresponding to the principal value of $\arg z$ is called the *principal branch* of $\log z$. From now on, the principal branch of the complex logarithmic function is denoted by $\mathrm{Log}\, z$, that is,

$$\mathrm{Log}\, z = \ln |z| + i\, \mathrm{Arg}\, z, \quad -\pi < \mathrm{Arg}\, z \le \pi. \qquad (3.3.5)$$

One may write

$$\log z = \mathrm{Log}\, z + 2k\pi i, \quad k \text{ is any integer.} \qquad (3.3.6)$$

For example, $\mathrm{Log}\, i = \frac{\pi}{2}i$ and $\log i = \frac{\pi}{2}i + 2k\pi i$, k is any integer.

By definition, $z = e^{\log z}$, however it would be incorrect to write $z = \log e^z$ since the logarithmic function is multi-valued.

Given two non-zero complex numbers z_1 and z_2, we have

$$\ln |z_1 z_2| = \ln |z_1| + \ln |z_2|,$$
$$\arg(z_1 z_2) = \arg z_1 + \arg z_2,$$

from which it can be deduced that

$$\log(z_1 z_2) = \log z_1 + \log z_2. \qquad (3.3.7)$$

The equality sign in eq. (3.3.7) actually means that any value of $\log(z_1 z_2)$ equals some value of $\log z_1$ plus some value of $\log z_2$.

From eq. (3.3.5), the real and imaginary parts of $\mathrm{Log}\, z$ are seen to be $\ln |z|$ and $\mathrm{Arg}\, z$, respectively. One then deduces that $\ln |z|$ is harmonic everywhere except at $z = 0$, and $\mathrm{Arg}\, z$ is harmonic inside the domain $-\pi < \mathrm{Arg}\, z < \pi$.

Example 3.3.1 Show from the first principles that

$$\frac{d}{dz} \log z = \frac{1}{z}, \quad z \ne 0, \infty.$$

Solution   Supposing we take differentiation along the x-axis, the derivative of $\log z$ becomes

$$\frac{d}{dz} \log z = \frac{\partial}{\partial x} \ln r + i \frac{\partial}{\partial x}(\theta + 2k\pi), \quad z \neq 0, \infty,$$

where $r = \sqrt{x^2 + y^2}$, $\theta = \tan^{-1} \frac{y}{x}$ and k is any integer. The above expression can be simplified to

$$\begin{aligned}
\frac{d}{dz} \log z &= \frac{1}{r} \frac{\partial r}{\partial x} + i \frac{\partial \theta}{\partial x} \\
&= \frac{1}{\sqrt{x^2 + y^2}} \frac{x}{\sqrt{x^2 + y^2}} + i \frac{-y}{x^2 + y^2} \\
&= \frac{x - iy}{x^2 + y^2} = \frac{\bar{z}}{z\bar{z}} = \frac{1}{z}, \quad z \neq 0, \infty.
\end{aligned}$$

3.3.1 Heat source

Consider the complex function

$$T(z) = -\lambda \operatorname{Re}(\operatorname{Log} z) + A = \lambda \ln \frac{1}{r} + A, \quad \lambda > 0, \ z = re^{i\theta}, \quad (3.3.8)$$

where A is a constant and $z \neq 0$. One argues that this function is a feasible steady state temperature function since it is harmonic everywhere except at $z = 0$. Since the temperature at the origin becomes infinite, $T(z)$ is not analytic at $z = 0$. The temperature field has radial symmetry since $T(z)$ is independent of θ.

In polar coordinates, the local heat flux across a curve Γ is given by (see Section 2.6)

$$Q = -K \nabla T \cdot \mathbf{n},$$

where $\mathbf{n}$ is the local normal vector to the curve Γ, and K is the thermal conductivity of the material. When the steady state temperature field has radial symmetry, the normal gradient $\nabla T \cdot \mathbf{n}$ is simply dT/dr. For the temperature field defined in eq. (3.3.8), the net rate of heat energy flowing across the circle $|z| = R$ is then given by

$$\bar{Q}\Big|_{r=R} = -K \int_0^{2\pi} \frac{dT}{dr}\Big|_{r=R} R \, d\theta = K \int_0^{2\pi} \lambda \frac{1}{R} R \, d\theta = 2\pi\lambda K, \quad (3.3.9)$$

which is constant and independent of R. Since the net heat flux is positive, this indicates that heat is flowing radially outward from the origin. The independence of R in the above expression for the heat flux reveals that the same amount of heat energy flows across every circle centered at the origin when the steady state is attained. Let m denote the amount of heat energy flowing out from the origin per unit time. Then λ and m are related by

$$m = 2\pi\lambda K \quad \text{or} \quad \lambda = \frac{m}{2\pi K}.$$

Combining the above results, we deduce that $T(z)$ as defined in eq. (3.3.8) refers to the steady state temperature distribution due to a *heat source* of intensity $2\pi\lambda K$ placed at the origin. In particular, the temperature along the circumference of the unit circle $|z| = 1$ is equal to A. The steady state condition can only be attained when there is no net heat energy accumulated within any pair of concentric circles centered at the source. The amount of heat flux flowing across any circle centered at the heat source is exactly equal to the heat energy generated at the heat source per unit time. In terms of source intensity m, thermal conductivity K and temperature value along the unit circle $T(1)$, the temperature function due to a heat source at the origin can be expressed as

$$T(r) = \frac{m}{2\pi K} \ln\frac{1}{r} + T(1). \tag{3.3.10}$$

Remark In the above derivation, we start with a feasible temperature function and attempt to derive a physical interpretation of the function. Conversely, one may start from the laws of physics and derive the required form of the steady state temperature distribution. This approach is illustrated in the following example.

Example 3.3.2 Suppose a heat source of intensity m is placed at the origin so that heat flows radially outward uniformly in all directions. When the steady state condition prevails, the net rate of heat energy flowing across any circle $|z| = r$ should be the same, independent of r. Otherwise, there will be net heat accumulation. Show that the temperature function must be of the form

$$T(z) = \alpha \ln|z| + \beta, \quad \alpha < 0 \text{ and } \beta \text{ is real}.$$

Give a physical interpretation to the parameters α and β.

Solution In order that the net rate of heat energy flowing across any circle $|z| = r$ is independent of r, we deduce from eq. (3.3.9) that the temperature gradient must be of the form

$$\frac{dT}{dr} = \frac{\alpha}{r}, \quad \alpha \text{ is some real constant}, \tag{i}$$

so that the source intensity

$$m = -K \int_0^{2\pi} \frac{dT}{dr} r \, d\theta = -K \int_0^{2\pi} \alpha \, d\theta = -2\pi K \alpha \qquad \text{(ii)}$$

is a constant. The temperature function can be obtained by integrating eq. (i) to give

$$T(r) = \alpha \ln r + \beta, \quad \beta \text{ is an arbitrary constant.}$$

The constant β is seen to be equal to $T(1)$, which is a real quantity [see eq. (3.3.10)]. From eq. (ii), α is equal to $-\frac{m}{2\pi K}$, which is a negative quantity. The resulting formula agrees with the temperature function given in eq. (3.3.10). The sign of α is reversed when we deal with a heat sink (negative source). In terms of the complex variable z, where $z = re^{i\theta}$, the temperature function takes the form

$$T(z) = \alpha \ln |z| + \beta, \quad \alpha < 0 \text{ and } \beta \text{ is real.}$$

3.3.2 *Temperature distribution in the upper half-plane*

We would like to illustrate that the solution to the steady state temperature distribution in the upper half-plane, Im $z > 0$, can be found readily if the boundary temperature values along the x-axis assume some discrete constant values. First, let $T(x, 0)$ be given by

$$T(x, 0) = \begin{cases} 0 & x > 0 \\ U & x < 0 \end{cases} \qquad (3.3.11)$$

which has a jump of discontinuity at $x = 0$. Consider the function Arg $z =$ Im(Log z) which is harmonic in the upper half-plane since it is the imaginary part of the analytic function Log z in the same domain. Furthermore,

$$\text{Arg } x = \begin{cases} 0 & x > 0 \\ \pi & x < 0 \end{cases}.$$

Hence, one deduces that the function

$$T(z) = \frac{U}{\pi} \text{Arg } z = \frac{U}{\pi} \text{Im(Log } z), \quad \text{Im } z > 0, \qquad (3.3.12)$$

is a solution to this temperature distribution problem since $T(z)$ is harmonic in the domain Im $z > 0$ and satisfies the prescribed boundary temperature values. Indeed, by the uniqueness property of solutions to the Laplace equation that governs $T(x, y)$ (see Subsection 7.1.1), this is the *unique* solution to the present temperature distribution problem.

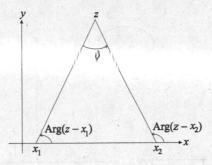

Figure 3.3. The angle θ is facing the segment of the x-axis with boundary temperature value U.

We note that Arg x is related to the well-known Heaviside function $H(x)$ defined by

$$H(x) = \begin{cases} 1 & x > 0 \\ 0 & x < 0 \end{cases}. \tag{3.3.13}$$

In terms of $H(x)$, we observe that

$$T(x, 0) = U[1 - H(x)] \quad \text{and} \quad \text{Arg } x = \pi[1 - H(x)].$$

The above solution method can be easily generalized to the following form of boundary temperature distribution along the x-axis:

$$
\begin{aligned}
T(x, 0) &= \begin{cases} U & x_1 < x < x_2 \\ 0 & x < x_1 \text{ or } x > x_2 \end{cases} \\
&= U[H(x - x_1) - H(x - x_2)]. \tag{3.3.14}
\end{aligned}
$$

Since the Laplace equation is linear, the superposition of solutions is also a solution. The solution to this new problem is seen to be

$$T(z) = \frac{U}{\pi}[\text{Arg}(z - x_2) - \text{Arg}(z - x_1)], \quad \text{Im } z > 0. \tag{3.3.15}$$

A simple geometrical visualization may be helpful to recognize the form of the above solution. In Figure 3.3, we set θ to be the angle subtended at z and bounded by the rays joining z to the points $z = x_1$ and $z = x_2$ on the x-axis. Geometrically, the subtended angle faces the segment of the x-axis with boundary temperature value U. The value of θ is seen to be $\text{Arg}(z - x_2) - \text{Arg}(z - x_1)$, and so

$$T(z) = \frac{U}{\pi}\theta, \quad \text{Im } z > 0. \tag{3.3.16}$$

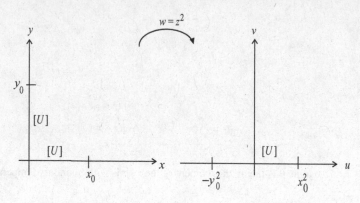

Figure 3.4. The mapping $w = z^2$ maps the first quadrant in the z-plane onto the upper half w-plane. The boundary temperature value along the u-axis becomes $T(u, 0) = U[H(u + y_0^2) - H(u - x_0^2)]$.

For points at close proximity to the two points of discontinuity $z = x_1$ and $z = x_2$, the temperature function can assume any value ranging from 0 to U. Consider the case where $z \to x_2$. It is seen that $\text{Arg}(z - x_1) \approx 0$ and so $T(z) \approx (U/\pi) \text{Arg}(z - x_2)$. For $\text{Im } z > 0$, $\text{Arg}(z - x_2)$ takes values from 0 to π, depending on the direction of approach of the point z to x_2. For example, suppose z approaches x_2 along the vertical direction such that $\text{Arg}(z - x_2) = \pi/2$. Then

$$\lim_{\substack{z \to x_2 \\ \text{Arg}(z - x_2) = \pi/2}} T(z) = \frac{U}{2}.$$

Readers are invited to seek the generalization of the present problem to a boundary temperature distribution which has a finite number of discrete constant values along the x-axis (see Problem 3.25).

Example 3.3.3   Find the steady state temperature distribution inside the first quadrant $x > 0$, $y > 0$, where the boundary temperature values are given by

$$T(x, 0) = \begin{cases} U & 0 \leq x < x_0 \\ 0 & x > x_0 \end{cases} \quad \text{and} \quad T(0, y) = \begin{cases} U & 0 \leq y < y_0 \\ 0 & y > y_0 \end{cases}.$$

Solution   The mapping $w = z^2$ maps the first quadrant in the z-plane, $z = x + iy$, onto the upper half w-plane, $w = u + iv$ (see Figure 3.4).

In the w-plane, the boundary temperature value along the u-axis is given by $T(u, 0) = U[H(u + y_0^2) - H(u - x_0^2)]$. Using eq. (3.3.15), the temperature function in the upper half w-plane is found to be

$$\phi(w) = \frac{U}{\pi}[\text{Arg}(w - r_0^2) - \text{Arg}(w + y_0^2)], \quad \text{Im } w > 0.$$

Transforming back to the z-plane, since $w = z^2$, the temperature function in the first quadrant in the z-plane is given by

$$T(z) = \frac{U}{\pi}\left[\text{Arg}(z^2 - x_0^2) - \text{Arg}(z^2 + y_0^2)\right], \quad \text{Re } z > 0 \text{ and Im } z > 0.$$

This is the solution to the given problem since each term in the above solution is harmonic, and $T(z)$ satisfies the prescribed boundary conditions along the boundary of the first quadrant.

Remark The above solution technique requires the following invariance property of the Laplace equation: if $\phi(x, y)$ is harmonic in a certain domain $\mathcal{D}$ in the z-plane, with $z = x + iy$, and if $w = f(z)$ is an analytic function which maps $\mathcal{D}$ onto a domain $\mathcal{D}'$ in the w-plane, where $w = u + iv$, then $\phi(u, v)$ is harmonic in $\mathcal{D}'$. This invariance property is discussed in detail in Subsection 8.1.1.

3.4 Inverse trigonometric and hyperbolic functions

Since the complex trigonometric and hyperbolic functions are defined in terms of the complex exponentials, we would expect that the inverses of these functions will be expressible in terms of the complex logarithms. Similarly to their real counterparts, the complex inverse trigonometric and hyperbolic functions are multi-valued.

First, we consider the inverse sine function and write

$$w = \sin^{-1} z, \tag{3.4.1}$$

or equivalently,

$$z = \sin w = \frac{e^{iw} - e^{-iw}}{2i}. \tag{3.4.2}$$

Equation (3.4.2) can be considered as a quadratic equation in e^{iw}, that is,

$$e^{2iw} - 2ize^{iw} - 1 = 0.$$

Solving this gives

$$e^{iw} = iz + (1 - z^2)^{\frac{1}{2}}.$$

Taking the logarithm of both sides of the above equation, it then follows that

$$w = \sin^{-1} z = \frac{1}{i} \log\left(iz + (1 - z^2)^{\frac{1}{2}}\right). \tag{3.4.3}$$

When $z \neq \pm 1$, the quantity $(1 - z^2)^{\frac{1}{2}}$ has two possible values. For each value, the logarithm generates infinitely many values. Therefore, $\sin^{-1} z$ has two sets of an infinite number of values. For example, consider

$$\sin^{-1} \frac{1}{2} = \frac{1}{i} \log\left(\frac{i}{2} \pm \frac{\sqrt{3}}{2}\right)$$

$$= \frac{1}{i} \left[\ln 1 + i \left(\frac{\pi}{6} + 2k\pi\right)\right] \text{ or } \frac{1}{i} \left[\ln 1 + i \left(\frac{5\pi}{6} + 2k\pi\right)\right]$$

$$= \frac{\pi}{6} + 2k\pi \quad \text{or} \quad \frac{5\pi}{6} + 2k\pi, \quad k \text{ is any integer.}$$

In a similar manner, we can derive the following formulas for the other inverse trigonometric and hyperbolic functions:

$$\cos^{-1} z = \frac{1}{i} \log(z + (z^2 - 1)^{\frac{1}{2}}), \tag{3.4.4a}$$

$$\tan^{-1} z = \frac{1}{2i} \log \frac{1 + iz}{1 - iz} \cot^{-1} z = \frac{1}{2i} \log \frac{z + i}{z - i}, \tag{3.4.4b}$$

$$\sinh^{-1} z = \log(z + (1 + z^2)^{\frac{1}{2}}), \quad \cosh^{-1} z = \log(z + (z^2 - 1)^{\frac{1}{2}}), \tag{3.4.4c}$$

$$\tanh^{-1} z = \frac{1}{2} \log \frac{1 + z}{1 - z}, \quad \coth^{-1} z = \frac{1}{2} \log \frac{z + 1}{z - 1}, \text{ etc.} \tag{3.4.4d}$$

The derivative formulas for the inverse trigonometric functions are

$$\frac{d}{dz} \sin^{-1} z = \frac{1}{(1 - z^2)^{\frac{1}{2}}}, \quad \frac{d}{dz} \cos^{-1} z = -\frac{1}{(1 - z^2)^{\frac{1}{2}}},$$

$$\frac{d}{dz} \tan^{-1} z = \frac{1}{1 + z^2}, \quad \text{and so forth.} \tag{3.4.5}$$

Example 3.4.1   Find explicitly all values of $\tan^{-1}(i - 2)$.

Solution   Applying formula (3.4.4b), we have

$$\tan^{-1}(i - 2) = \frac{1}{2i} \log \frac{1 + i(i - 2)}{1 - i(i - 2)}$$

$$= \frac{1}{2i} \log \frac{-2i}{2(1 + i)}$$

$$= n\pi - \frac{3\pi}{8} + \frac{i}{4} \ln 2, \quad n = 0, \pm 1, \pm 2, \ldots .$$

Example 3.4.2 Show that if θ is real and $\sin\theta \sin\phi = 1$, then

$$\phi = \left(n + \frac{1}{2}\right)\pi \pm i\ln\left|\tan\frac{\theta}{2}\right|,$$

where n is an integer, even or odd, according to whether $\sin\theta > 0$ or $\sin\theta < 0$.

Solution Let $\phi = \alpha + i\beta$. The given equation then takes the form

$$\sin(\alpha + i\beta) = \operatorname{cosec}\theta.$$

By equating the real and imaginary parts of $\sin(\alpha + i\beta)$ to those of $\operatorname{cosec}\theta$, we obtain

$$\sin\alpha\,\cosh\beta = \operatorname{cosec}\theta \quad \text{and} \quad \cos\alpha\,\sinh\beta = 0.$$

The second equation implies either $\cos\alpha = 0$ or $\sinh\beta = 0$. However, when $\sinh\beta = 0$, we have $\cosh\beta = 1$. Substituting back into the first equation, we get

$$\sin\alpha = \operatorname{cosec}\theta.$$

There will be no solution for real α except when $\theta = \left(n + \frac{1}{2}\right)\pi$, where n is any integer. In this case, $\phi = \theta = \left(n + \frac{1}{2}\right)\pi$, which is just trivial. Excluding this trivial case, we are then left with $\cos\alpha = 0$. This gives $\alpha = \left(n + \frac{1}{2}\right)\pi$, where n is any integer, and β is determined from

$$\cosh\beta = \frac{\operatorname{cosec}\theta}{\sin\left(n + \dfrac{1}{2}\right)\pi} = \begin{cases} \operatorname{cosec}\theta & \text{if } n \text{ is even} \\ -\operatorname{cosec}\theta & \text{if } n \text{ is odd} \end{cases}.$$

Since $\cosh\beta > 0$ for all β, we have to choose n to be even when $\sin\theta > 0$ or n to be odd when $\sin\theta < 0$. The above results can be represented in the following succinct form:

$$\cosh\beta = \frac{e^\beta + e^{-\beta}}{2} = \operatorname{cosec}(\theta + n\pi).$$

The above equation can be considered as a quadratic equation in e^β. On solving for e^β, we obtain

$$e^\beta = \frac{1 \pm \cos(\theta + n\pi)}{\sin(\theta + n\pi)}$$

$$= \tan\frac{\theta + n\pi}{2} \quad \text{or} \quad \cot\frac{\theta + n\pi}{2}.$$

Combining these results, we observe that

(i) when $\sin\theta > 0$, n has to be even (write $n = 2k$, k is an integer), so

$$\tan\frac{\theta + n\pi}{2} = \tan\left(\frac{\theta}{2} + k\pi\right) = \left|\tan\frac{\theta}{2}\right|;$$

(ii) when $\sin\theta < 0$, n has to be odd (write $n = 2k + 1$, k is an integer), so

$$\tan\frac{\theta + n\pi}{2} = \tan\left(\frac{\theta + \pi}{2} + k\pi\right) = -\cot\frac{\theta}{2} = \left|\cot\frac{\theta}{2}\right|.$$

By following a similar argument, we obtain

$$\cot\frac{\theta + n\pi}{2} = \begin{cases} \left|\cot\frac{\theta}{2}\right| & \text{when } \sin\theta > 0 \\ \left|\tan\frac{\theta}{2}\right| & \text{when } \sin\theta < 0 \end{cases}.$$

Therefore, the possible solutions to β are given by

$$\beta = \ln\left|\tan\frac{\theta}{2}\right| \quad \text{or} \quad \beta = \ln\left|\cot\frac{\theta}{2}\right| = -\ln\left|\tan\frac{\theta}{2}\right|.$$

Summing up the results, the solutions for ϕ can be represented by

$$\phi = \alpha + i\beta = \left(n + \frac{1}{2}\right)\pi \pm i\ln\left|\tan\frac{\theta}{2}\right|,$$

where n is an integer, even or odd according to $\sin\theta > 0$ or $\sin\theta < 0$, respectively.

As a numerical example, consider the solution to

$$\sin\phi = 2 = \frac{1}{\sin\frac{\pi}{6}}.$$

Here $\theta = \frac{\pi}{6}$ and $\sin\theta > 0$, so we obtain

$$\phi = \left(2k + \frac{1}{2}\right)\pi \pm i\ln\left(\tan\frac{\pi}{12}\right), \quad k \text{ is any integer.}$$

Similarly, the solution to

$$\sin\phi = -2$$

is given by

$$\phi = \left(2k - \frac{1}{2}\right)\pi \pm i\ln\left(\tan\frac{\pi}{12}\right), \quad k \text{ is any integer.}$$

Referring to the mapping of $w = \sin z$ (see Figure 3.2a), the pair of points $z = \frac{\pi}{2} \pm i\ln\left(\tan\frac{\pi}{12}\right)$ in the z-plane are mapped to the same point $w = 2$. This is consistent with the mapping property that the vertical line $\text{Re}\,z = \frac{\pi}{2}$ in the z-plane is mapped onto the semi-infinite line $\text{Re}\,w \geq 1$ and $\text{Im}\,w = 0$ in the w-plane.

In general, given $w = \alpha_0$, where $\alpha_0 > 1$, we find $\theta_0, 0 < \theta_0 < \frac{\pi}{2}$, such that $\sin \theta_0 = \frac{1}{\alpha_0}$. The pair of preimages of $w = \alpha_0$ within the strip $-\frac{\pi}{2} \le \operatorname{Re} z \le \frac{\pi}{2}$ under the mapping $w = \sin z$ are

$$z = \frac{\pi}{2} \pm i \ln \left(\tan \frac{\theta_0}{2} \right).$$

Similarly, the pair of preimages of $w = -\alpha_0$ within the same strip under the same mapping are

$$z = -\frac{\pi}{2} \pm i \ln \left(\tan \frac{\theta_0}{2} \right).$$

3.5 Generalized exponential, logarithmic and power functions

In this section, we would like to give definitions of the following quantities: i^z, z^i, $\log_i z$, etc. First, consider the *generalized exponential function*

$$f(z) = a^z, \tag{3.5.1}$$

where a is complex in general and $z = x + iy$ is a complex variable. Suppose we write

$$\log a = \ln |a| + i(\operatorname{Arg}\, a + 2k\pi), \quad k = 0, \pm 1, \pm 2, \ldots;$$

since $a = e^{\log a}$, we then have

$$a^z = e^{[\ln |a| + i(\operatorname{Arg}\, a + 2k\pi)]\,(x+iy)}$$

$$= e^{[x \ln |a| - y(\operatorname{Arg}\, a + 2k\pi)]}\, e^{i[y \ln |a| + x(\operatorname{Arg}\, a + 2k\pi)]}$$

$$= |a|^x e^{-y(\operatorname{Arg}\, a + 2k\pi)} [\cos(y \ln |a| + x(\operatorname{Arg}\, a + 2k\pi))$$

$$+ i \sin(y \ln |a| + x(\operatorname{Arg}\, a + 2k\pi))], \quad k = 0, \pm 1, \pm 2, \ldots. \tag{3.5.2}$$

The function is not multi-valued, though apparently we have infinitely many choices for k. The value k is related to the choice of the value for the argument of a and it has nothing to do with the complex variable z. Each choice of k corresponds to a separate function but not a particular branch of a multi-valued function. We usually take $k = 0$ as convention so that when a is real, the expression for a^z in eq. (3.5.2) reduces to

$$a^z = e^{(\ln a)z}. \tag{3.5.3}$$

The inverse of the generalized exponential function a^z is the logarithm function to the complex base a, denoted by $\log_a z$. Suppose we write $w = \log_a z$; then $z = a^w$. Similarly, we choose a branch of a^w as above, that is, fix the value of k in $\log a$, where $\log a = \operatorname{Log} a + 2k\pi i$. Once the branch is fixed (say $k = k_0$ for some chosen k_0), then $z = e^{\lambda w}$, where $\lambda = \text{Log } a + 2k_0\pi i$. This leads to

$$w = \log_a z = \frac{\log z}{\lambda} = \frac{\log z}{\text{Log } a + 2k_0\pi i}. \tag{3.5.4}$$

In the denominator, λ is one of the infinitely many possible values of $\log a$. Note that $\log_a z$ remains a multi-valued function, like other logarithm functions with a real base. For example, suppose we choose k_0 to be 2. Then

$$\text{Log}_{1+i}(1 - i) = \frac{\text{Log}(1 - i)}{\frac{1}{2}\ln 2 + i\left(4\pi + \frac{\pi}{4}\right)} = \frac{\frac{1}{2}\ln 2 - \frac{\pi}{4}i}{\frac{1}{2}\ln 2 + \frac{17\pi}{4}i}.$$

Lastly, we consider the *generalized power function*

$$f(z) = z^a, \tag{3.5.5}$$

where a is complex in general and

$$z = x + iy = re^{i\theta} = |z|e^{i(\text{Arg } z + 2k\pi)}$$

is a complex variable. Consider the following cases:

(i) When $a = n$, n is an integer,

$$z^n = |z|^n \, e^{in\text{Arg } z}.$$

(ii) When a is rational, $a = m/n$ where m, n are irreducible integers, we have

$$z^{\frac{m}{n}} = e^{\frac{m}{n}\log z}$$
$$= |z|^{\frac{m}{n}} \, e^{i\frac{m}{n}\text{Arg } z} \, e^{2k\frac{m}{n}\pi i}, \qquad k = 0, 1, \dots, n-1.$$

The factor $e^{2k\frac{m}{n}\pi i}$ takes on n different values for $k = 0, 1, \dots, n-1$, then repeats itself with period n if k continues to increase through the integers. This is precisely the same result as that of De Moivre's theorem (see Subsection 1.2.1). The power function has n different branches, corresponding to the different values of k.

(iii) When $a = \alpha + i\beta$,

$$z^a = e^{(\alpha + i\beta)[\ln|z| + i(\text{Arg } z + 2k\pi)]}$$
$$= e^{\alpha \ln|z| - \beta(\text{Arg } z + 2k\pi)} \, e^{i[\beta \ln|z| + \alpha(\text{Arg } z + 2k\pi)]}$$
$$= |z|^\alpha e^{-\beta(\text{Arg } z + 2k\pi)}[\cos(\beta \ln|z| + \alpha(\text{Arg } z + 2k\pi))$$
$$+ i\sin(\beta \ln|z| + \alpha(\text{Arg } z + 2k\pi))], \quad k = 0, \pm 1, \pm 2, \dots.$$

In this case, z^a has infinitely many branches.

Example 3.5.1 Find the principal value of each of the following complex quantities:

(a) $(1 - i)^{1+i}$; (b) 3^{3-i}; (c) 2^{2i}.

Solution

(a) Principal value of $(1 - i)^{1+i} = e^{(1+i)\text{Log}(1-i)} = e^{(1+i)(\ln\sqrt{2}-\frac{\pi}{4}i)}$

$$= e^{\left(\ln\sqrt{2}+\frac{\pi}{4}\right)+i\left(\ln\sqrt{2}-\frac{\pi}{4}\right)}$$

$$= \sqrt{2}e^{\frac{\pi}{4}}\left[\cos\left(\ln\sqrt{2}-\frac{\pi}{4}\right)\right.$$
$$\left. +i\sin\left(\ln\sqrt{2}-\frac{\pi}{4}\right)\right].$$

(b) Principal value of $3^{3-i} = e^{(3-i)\text{Log}\,3} = e^{3\ln 3 - i\ln 3}$

$$= 27\left[\cos(\ln 3) - i\sin(\ln 3)\right].$$

(c) Principal value of $2^{2i} = e^{2i\ln 2} = \cos(\ln 4) + i\sin(\ln 4)$.

Example 3.5.2 The power function

$$w = f(z) = [z(z - 1)(z - 2)]^{1/2}$$

has two branches. Show that $f(-1)$ can be either $-\sqrt{6}i$ or $\sqrt{6}i$. Suppose the branch that corresponds to $f(-1) = -\sqrt{6}i$ is chosen; find the value of the function at $z = i$.

Solution The given power function can be expressed as

$$w = f(z) = |z(z - 1)(z - 2)|^{1/2}\, e^{i[\text{Arg}\,z\, +\, \text{Arg}(z-1)\, +\, \text{Arg}(z-2)]/2}\, e^{ik\pi}, \quad k = 0, 1,$$

where the two possible values of k correspond to the two branches of the double-valued power function. Note that at $z = -1$,

$$\text{Arg}\,z = \text{Arg}(z - 1) = \text{Arg}(z - 2) = \pi \quad \text{and} \quad |z(z - 1)(z - 2)|^{1/2} = \sqrt{6},$$

so $f(-1)$ can be either $\sqrt{6}e^{i3\pi/2} = -\sqrt{6}i$ or $\sqrt{6}e^{i3\pi/2}e^{i\pi} = \sqrt{6}i$. The branch that gives $f(-1) = -\sqrt{6}i$ corresponds to $k = 0$. With the choice of that branch, we have

$$f(i) = |i(i-1)(i-2)|^{1/2} \, e^{i[\text{Arg } i \, + \, \text{Arg}(i-1) \, + \, \text{Arg}(i-2)]/2}$$
$$= (\sqrt{2}\sqrt{5})^{1/2} \, e^{i(\frac{\pi}{2}+\frac{3\pi}{4}+\pi-\tan^{-1}\frac{1}{2})/2}$$
$$= (10)^{1/4}e^{i\pi}e^{\frac{i}{2}(\frac{\pi}{4}-\tan^{-1}\frac{1}{2})}$$
$$= -(10)^{1/4}e^{\frac{i}{2}(\tan^{-1}1-\tan^{-1}\frac{1}{2})}$$
$$= -(10)^{1/4}e^{\frac{i}{2}\left(\tan^{-1}\frac{1-\frac{1}{2}}{1+\frac{1}{2}}\right)}$$
$$= -(10)^{1/4}e^{\frac{i}{2}\tan^{-1}\frac{1}{3}}.$$

3.6 Branch points, branch cuts and Riemann surfaces

We have discussed the multi-valuedness of power functions and logarithmic functions in previous sections. The mapping property of a complex function would be much easier to visualize if the mapping were one-to-one. A function $f(z)$ is said to be *univalent* in a domain $\mathcal{D}$ if it is one-to-one and analytic in $\mathcal{D}$. Correspondingly, $\mathcal{D}$ is called a *domain of univalence* for $f(z)$. For example, the function $w = z^n$, n is a positive integer, is univalent in the sectoral domain $\{z : \theta_0 < \arg z < \theta_0 + \frac{2\pi}{n}, \; \theta_0 \text{ is any value}\}$.

The inverse of the above function is $w = z^{1/n}$, known to be multi-valued. Given any value of z, there are n distinct nth roots of z. Each root corresponds to a specific branch of the multi-valued function. Every branch of a multi-valued function has the same domain. How do we separate the branches so that every branch has its own copy of the domain? Each branch of the function now becomes single-valued in its own domain.

In this section, we would like to discuss an ingenious construction, known as the Riemann surfaces, in order to achieve the above objective. A *Riemann surface* consists of overlapping sheets (the number of sheets can be finite or infinite) and these sheets are connected by *branch cuts*. The end points of a branch cut are called the *branch points*. The concepts of branch points, branch cuts and Riemann surfaces for multi-valued functions are exemplified below.

First, consider the simple example $w = z^{1/2}$. For a given $z_0 = r_0 e^{i\theta_0}$ ($0 \le \theta_0 < 2\pi$), the two roots are $\sqrt{r_0}e^{i\theta_0/2}$ and $\sqrt{r_0}e^{i(\theta_0+2\pi)/2} = -\sqrt{r_0}e^{i\theta_0/2}$. Suppose we start from $r_0 e^{i\theta_0}$ in the z-plane and follow the path of z which traverses one complete loop around the origin and back to the starting point; the argument of z increases from θ_0 to $\theta_0 + 2\pi$. The corresponding image point in the w-plane changes from $\sqrt{r_0}e^{i\theta_0/2}$ to $\sqrt{r_0}e^{i(\theta_0+2\pi)/2} = -\sqrt{r_0}e^{i\theta_0/2}$. Note that the two image points lie in different branches of the double-valued function $w = z^{1/2}$. Suppose the path of z traverses once more around $z = 0$ in the

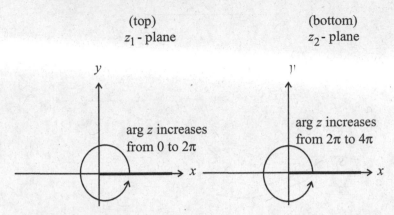

Figure 3.5. The Riemann surface of $w = z^{1/2}$ consists of two sheets: z_1-plane (top) and z_2-plane (bottom). The branch points are $z = 0$ and $z = \infty$, and the branch cut is taken to be along the positive real axis on each sheet. The path of z traversing a closed circuit around the origin moves from one sheet to the other sheet.

anticlockwise direction and back to the same starting point; the argument of z now becomes $\theta_0 + 4\pi$ and leads $w = z^{1/2}$ to return to the value $r_0 e^{i\theta_0/2}$. Note that the value $w = z^{1/2}$ remains unchanged if the complete circuit in the z-plane does not include the point $z = 0$.

Suppose any closed loop around a point always carries every branch of a given multi-valued function into another branch. Then the enclosed point is called a *branch point* of the function. The branch point is said to be of order $n - 1$ (a finite positive integer) if n complete circuits in the same direction around the point carry every branch of the function back to itself. For example, $z = 0$ is a branch point of order one of the function $w = z^{1/2}$.

Is the point at infinity a branch point of $w = z^{1/2}$? To answer this question, we need to examine the case of a complete loop traversing around the complex infinity. Recall that the complex infinity in the extended z-plane corresponds to the north pole on the Riemann sphere, and so a small closed curve around the north pole on the Riemann sphere corresponds to a large closed curve in the complex plane. When the path of z moves around a large closed loop in the complex plane, $w = z^{1/2}$ is seen to move to a new branch. Hence, $z = \infty$ is also a branch point of $w = z^{1/2}$.

A multi-valued function may be regarded as single-valued if we suitably generalize its domain of definition. For the function $w = z^{1/2}$, suppose we take two copies of the z-plane superimposed upon each other. The argument of z on the top sheet (called it the z_1-plane) ranges from 0 to 2π, while that on the bottom sheet (called it the z_2-plane) ranges from 2π to 4π (see Figure 3.5). In order that the path of z that traverses a complete circuit around the origin moves

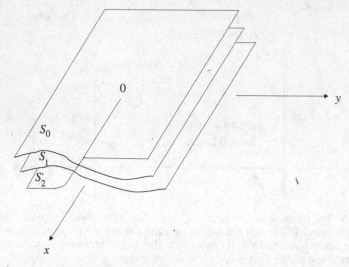

Figure 3.6. Sketch of the three-sheeted structure of the Riemann surface for the multi-valued function $w = z^{1/3}$. The bottom sheet S_2 is joined back to the top sheet S_0. The branch cuts are chosen to be along the positive real axis.

from one sheet to the other sheet, we place a cut along the positive real axis of each sheet. This cut is called the *branch cut*, and it links the two branches of the multi-valued function. Naturally, the two ends of this branch cut are the two branch points of the function. The lower edge along the branch cut of the top sheet (the ray defined by arg $z = 2\pi^-$) is joined to the upper edge along the branch cut of the bottom sheet (the ray defined by arg $z = 2\pi^+$). Likewise, the lower edge of the cut on the bottom sheet (the ray defined by arg $z = 4\pi^-$) is joined to the upper edge of the cut on the top sheet (the ray defined by arg $z = 0^+$). These two sheets together with the branch cuts constitute the *Riemann surface* of the double-valued function $w = z^{1/2}$.

The construction of the Riemann surface can be easily generalized to $w = z^{1/n}$, where n is a positive integer. There will be n sheets in the corresponding Riemann surface. The branch points are $z = 0$ and $z = \infty$, and the order is equal to $n - 1$. A sketch of the three-sheeted structure of the Riemann surface for $w = z^{1/3}$ is illustrated in Figure 3.6.

Logarithmic function

The logarithmic function $w = \log z$ has infinitely many values for each z, so we expect its Riemann surface to consist of infinitely many sheets. The sheets are joined together in a similar manner to those of $w = z^{1/2}$, that is, the lower edge of the cut on the z_k-sheet is joined to the upper edge of the cut on the z_{k+1}-sheet. However, it is *not* necessary to join the first and the last sheet, unlike those of $w = z^{1/n}$. Indeed it is impossible to define which sheet is the first and which is the last when the number of sheets is infinite. The branch points can be deduced to be $z = 0$ and $z = \infty$, using the same technique of observing the change of branch when z moves around a closed loop containing the branch point.

Where do we place the branch cut? Supposing the principal branch of the logarithmic function is chosen such that Arg z is lying between $-\pi$ and π, the choice of the branch cut is then taken along the negative real axis of each Riemann sheet.

Example 3.6.1   Consider the multi-valued function

$$w = f(z) = (z^2 - 1)^{1/2}.$$

Find the two branch points of the function. Describe the possible branch cut and the Riemann surface of the function.

Solution   We let

$$z - 1 = r_1 e^{i\theta_1} \qquad \text{and} \qquad z + 1 = r_2 e^{i\theta_2}$$

so that

$$w = [r_1 r_2 e^{i(\theta_1 + \theta_2)}]^{\frac{1}{2}} = \sqrt{r_1 r_2} \, e^{\frac{i\theta_1}{2}} e^{\frac{i\theta_2}{2}}.$$

Suppose we start with a particular point z_0 such that $\theta_1 = \alpha_1$ and $\theta_2 = \alpha_2$. When z moves in the anticlockwise sense once around $z = 1$ but not $z = -1$, the value of θ_1 increases from α_1 to $\alpha_1 + 2\pi$ but θ_2 remains the same value. The new value of w becomes

$$w = \sqrt{r_1 r_2} \, e^{\frac{i(\alpha_1 + 2\pi)}{2}} e^{\frac{i\alpha_2}{2}} = -\sqrt{r_1 r_2} \, e^{\frac{i\alpha_1}{2}} e^{\frac{i\alpha_2}{2}},$$

which is different from the original value. This signifies a change in branch and therefore $z = 1$ is a branch point of $f(z)$. Similarly, if we consider a closed path around $z = -1$ but not $z = 1$, θ_1 remains the same value but θ_2 increases from α_2 to $\alpha_2 + 2\pi$. This causes a change in the value of w. Therefore, $z = -1$ is the other branch point of $f(z)$. If z moves in the anticlockwise sense around a sufficiently large circuit that includes both branch points $z = \pm 1$, then θ_1 increases from α_1 to $\alpha_1 + 2\pi$ and θ_2 also increases from α_2 to $\alpha_2 + 2\pi$. Now, the value of w becomes

$$w = \sqrt{r_1 r_2} e^{\frac{i(\alpha_1 + 2\pi)}{2}} e^{\frac{i(\alpha_2 + 2\pi)}{2}} = \sqrt{r_1 r_2} \, e^{\frac{i\alpha_1}{2}} e^{\frac{i\alpha_2}{2}},$$

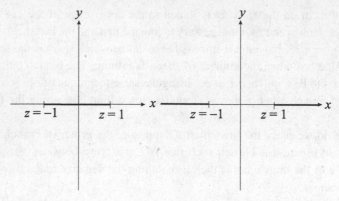

Figure 3.7. One choice of the branch cut of the multi-valued function $f(z) = (z^2 - 1)^{1/2}$ is along the line segment joining $z = -1$ and $z = 1$. A new branch of the function is encountered when the path of z traverses across the branch cut. Another possible choice of the branch cut is the union of the two line segments: $z = -1$ to $z = \infty$ along the negative real axis and $z = 1$ to $z = \infty$ along the positive real axis. In this case, one may visualize the two branch points as being joined by a line segment going through the complex infinity.

which is the same as the original value. There is no change in branch, so $z = \infty$ is not a branch point.

The branch cut can be taken to be either (i) a cut between the branch points $z = 1$ and $z = -1$, or (ii) two cuts along the real axis, one from $z = 1$ to $z = \infty$ and the other from $z = -1$ to $z = \infty$ (see Figure 3.7). The single-valuedness of the function is ensured when the path of z does not cross these branch cuts. The Riemann surface of the function consists of two sheets superimposed on each other and they are joined along the branch cuts. The end points of the cuts are $z = -1$ and $z = 1$. On making one complete loop around either $z = -1$ or $z = 1$ but not both, we start on one sheet and wind up on another sheet of the Riemann surface.

Example 3.6.2 The inverse cosine function is related to the logarithm function by [see eq. (3.4.4a)]

$$w = \cos^{-1} z = \frac{1}{i} \log(z + (z^2 - 1)^{1/2}).$$

Find the branch points and branch cuts of this multi-valued function.

Solution Since $z = -1$ and $z = 1$ are known to be the branch points of the double-valued function $(z^2 - 1)^{1/2}$, the same is true for the inverse cosine function. Though $z = \infty$ is not a branch point of $(z^2 - 1)^{1/2}$, the argument of $z + (z^2 - 1)^{1/2}$ increases by 2π when z traverses in a simple closed curve encircling both branch points $z = -1$ and $z = 1$. Therefore, $z = \infty$ is seen to be a branch point of $\frac{1}{i} \log(z + (z^2 - 1)^{1/2})$. Since $z = \infty$ is a branch point of $\cos^{-1} z$, the branch cuts are chosen to be semi-infinite line segments emanating from the two branch points $z = -1$ and $z = 1$ and extending to infinity.

A convenient choice of the branch cuts would be the two line segments along the real axis: $-\infty < x < -1$ and $1 < x < \infty$. Let $\mathcal{D}$ be the domain of definition of the branch with respect to these two branch cuts. Each branch of the inverse cosine function maps $\mathcal{D}$ in the z-plane onto a vertical infinite strip of width 2π in the w-plane. The single-valued branch function takes the upper half-plane of $\mathcal{D}$ onto a vertical infinite strip of width π and the lower half-plane of $\mathcal{D}$ onto a neighboring vertical infinite strip of the same width.

3.6.1 Joukowski mapping

The complex function

$$w = f(z) = \frac{1}{2}\left(z + \frac{1}{z}\right), \quad z \neq 0, \tag{3.6.1}$$

is called the Joukowski function. This function is analytic everywhere except at $z = 0$. Since $w = f(z) = f\left(\frac{1}{z}\right)$, the pair of points z_1 and z_2 in the z-plane which satisfy $z_1 z_2 = 1$ are mapped onto the same image point in the w-plane. Hence, the Joukowski mapping is univalent in a domain $\mathcal{D}$ if and only if there are no two points z_1 and z_2 in $\mathcal{D}$ which satisfy $z_1 z_2 = 1$. For example, the Joukowski mapping is univalent in the following domains: (i) $|z| < 1$, (ii) $|z| > 1$, (iii) Im $z > 0$, (iv) Im $z < 0$.

The Joukowski mapping is a two-to-one mapping. To examine its mapping properties, we write $z = re^{i\theta}$ and $w = u + iv$, and consider

$$w = u + iv = \frac{1}{2}\left(re^{i\theta} + \frac{1}{r}e^{-i\theta}\right)$$

$$= \frac{1}{2}\left(r + \frac{1}{r}\right)\cos\theta + \frac{i}{2}\left(r - \frac{1}{r}\right)\sin\theta,$$

so that

$$u = \frac{1}{2}\left(r + \frac{1}{r}\right)\cos\theta \quad \text{and} \quad v = \frac{1}{2}\left(r - \frac{1}{r}\right)\sin\theta.$$

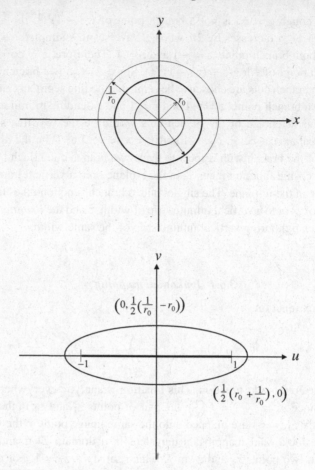

Figure 3.8. The Joukowski mapping takes concentric circles centered at the origin in the z-plane onto cofocal ellipses in the w-plane. Any two circles with the product of their radii equal to one are mapped onto the same ellipse. In particular, the unit circle $|z| = 1$ is mapped onto the line segment $[-1, 1]$ on the u-axis (considered as a degenerated ellipse).

The circle $|z| = r_0 < 1$ is mapped onto the ellipse defined by (see Figure 3.8)

$$u = \frac{1}{2}\left(r_0 + \frac{1}{r_0}\right)\cos\theta \quad \text{and} \quad v = \frac{1}{2}\left(r_0 - \frac{1}{r_0}\right)\sin\theta.$$

The semi-major and semi-minor axes of the ellipse are seen to be $\frac{1}{2}\left(r_0 + \frac{1}{r_0}\right)$ and $\frac{1}{2}\left(\frac{1}{r_0} - r_0\right)$, respectively. The ellipse degenerates into the line segment $[-1, 1]$ on the u-axis when $r_0 \to 1$.

The domain $\{z: |z| < 1\}$ in the z-plane is mapped onto the whole w-plane minus the line segment $[-1, 1]$ along the real axis. By virtue of the reciprocity property $f(z) = f\left(\frac{1}{z}\right)$, the domain exterior to the unit circle in the z-plane is also mapped onto the whole w-plane (see Figure 3.8).

The inverse of the Joukowski function is found to be

$$w = z + (z^2 - 1)^{1/2}, \tag{3.6.2}$$

which is a double-valued function. Following similar techniques used in Example 3.6.1, one can check that $z = 1$ and $z = -1$ are branch points of order one, but $z = \infty$ is not a branch point of the Riemann surface of the inverse Joukowski function.

Suppose we choose the domain $\mathcal{D}$ to be the whole z-plane minus the line segment $[-1, 1]$ on the x-axis. Then the two branches of the inverse Joukowski function are given by

$$w_1 = z + \sqrt{z^2 - 1} \quad \text{with} \quad w_1(i) = (1 + \sqrt{2})i, \tag{3.6.3a}$$

$$w_2 = z - \sqrt{z^2 - 1} \quad \text{with} \quad w_2(i) = (1 - \sqrt{2})i. \tag{3.6.3b}$$

They map the domain $\mathcal{D}$ onto the exterior and interior of the unit circle $|w| = 1$ in the w-plane, respectively (see also Problem 3.37).

Example 3.6.3 Find an analytic function $w = f(z)$ which maps the interior of the unit circle $|z| < 1$ minus the segments $(-1, -1 + h]$ and $[1 - h, 1)$, where $0 < h < 1$, on the real axis onto the interior of the unit circle $|w| < 1$.

Solution Recall that the Joukowski mapping

$$w_1 = \frac{1}{2}\left(z + \frac{1}{z}\right)$$

takes the domain $\{z: |z| < 1\}$ to the whole w_1-plane minus the segment $[-1, 1]$ along the real axis. Also, the segments $(-1, -1 + h]$ and $[1 - h, 1)$ along the real axis in the z-plane are mapped onto the segments $[-\delta, -1)$ and $(1, \delta]$ along the real axis in the w_1-plane, where $\delta = \frac{1}{2}\left(1 - h + \frac{1}{1-h}\right)$. Note that $[-\delta, \delta]$ is the union of the three segments: $[-\delta, -1)$, $[-1, 1]$ and $(1, \delta]$. Hence, the above Joukowski mapping takes the given domain in the z-plane onto the whole w_1-plane minus the segment $[-\delta, \delta]$ along the real axis. Suppose we take

$$w_2 = \frac{w_1}{\delta};$$

then the segment $[-\delta, \delta]$ on the real axis in the w_1-plane becomes $[-1, 1]$ on the real axis in the w_2-plane. By virtue of eq. (3.6.3b), the inverse of the Joukowski mapping

$$w = w_2 - \sqrt{w_2^2 - 1}$$

takes the whole w_2-plane minus the segment $[-1, 1]$ on the real axis onto the interior of the unit circle $|w| < 1$. Combining the three transformations, the analytic function which effects the mapping is found to be

$$w = \frac{z + \frac{1}{z}}{1 - h + \frac{1}{1-h}} - \sqrt{\left(\frac{z + \frac{1}{z}}{1 - h + \frac{1}{1-h}} \right)^2 - 1}.$$

3.7 Problems

3.1. Verify the following identities:

    (a) $\cosh^2 z - \sinh^2 z = 1$;   (b) $\cosh^2 z + \sinh^2 z = \cosh 2z$;
    (c) $\sinh(z_1 + z_2) = \sinh z_1 \cosh z_2 + \cosh z_1 \sinh z_2$.

3.2. Show that

    (a) $\tan iz = i \tanh z$;   (b) $\cot iz = -i \coth z$;
    (c) $\coth z = \dfrac{\sinh 2x - i \sin 2y}{\cosh 2x - \cos 2y}$, $z = x + iy$.

3.3. Show that if $\cos(z + w) = \cos z$ for any z in the complex plane, then $w = 2k\pi$, where k is any integer.

3.4. Find the general solution for each of the following equations:

    (a) $e^z = 2i$; (b) $\sin z = \cosh 3$.

3.5. Show that

    (a) $|e^z|$ is bounded when $\operatorname{Re} z \leq \alpha$;
    (b) $|\cos(x + iy)|$ is unbounded as $y \to \infty$.

3.6. For each of the functions

    (a) $\sin z$, (b) $\tan z$, (c) $\coth z$,
       find the set of points z such that the function assumes
       (i) real value,
       (ii) purely imaginary value.

3.7. Let $z = x + iy$; express the following modulus quantities in terms of x and y,

(a) $|\tan z|$;   (b) $|\tanh z|$.

3.8. For $0 < |z| < 1$, show that

$$\frac{|z|}{4} < |e^z - 1| < \frac{7|z|}{4}.$$

Hint: Consider

$$|e^z - 1| \le |z| \left(1 + \frac{|z|}{2!} + \frac{|z|^2}{3!} + \cdots \right)$$

$$\le |z| \left[1 + \frac{1}{2} \left(1 + \frac{1}{3} + \frac{1}{3^2} + \cdots \right) \right] \text{ since } 0 < |z| < 1.$$

In a similar manner, we obtain

$$|e^z - 1| \ge |z| \left[1 - \left(\frac{1}{2!} + \frac{1}{3!} + \cdots \right) \right].$$

3.9. Prove that

(a) $|e^z - 1| \le e^{|z|} - 1 \le |z| \, e^{|z|}$;
(b) $|\text{Im } z| \le |\sin z| \le e^{|\text{Im } z|}$.

Hint: To prove the left-hand side inequality of part (a), consider

$$|e^z - 1| = |e^{z/2}||e^{z/2} - e^{-z/2}| = 2|e^{z/2}| \left| \sinh \frac{z}{2} \right|$$

$$e^{|z|} - 1 = e^{|z|/2}(e^{|z|/2} - e^{-|z|/2}) = 2e^{|z|/2} \sinh \frac{|z|}{2}.$$

In addition, try to establish

$$\left| \sinh \frac{z}{2} \right| \le \sinh \frac{|z|}{2}.$$

3.10. Find a complex number z that satisfies

$$|\sin z| > 1 \quad \text{and} \quad |\cos z| > 1.$$

3.11. Show that

$$|\log z| = |\log(1 - z)| \quad \text{when Re } z = \frac{1}{2}.$$

3.12. Suppose $|z| \leq R$; show that

$$|\sin z| \leq \cosh R \quad \text{and} \quad |\cos z| \leq \cosh R.$$

3.13. Show that

$$(1+i)\cot(\alpha+i\beta)+(1-i)\cot(\alpha-i\beta) = 2\frac{\sin 2\alpha + \sinh 2\beta}{\cosh 2\beta - \cos 2\alpha}.$$

3.14. Let $z = x + iy$; show that

$$\lim_{n\to\infty}\left|\left(1+\frac{z}{n}\right)^n\right| = e^x \quad \text{and} \quad \lim_{n\to\infty}\text{Arg}\left(1+\frac{z}{n}\right)^n = y.$$

Hint: $\lim_{n\to\infty}\dfrac{\tan^{-1}\frac{y}{n}}{\frac{y}{n}} = 1$

3.15. Find all roots of each of the following equations:

(a) $\sin z + \cos z = 2$;   (b) $\sin z - \cos z = 3$;   (c) $\sin z - \cos z = i$;
(d) $\cosh z - \sinh z = 1$;   (e) $\sinh z - \cosh z = 2i$;
(f) $2\cosh z + \sinh z = i$;   (g) $\cos z = \cosh z$;
(h) $\sin z = i \sinh z$;   (i) $\cos z = i \sinh 2z$.

3.16. (a) Suppose z moves along the parabola $y = x^2$; find

$$\lim_{z\to\infty} e^z.$$

(b) Suppose z moves along the imaginary axis; describe the behavior of the following functions: (i) $\sin z$, (ii) $\cosh z$.
(c) Show that $|\sin z|$ is bounded when $\text{Im } z = \alpha$.

3.17. Evaluate the following quantities:

(a) $\text{Log}(2 - 3i)$;   (b) $\log(-2 + 3i)$;
(c) $\cos(2 + i)$;   (d) $\coth(2 + i)$;
(e) $\tanh\left(\ln 3 + \dfrac{\pi i}{4}\right)$;   (f) $\cos^{-1} i$;   (g) $\sin^{-1} i$;   (h) $\tan^{-1} i$.

3.18. For each of the following functions, find the domain of analyticity:

(a) $\dfrac{e^z}{z\cos z}$;   (b) $\dfrac{e^z}{\sin z + \cos z}$;   (c) $\dfrac{1}{\cosh z + e^z}$.

3.19. Show that

$$\left(\frac{ia-1}{ia+1}\right)^{ib} = \exp(-2b\,\cot^{-1} a), \quad \text{where } a \text{ and } b \text{ are real.}$$

3.20. Show that

$$\coth^{-1}\frac{z}{a} = \frac{1}{2}\ln\frac{z+a}{z-a}$$

$$= \frac{1}{4}\ln\frac{(x+a)^2+y^2}{(x-a)^2+y^2} + \frac{i}{2}\left(\tan^{-1}\frac{2ay}{a^2-x^2-y^2} + 2k\pi\right),$$

where a is positive real and k is any integer.

3.21. Find the image of the infinite horizontal strip $\{z: -1 < \operatorname{Im} z \le 1\}$ under the mapping $w = f(z) = e^{2\pi z}$. Find the corresponding inverse function for $f(z)$ whose image is the above horizontal strip.

3.22. Show that the mapping function

$$w = \cosh z$$

maps the semi-infinite strip $\{z = x + iy : x \ge 0 \text{ and } 0 \le y \le \frac{\pi}{2}\}$ in the z-plane onto the first quadrant of the w-plane.

3.23. Detect any fault in the following argument: using the fact that $(-z)^2 = z^2$, we obtain

$$2\operatorname{Log}(-z) = 2\operatorname{Log} z,$$

and so

$$\operatorname{Log}(-z) = \operatorname{Log} z, \quad \text{for all } z \text{ in } \mathbb{C}\backslash\{0\}.$$

3.24. Consider the temperature function

$$T(z) = \ln|z - z_1| - \ln|z - z_2|,$$

where z_1 and z_2 are distinct points in the complex plane. Explain why $T(z)$ is harmonic everywhere except at z_1 and z_2. Where are the locations of the heat source and heat sink in the temperature field defined by this temperature function? Find the set of points at which the temperature value equals zero.

3.25. Find the steady state temperature distribution in the upper half-plane, where the boundary temperature values along the x-axis are given by

$$T(x, 0) = \begin{cases} U_1 & x < x_1 \\ U_2 & x_1 < x < x_2 \\ \vdots & \vdots \\ U_n & x_{n-1} < x < x_n \\ U_{n+1} & x_n < x \end{cases}$$

3.26. Find the steady state temperature distribution $T(z)$ in the semi-infinite vertical strip $\{z: -\frac{\pi}{2} < \operatorname{Re} z < \frac{\pi}{2} \text{ and } \operatorname{Im} z > 0\}$, where the boundary temperature values are

$$\begin{cases} T\left(\frac{\pi}{2}+iy\right) = T\left(-\frac{\pi}{2}+iy\right) = 0, & y > 0 \\ T(x) = U, & -\frac{\pi}{2} < x < \frac{\pi}{2}, \qquad U \text{ is real positive} \end{cases}.$$

Hint: The mapping $w = \sin z$ maps the given semi-infinite strip in the z-plane onto the upper half w-plane.

3.27. Consider the multi-valued function

$$w = f(z) = z^{1/3}, \quad z \in \mathbb{C}\backslash[0, \infty).$$

Supposing we choose the branch such that $f(i) = e^{\frac{5}{6}\pi i}$, compute $f(-2)$.

3.28. Consider the multi-valued function

$$f(z) = [(1-z)^3 z]^{1/4}.$$

Show that $z = 0$ and $z = 1$ are branch points of the function, and find their order. Is $z = \infty$ a branch point? Describe the Riemann surface of the function.

3.29. Consider the logarithmic function

$$w = \log(z - \alpha), \quad \alpha \text{ is complex.}$$

Show that $z = \alpha$ and $z = \infty$ are branch points of the function. Explain why any simple curve that starts at $z = \alpha$ and ends at ∞ can be a branch cut of the Riemann surface of the function.

3.30. The inverse tangent function can be expressed as

$$w = \tan^{-1} z = \frac{1}{2i} \log \frac{i-z}{i+z}.$$

Examine whether the points $z = i$, $z = -i$ and $z = \infty$ are branch points of the function. Compute $\tan^{-1} i$, $\tan^{-1}(-i)$ and $\tan^{-1}\infty$.

3.31. Recall that

$$\tanh z = \frac{\sinh z}{\cosh z} = \frac{e^z - e^{-z}}{e^z + e^{-z}}.$$

If we choose the principal branch of $\tanh^{-1} z$ to be that for which $\tanh^{-1} 0 = 0$, prove that

$$\tanh^{-1} z = \frac{1}{2}\text{Log}\left(\frac{1+z}{1-z}\right).$$

Find the derivative of $\tanh^{-1} z$.

3.32. Consider the multi-valued function

$$f(z) = \log(1 - z^2)$$

where the domain $\mathcal{D}$ is defined to be the whole complex plane minus the following three branch cuts (see the figure):

  (i) line segment (including the end points) joining -1 and i,
 (ii) line segment (including the end points) joining 1 and i,
(iii) semi-infinite line segment: $x = 0$, $y \geq 1$.

  (a) Show that the function f can be separated into single-valued branches in the above domain $\mathcal{D}$.
  (b) Supposing we choose the branch where $f(0) = 0$, find the corresponding value of this branch of the function at $z = 2$.
  (c) Is the choice of the point $z = i$ (starting point of the semi-infinite branch cut) unique?

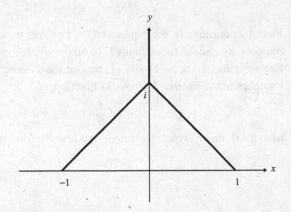

3.33. Is $[z(z + 1)]^{1/2}$ the same function as $z^{1/2}(z + 1)^{1/2}$?

3.34. Consider the many-to-one mapping

$$f(z) = (z - \alpha)(z - \beta),$$

and let ℓ be any line through the point $(\alpha + \beta)/2$ which divides the z-plane into two half-planes. Show that each of the open half-planes determined by ℓ is a domain of univalence for $f(z)$.

3.35. Show that the necessary and sufficient condition for the function

$$f(z) = e^{\alpha z}, \quad \alpha = a + ib \neq 0,$$

to be single-valued inside the infinite strip $\left\{z: -\frac{\pi}{2} < \text{Im } z < \frac{\pi}{2}\right\}$ is given by

$$a^2 + b^2 \leq 2|a|.$$

3.36. Consider the multi-valued function

$$w = f(z) = [z(1-z)^3]^{1/4}.$$

Supposing we choose the branch where

$$f(-1) = 8^{1/4}e^{i\pi/4}e^{i\pi/2} = 2^{1/4}(-1+i),$$

show that the value of w corresponding to this branch along the segment $(0, 1)$ on the x-axis is given by

$$f(x) = i\sqrt[4]{x(1-x)^3}, \quad x \in (0, 1).$$

3.37. Consider the mapping represented by the inverse Joukowski function

$$w = f(z) = z + (z^2 - 1)^{1/2}.$$

Find the preimage in the z-plane of the ray Arg $w = \theta_0$ in the w-plane. Suppose we choose the domain $\mathcal{D}$ to be the whole z-plane minus the two line segments $(-\infty, -1)$ and $(1, \infty)$ on the x-axis. Show that the two branches of the inverse Joukowski function

$$w_1 = z + \sqrt{z^2 - 1} \quad \text{and} \quad w_2 = z - \sqrt{z^2 - 1}$$

map the domain $\mathcal{D}$ onto the upper and lower half w-plane, respectively.

4

Complex Integration

The methods of integration of complex functions and their underlying theories are discussed in this chapter. The cornerstones in complex integration are the Cauchy–Goursat theorem and the Cauchy integral formula. A fascinating result deduced from the Cauchy integral formula is that if a complex function is analytic at a point, then its derivatives of all orders exist and these derivatives are analytic at that point. Other important theorems include Gauss' mean value theorem, Liouville's theorem, and the maximum modulus theorem.

Many properties of the complex integrals are very similar to those of the real line integrals. For example, when the integrand satisfies certain conditions, the integral can be computed by finding the primitive function of the integrand and evaluating the primitive function at the two end points of the integration path. However, there are other properties that are unique to integration in the complex plane.

In the last section, we link the study of conservative fields in physics with the mathematical theory of analytic functions and complex integration. The prototype conservative fields considered include the gravitational potential fields, electrostatic fields and potential fluid flow fields. The potential functions in these physical models are governed by the Laplace equation, and so their solutions are harmonic functions. Complex variables techniques are seen to be effective analytical tools for solving these physical models.

4.1 Formulations of complex integration

The integration of complex functions in the complex plane is seen to resemble closely the integration of real functions in the two-dimensional plane. In this section, we first consider the formulation of the integration of an *arbitrary* complex function following a similar approach to that for a real line integral. The celebrated Cauchy integral theory for dealing with complex integrals involving analytic integrand functions will be discussed in the next section.

4.1.1 Definite integral of a complex-valued function of a real variable

Consider a complex-valued function $f(t)$ of a real variable t:

$$f(t) = u(t) + iv(t), \tag{4.1.1}$$

which is assumed to be a piecewise continuous function defined in the closed interval $a \leq t \leq b$. To integrate $f(t)$ from $t = a$ to $t = b$, we use the natural definition

$$\int_a^b f(t)\, dt = \int_a^b u(t)\, dt + i \int_a^b v(t)\, dt. \tag{4.1.2}$$

Properties of a complex integral with real variable of integration

1. $$\operatorname{Re} \int_a^b f(t)\, dt = \int_a^b \operatorname{Re} f(t)\, dt = \int_a^b u(t)\, dt. \tag{4.1.3a}$$

2. $$\operatorname{Im} \int_a^b f(t)\, dt = \int_a^b \operatorname{Im} f(t)\, dt = \int_a^b v(t)\, dt. \tag{4.1.3b}$$

3. $$\int_a^b [\gamma_1 f_1(t) + \gamma_2 f_2(t)]\, dt = \gamma_1 \int_a^b f_1(t)\, dt + \gamma_2 \int_a^b f_2(t)\, dt, \tag{4.1.3c}$$

   where γ_1 and γ_2 are any complex constants.

4. $$\left| \int_a^b f(t)\, dt \right| \leq \int_a^b |f(t)|\, dt. \tag{4.1.3d}$$

The proofs of the first three properties are obvious. The last property can be shown using the following argument. We consider

$$\left| \int_a^b f(t)\, dt \right| = e^{-i\phi} \int_a^b f(t)\, dt = \int_a^b e^{-i\phi} f(t)\, dt,$$

where $\phi = \operatorname{Arg} \left(\int_a^b f(t)\, dt \right)$. Since $\left| \int_a^b f(t)\, dt \right|$ is real, we deduce that

$$\left| \int_a^b f(t)\, dt \right| = \operatorname{Re} \int_a^b e^{-i\phi} f(t)\, dt = \int_a^b \operatorname{Re} [e^{-i\phi} f(t)]\, dt$$

$$\leq \int_a^b |e^{-i\phi} f(t)|\, dt = \int_a^b |f(t)|\, dt.$$

Example 4.1.1   Suppose α is real. Show that

$$|e^{2\alpha\pi i} - 1| \leq 2\pi |\alpha|$$

Solution   Let $f(t) = e^{i\alpha t}$, where α and t are real. Substituting the function into eq. (4.1.3d), we obtain

$$\left| \int_0^{2\pi} e^{i\alpha t}\, dt \right| \leq \int_0^{2\pi} |e^{i\alpha t}|\, dt = 2\pi.$$

The left-hand side of the above inequality is equal to

$$\left| \int_0^{2\pi} e^{i\alpha t}\, dt \right| = \left| \frac{e^{i\alpha t}}{i\alpha} \right|_0^{2\pi} = \frac{|e^{2\alpha\pi i} - 1|}{|\alpha|}.$$

Combining the results, we obtain

$$|e^{2\alpha\pi i} - 1| \leq 2\pi |\alpha|, \quad \alpha \text{ is real.}$$

4.1.2 Complex integrals as line integrals

Consider a curve C which is a set of points $z = (x, y)$ in the complex plane defined by

$$x = x(t), \quad y = y(t), \quad a \leq t \leq b,$$

where $x(t)$ and $y(t)$ are continuous functions of the real parameter t. One may write

$$z(t) = x(t) + iy(t) \text{ and } \quad a \leq t \leq b.$$

Recall that $\frac{z'(t)}{|z'(t)|}$ gives the unit tangent vector to the curve at the point t, which is well defined provided that $z'(t) \neq 0$ in the open interval $a < t < b$. The curve is said to be *smooth* if $z(t)$ has a continuous derivative in the closed interval $a \leq t \leq b$ and nonzero in the open interval $a < t < b$. A *contour* is defined as a curve consisting of a finite number of smooth curves joined end to end. A contour is said to be a *simple closed contour* if only the initial and final values of $z(t)$ are the same.

Let $f(z)$ be any complex function defined in a domain $\mathcal{D}$ in the complex plane and let C be any contour contained in $\mathcal{D}$ with initial point z_0 and terminal point z. We divide the contour C into n subarcs by the discrete points $z_0, z_1, z_2, \ldots, z_{n-1}, z_n = z$ arranged consecutively along the direction of increasing t. Let ζ_k be an arbitrary point on the subarc $z_k z_{k+1}$ (see Figure 4.1)

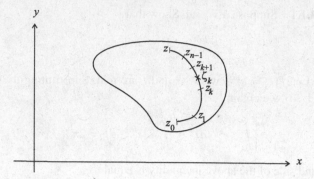

Figure 4.1. Subdivision of the contour into n subarcs by the discrete points $z_0, z_1, \ldots, z_{n-1}, z_n = z$.

and form the sum

$$\sum_{k=0}^{n-1} f(\zeta_k)(z_{k+1} - z_k).$$

We write $\Delta z_k = z_{k+1} - z_k$. Let $\lambda = \max_k |\Delta z_k|$ and take the limit

$$\lim_{\substack{\lambda \to 0 \\ n \to \infty}} \sum_{k=0}^{n-1} f(\zeta_k) \, \Delta z_k.$$

The above limit is defined to be the *contour integral* of $f(z)$ along the contour C. Symbolically, we write

$$\int_C f(z) \, dz = \lim_{\substack{\lambda \to 0 \\ n \to \infty}} \sum_{k=0}^{n-1} f(\zeta_k) \, \Delta z_k. \tag{4.1.4}$$

If the above limit exists, then the function $f(z)$ is said to be integrable along the contour C.

The contour integral defined in eq. (4.1.4) can be related to the integral of a complex function of a real variable. If we write

$$\frac{dz(t)}{dt} = \frac{dx(t)}{dt} + i \frac{dy(t)}{dt}, \quad a \le t \le b,$$

then

$$\int_C f(z) \, dz = \int_a^b f(z(t)) \, \frac{dz(t)}{dt} \, dt.$$

Writing $f(z) = u(x, y) + i v(x, y)$ and $dz = dx + i\, dy$, we have

$$\int_C f(z)\, dz = \int_C u\, dx - v\, dy + i \int_C u\, dy + v\, dx$$

$$= \int_a^b \left[u(x(t), y(t)) \frac{dx(t)}{dt} - v(x(t), y(t)) \frac{dy(t)}{dt} \right] dt \quad (4.1.5)$$

$$+ i \int_a^b \left[u(x(t), y(t)) \frac{dy(t)}{dt} + v(x(t), y(t)) \frac{dx(t)}{dt} \right] dt.$$

Since a contour integral can be defined in terms of real line integrals, the usual properties of real line integrals are carried over to their complex counterparts. Some of these properties are included below.

(i) $\int_C f(z)\, dz$ is independent of the parametrization of C.

(ii) $\int_{-C} f(z)\, dz = -\int_C f(z)\, dz$, where $-C$ is the opposite curve of C.

(iii) The integral of $f(z)$ along a string of contours is equal to the sum of the integrals of $f(z)$ along each of these contours.

Estimation of the modulus value of a complex integral The upper bound of the modulus value of a complex integral can be related to the arc length of the contour C and the maximum value of $|f(z)|$ along C. In fact,

$$\left| \int_C f(z)\, dz \right| \leq ML, \quad (4.1.6)$$

where M is the upper bound of $|f(z)|$ along C and L is the arc length of the contour C. To show the above *modulus inequality*, we consider

$$\left| \int_C f(z)\, dz \right| = \left| \int_a^b f(z(t)) \frac{dz(t)}{dt}\, dt \right|$$

$$\leq \int_a^b |f(z(t))| \left| \frac{dz(t)}{dt} \right| dt$$

$$\leq \int_a^b M \left| \frac{dz(t)}{dt} \right| dt$$

$$= M \int_a^b \sqrt{\left(\frac{dx(t)}{dt} \right)^2 + \left(\frac{dy(t)}{dt} \right)^2}\, dt = ML.$$

Example 4.1.2 Evaluate the integral

$$\oint_C \frac{1}{z - z_0}\, dz,$$

where C is a circle centered at z_0 and of any radius. The path is traced out once in the anticlockwise direction.

Solution   The circle C can be parametrized by

$$z(t) = z_0 + re^{it}, \quad 0 \le t \le 2\pi,$$

where r is any positive real number. The contour integral becomes

$$\oint_C \frac{1}{z - z_0}\, dz = \int_0^{2\pi} \frac{1}{z(t) - z_0}\, \frac{dz(t)}{dt}\, dt = \int_0^{2\pi} \frac{ire^{it}}{re^{it}}\, dt = 2\pi i.$$

Interestingly, the value of the integral is independent of the radius of the circle.

Example 4.1.3   Evaluate the integral

$$\int_C \frac{1}{z^2}\, dz$$

along the following two paths:

(a) the straight line segment joining 1 and $2 + i$;
(b) the horizontal line from 1 to 2, then the vertical line from 2 to $2 + i$.

Solution

(a) The parametric form of the straight line segment joining 1 and $2 + i$ is given by $z(t) = 1 + (1 + i)t, \ 0 \le t \le 1$. The contour integral can be expressed as

$$\int_0^1 \frac{1}{[1 + (1 + i)t]^2}\, (1 + i)\, dt = 1 - \frac{1}{2 + i} = \frac{3 + i}{5}.$$

(b) For the second path, the contour integral can be expressed as

$$\int_1^2 \frac{1}{x^2}\, dx + \int_0^1 \frac{1}{(2 + iy)^2}\, i\, dy = \left(1 - \frac{1}{2}\right) + \left(\frac{1}{2} - \frac{1}{2 + i}\right) = \frac{3 + i}{5}.$$

Remark   Apparently, the value of the integral is independent of the path of integration and can be found directly from

$$\int_1^{2+i} \frac{1}{z^2}\, dz = -\frac{1}{z}\Big|_1^{2+i} = \frac{3 + i}{5},$$

where $-\frac{1}{z}$ is a primitive function of $\frac{1}{z^2}$.

Example 4.1.4 Evaluate the integral

$$\int_C |z|^2 \, dz,$$

where the contour C is

(a) the line segment with initial point -1 and final point i;
(b) the arc of the unit circle $|z| = 1$ traversed in the clockwise direction with initial point -1 and final point i.

Do the two results agree?

Solution

(a) Parametrize the line segment by

$$z = -1 + (1 + i)t, \quad 0 \le t \le 1,$$

so that

$$|z|^2 = (-1 + t)^2 + t^2 \quad \text{and} \quad dz = (1 + i) \, dt.$$

The value of the contour integral becomes

$$\int_C |z|^2 \, dz = \int_0^1 (2t^2 - 2t + 1)(1 + i) \, dt = \frac{2}{3}(1 + i).$$

(b) Along the unit circle $|z| = 1$, we have $z = e^{i\theta}$ and $dz = ie^{i\theta} d\theta$. The initial and final points of the path correspond to $\theta = \pi$ and $\theta = \pi/2$, respectively. The contour integral can be evaluated as

$$\int_C |z|^2 \, dz = \int_\pi^{\frac{\pi}{2}} ie^{i\theta} \, d\theta = e^{i\theta} \Big|_\pi^{\frac{\pi}{2}} = 1 + i.$$

The results in (a) and (b) do not agree. Hence, the value of this contour integral does depend on the path of integration.

Example 4.1.5 Let C be the closed contour consisting of four straight line segments, Re $z = \pm a$ and Im $z = \pm a$ ($a > 0$), oriented in the anticlockwise direction (C represents a square of sides $2a$). Evaluate the integral

$$\oint_C \text{Re } z \, dz.$$

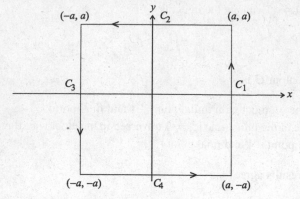

Figure 4.2. The contour is a closed square of sides $2a$.

Solution   As shown in Figure 4.2, the closed contour C consists of the four line segments

$$C_1 = \{z : z = a + it, \ -a \leq t \leq a\}, \ dz = i \, dt,$$
$$C_2 = \{z : z = -t + ia, \ -a \leq t \leq a\}, \ dz = -dt,$$
$$C_3 = \{z : z = -a - it, \ -a \leq t \leq a\}, \ dz = -i \, dt,$$
$$C_4 = \{z : z = t - ia, \ -a \leq t \leq a\}, \ dz = dt,$$

where $C = C_1 \cup C_2 \cup C_3 \cup C_4$. The respective contour integrals along the line segments are found to be

$$\int_{C_1} \operatorname{Re} z \, dz = \int_{-a}^{a} ai \, dt = 2a^2 i$$

$$\int_{C_2} \operatorname{Re} z \, dz = \int_{-a}^{a} (-t)(-dt) = 0$$

$$\int_{C_3} \operatorname{Re} z \, dz = \int_{-a}^{a} (-a)(-i \, dt) = 2a^2 i$$

$$\int_{C_4} \operatorname{Re} z \, dz = \int_{-a}^{a} t \, dt = 0.$$

The contour integral around the closed square C is given by the sum of the contour integrals along C_1, C_2, C_3 and C_4. Adding the four contour integrals together, we obtain

$$\oint_C \operatorname{Re} z \, dz = \int_{C_1} \operatorname{Re} z \, dz + \int_{C_2} \operatorname{Re} z \, dz + \int_{C_3} \operatorname{Re} z \, dz + \int_{C_4} \operatorname{Re} z \, dz$$
$$= 4a^2 i.$$

Example 4.1.6 Show that

(a) $\left| \int_C (x^2 + iy^2)\, dz \right| \le 2$, where C is the line segment joining $-i$ to i;

(b) $\left| \int_C (x^2 + iy^2)\, dz \right| \le \pi$, where C is the right half-circle $|z| = 1$ and Re $z \ge 0$;

(c) $\left| \int_C \frac{1}{z^2}\, dz \right| \le 2$, where C is the line segment joining $-1 + i$ and $1 + i$.

Solution

(a) Using modulus inequality (4.1.6), we obtain

$$\left| \int_C (x^2 + iy^2)\, dz \right| \le \int_C |x^2 + iy^2|\, |dz|, \quad z = x + iy.$$

By the triangle inequality, we deduce that

$$|x^2 + iy^2| \le |x^2| + |iy^2| = |x|^2 + |y|^2.$$

The line segment joining $-i$ to i can be defined as $z = iy$, $-1 \le y \le 1$. Along the line segment, we have $|x|^2 + |y|^2 \le 1$. Combining the results, we obtain

$$\left| \int_C (x^2 + iy^2)\, dz \right| \le \int_C |dz| = \int_{-1}^{1} |i\, dy| = 2.$$

(b) Similarly, we use modulus inequality (4.1.6) to establish

$$\left| \int_C (x^2 + iy^2)\, dz \right| \le \int_C (|x|^2 + |y|^2)\, |dz| = \int_C |z|^2\, |dz|.$$

The contour C is the right half unit circle, $|z| = 1$ and Re $z \ge 0$. We then have

$$\int_C |z|^2\, |dz| = \int_C |dz| = \int_{-\pi/2}^{\pi/2} |ie^{i\theta}|\, d\theta = \pi,$$

and so

$$\left| \int_C (x^2 + iy^2)\, dz \right| \le \pi.$$

(c) Along the contour C, we have $z = x + i$, $-1 \le x \le 1$, so that $1 \le |z| \le \sqrt{2}$ and, correspondingly, $\frac{1}{2} \le \frac{1}{|z|^2} \le 1$. Here, $M = \max_{z \in C} \frac{1}{|z|^2} = 1$ and the arc length $L = 2$. Using modulus inequality (4.1.6), we have

$$\left| \int_C \frac{1}{z^2} \, dz \right| \le ML = 2.$$

4.2 Cauchy integral theorem

One is tempted to ask under what conditions does

$$\int_{C_1} f(z) \, dz = \int_{C_2} f(z) \, dz, \tag{4.2.1}$$

where C_1 and C_2 are two contours in a domain $\mathcal{D}$ with the same initial and final points and $f(z)$ is piecewise continuous inside $\mathcal{D}$. We observe that the property of path independence is valid for $f(z) = \frac{1}{z^2}$ in Example 4.1.3, but it fails when $f(z) = |z|^2$ in Example 4.1.4. The above query is equivalent to the question: when does

$$\oint_C f(z) \, dz = 0 \tag{4.2.2}$$

hold, where C is any closed contour lying completely inside $\mathcal{D}$? The equivalence of eqs. (4.2.1) and (4.2.2) is revealed if we treat C as $C_1 \cup -C_2$. In the above examples, we observe that $f(z) = \frac{1}{z^2}$ is analytic everywhere except at $z = 0$ but $f(z) = |z|^2$ is nowhere analytic. The observation suggests that analyticity of the integrand may play an important role in establishing the validity of eq. (4.2.2). The conjecture is confirmed by the renowned *Cauchy integral theorem*.

Theorem 4.2.1 (Cauchy integral theorem) *Let $f(z)$ be analytic on and inside a simple closed contour C and let $f'(z)$ be continuous on and inside C. Then*

$$\oint_C f(z) \, dz = 0.$$

Proof The proof of the Cauchy integral theorem requires Green's theorem from real calculus. Green's theorem can be stated as: given a positively oriented closed contour C, if the two real functions $P(x, y)$ and $Q(x, y)$ have continuous first-order partial derivatives throughout the closed region consisting of all points on and inside C, then

$$\oint_C P \, dx + Q \, dy = \iint_{\mathcal{D}} (Q_x - P_y) \, dx dy, \tag{4.2.3}$$

where $\mathcal{D}$ is the simply connected domain bounded by C.

From eq. (4.1.5), suppose we write

$$f'(z) = u(x, y) + i v(x, y), \quad z = x + iy;$$

we have

$$\oint_C f(z)\, dz = \oint_C u\, dx - v\, dy + i \oint_C v\, dx + u\, dy.$$

One can infer from the continuity of $f'(z)$ that $u(x, y)$ and $v(x, y)$ have continuous derivatives on and inside C. Using Green's theorem, the two real line integrals can be transformed into double integrals. This gives

$$\oint_C f(z)\, dz = \iint_D (-v_x - u_y)\, dxdy + i \iint_D (u_x - v_y)\, dxdy.$$

Both integrands in the double integrals are equal to zero due to the Cauchy–Riemann relations, hence the Cauchy integral theorem is established.

In 1903, Goursat was able to obtain the same result as in eq. (4.2.2) without assuming the continuity of $f'(z)$. This stronger version is called the *Goursat theorem*. The omission of the continuity assumption is important. Based on the Goursat theorem, we can show later that the derivative f' is also analytic without assuming continuity of f' (see Theorem 4.3.2).

Theorem 4.2.2 (Goursat theorem) *Given a simple closed contour C, let $f(z)$ be analytic on and inside C. Then*

$$\oint_C f(z)\, dz = 0.$$

We choose to omit the proof of the Goursat theorem here since the procedures are rather technical. Readers interested in the proof may consult some other texts in complex variables.[†]

Corollary 1 The integral of a function $f(z)$ that is analytic throughout a simply connected domain D depends on the end points and not on the particular contour taken. Suppose α and β are inside D, and C_1 and C_2 are any contours inside D joining α to β. We have

$$\int_{C_1} f(z)\, dz = \int_{C_2} f(z)\, dz. \tag{4.2.4}$$

[†] A detailed proof of the Goursat theorem can be found in *Complex Variables and Applications* by J.W. Brown and R.V. Churchill, 7th edition, McGraw-Hill, 2003.

The basic essence of this corollary has been discussed at the beginning of this section. As deduced from this corollary, the Goursat theorem can be stated in the following alternative form:

If a function $f(z)$ is analytic throughout a simply connected domain $\mathcal{D}$, then for *any* simple closed contour C lying completely inside $\mathcal{D}$, we have

$$\oint_C f(z)\, dz = 0.$$

Corollary 2 Let $f(z)$ be analytic throughout a simply connected domain $\mathcal{D}$. Consider a fixed point $z_0 \in \mathcal{D}$; by virtue of Corollary 1, we deduce that

$$F(z) = \int_{z_0}^{z} f(\zeta)\, d\zeta, \quad \text{for any } z \in \mathcal{D}, \tag{4.2.5}$$

is a well-defined function in $\mathcal{D}$. Moreover, it can be shown that (see Problem 4.9)

$$F'(z) = f(z), \quad \text{for any } z \in \mathcal{D}, \tag{4.2.6}$$

so $F(z)$ is analytic throughout $\mathcal{D}$. Here, $F(z)$ may be visualized as a primitive function of $f(z)$.

This corollary may be considered as the complex counterpart of the fundamental theorem of real calculus. If we integrate $f(z)$ along any contour joining α and β inside $\mathcal{D}$, then the value of the integral is given by

$$\int_{\alpha}^{\beta} f(z)\, dz = F(\beta) - F(\alpha), \quad \alpha \text{ and } \beta \in \mathcal{D}. \tag{4.2.7}$$

The above formula has been verified in Example 4.1.3.

Corollary 3 Let C, C_1, $C_2, \ldots, C_n$ be positively oriented closed contours, where C_1, $C_2, \ldots, C_n$ are all inside C. For C_1, $C_2, \ldots, C_n$, each of these contours lies outside of the other contours. Let int C_i denote the collection of all points bounded by C_i, $i = 1, 2, \ldots, n$. Let $f(z)$ be analytic on the set $S: C \cup \text{int } C \setminus \text{int } C_1 \setminus \text{int } C_2 \setminus \cdots \setminus \text{int } C_n$ (see the shaded area in Figure 4.3). Then

$$\oint_C f(z)\, dz = \sum_{k=1}^{n} \oint_{C_k} f(z)\, dz. \tag{4.2.8}$$

The proof for the case when $n = 2$ is presented below. The extension to the general case is straightforward.

Proof To each interior closed contour C_i, $i = 1, 2$, we place one cut which joins C_i with the exterior contour C. The cut lines provide the passage from

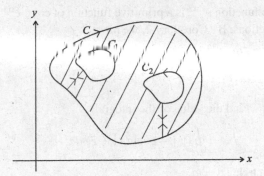

Figure 4.3. The shaded region becomes simply connected with the introduction of the cuts.

the exterior contour to the interior contours. We now construct the boundary curve for the multiply connected region: $C \cup \operatorname{int} C \setminus \operatorname{int} C_1 \setminus \operatorname{int} C_2$ (see the shaded area in Figure 4.3). The constructed boundary curve is composed of $C \cup -C_1 \cup -C_2$ together with the cut lines. Each cut line travels twice in opposite directions. To explain the negative signs in front of C_1 and C_2, we note that the interior contours traverse in the clockwise sense as parts of the positively oriented boundary curve. With the introduction of these cuts, the shaded region bounded within this constructed boundary curve becomes a simply connected set.

By the Cauchy–Goursat theorem, the integral of $f(z)$ along the above boundary curve vanishes. Also, since $f(z)$ is integrated along the cuts twice but in opposite directions, the various contributions to the line integral along the cuts are canceled off. We then have

$$\oint_C f(z)\, dz + \int_{-C_1} f(z)\, dz + \int_{-C_2} f(z)\, dz = 0,$$

so that

$$\oint_C f(z)\, dz = \int_{C_1} f(z)\, dz + \int_{C_2} f(z)\, dz.$$

Example 4.2.1 Evaluate the integral

$$\int_\alpha^\beta \cos z\, e^{\sin z}\, dz,$$

where α and β are any complex constants.

Solution   The function $e^{\sin z}$ is a primitive function of $\cos z\, e^{\sin z}$, and they are both entire functions. By Corollary 2, we have

$$\int_\alpha^\beta \cos z\, e^{\sin z}\, dz = e^{\sin z}\Big|_\alpha^\beta = e^{\sin \beta} - e^{\sin \alpha}.$$

Example 4.2.2   Find the value of the integral

$$\oint_C (|z| - e^z \sin z^2 + \overline{z})\, dz,$$

where C is the circle $|z| = a$.

Solution   The integral is split into three individual integrals:

$$\oint_C (|z| - e^z \sin z^2 + \overline{z})\, dz = \oint_C |z|\, dz - \oint_C e^z \sin z^2\, dz + \oint_C \overline{z}\, dz.$$

The second integral equals zero since the integrand $e^z \sin z^2$ is an entire function. The first integral is found to be

$$\oint_C |z|\, dz = a \oint_C dz = 0,$$

since C is the circle $|z| = a$. To compute the third integral, we write $z = ae^{i\theta}$ and substitute $\overline{z}$ with $ae^{-i\theta}$ and dz with $aie^{i\theta}\, d\theta$. This gives

$$\oint_C \overline{z}\, dz = \int_0^{2\pi} ae^{-i\theta} aie^{i\theta}\, d\theta = a^2 i \int_0^{2\pi} d\theta = 2\pi a^2 i.$$

Combining the results, the value of the integral is found to be $2\pi a^2 i$.

Example 4.2.3   Let $\mathcal{D}$ be the domain that contains the whole complex plane except the origin and the negative real axis. Let Γ be an arbitrary contour, lying completely inside $\mathcal{D}$, that starts from 1 and ends at a point α (see Figure 4.4). Show that

$$\int_\Gamma \frac{dz}{z} = \operatorname{Log} \alpha.$$

Solution   Let Γ_1 be the line segment from 1 to $|\alpha|$ along the real axis, and Γ_2 be the circular arc centered at the origin and of radius $|\alpha|$ which extends from $|\alpha|$ to α. The union $\Gamma_1 \cup \Gamma_2 \cup -\Gamma$ forms a closed contour (see Figure 4.4). Since the integrand $\frac{1}{z}$ is analytic everywhere inside $\mathcal{D}$, by the Cauchy–Goursat theorem, we have

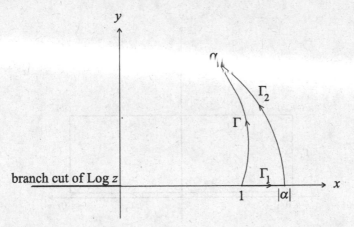

Figure 4.4. The contour Γ starts from $z = 1$ and ends at $z = \alpha$. The arc Γ_2 is part of the circle $|z| = |\alpha|$.

$$\int_\Gamma \frac{dz}{z} = \int_{\Gamma_1} \frac{dz}{z} + \int_{\Gamma_2} \frac{dz}{z}.$$

Since α does not lie on the negative real axis, $\operatorname{Arg}\alpha$ cannot assume the value π. If we write $\alpha = |\alpha|e^{i\operatorname{Arg}\alpha}$ $(-\pi < \operatorname{Arg}\alpha < \pi)$, then

$$\int_{\Gamma_1} \frac{dz}{z} = \int_1^{|\alpha|} \frac{dt}{t} = \ln|\alpha|$$

and

$$\int_{\Gamma_2} \frac{dz}{z} = \int_0^{\operatorname{Arg}\alpha} \frac{ire^{i\theta}}{re^{i\theta}} \, d\theta = i \operatorname{Arg}\alpha.$$

Combining the results, the integral is found to be

$$\int_\Gamma \frac{dz}{z} = \ln|\alpha| + i \operatorname{Arg}\alpha = \operatorname{Log}\alpha.$$

The primitive function of $\frac{1}{z}$ is $\operatorname{Log} z$. The above result can be obtained by applying the computational formula in eq. (4.2.7). Note that the given domain $\mathcal{D}$ is the domain of definition of $\operatorname{Log} z$, the principal branch of the complex logarithm function. The branch cut is taken to be the negative real axis and the branch points are $z = 0$ and $z = \infty$. By excluding the origin and the negative real axis in $\mathcal{D}$, we avoid multi-valuedness of the integral.

Example 4.2.4 Consider the integration of the function e^{-z^2} around the rectangular contour Γ with vertices $\pm a$, $\pm a + ib$ and oriented positively as shown

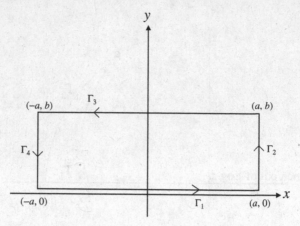

Figure 4.5. The configuration of the closed rectangular contour Γ.

in Figure 4.5. By letting $a \to \infty$ while keeping b fixed, show that

$$\int_{-\infty}^{\infty} e^{-x^2} e^{\pm 2ibx}\, dx = \int_{-\infty}^{\infty} e^{-x^2} \cos 2bx\, dx = e^{-b^2} \sqrt{\pi}.$$

The result is called the *Poisson integral.*

Solution   Since e^{-z^2} is an entire function, we have

$$\oint_{\Gamma} e^{-z^2}\, dz = 0,$$

by virtue of the Cauchy–Goursat theorem. The closed contour Γ consists of four line segments: $\Gamma = \Gamma_1 \cup \Gamma_2 \cup \Gamma_3 \cup \Gamma_4$, where (see Figure 4.5)

$$\begin{aligned}
\Gamma_1 &= \{x : -a \le x \le a\}, \\
\Gamma_2 &= \{a + iy : 0 \le y \le b\}, \\
\Gamma_3 &= \{x + ib : -a \le x \le a\}, \\
\Gamma_4 &= \{-a + iy : 0 \le y \le b\},
\end{aligned}$$

and Γ is oriented in the anticlockwise direction.

The contour integral can be split into four contour integrals, namely,

$$\oint_{\Gamma} e^{-z^2}\, dz = \int_{\Gamma_1} e^{-z^2}\, dz + \int_{\Gamma_2} e^{-z^2}\, dz + \int_{\Gamma_3} e^{-z^2}\, dz + \int_{\Gamma_4} e^{-z^2}\, dz.$$

The four contour integrals can be expressed as real integrals as follows:

$$\int_{\Gamma_1} e^{-z^2}\, dz = \int_{-a}^{a} e^{-x^2}\, dx,$$

$$\int_{\Gamma_2} e^{-z^2}\, dz = \int_{0}^{b} e^{-(a+iy)^2} i\, dy,$$

$$\int_{\Gamma_3} e^{-z^2}\, dz = \int_{a}^{-a} e^{-(x+ib)^2}\, dx,$$

$$= -e^{b^2}\left[\int_{-a}^{a} e^{-x^2} \cos 2bx\, dx - i\int_{-a}^{a} e^{-x^2} \sin 2bx\, dx\right],$$

$$\int_{\Gamma_4} e^{-z^2}\, dz = \int_{b}^{0} e^{-(-a+iy)^2} i\, dy.$$

First, we consider the bound on the modulus of the second integral. By the modulus inequality (4.1.6), we have

$$\left|\int_{\Gamma_2} e^{-z^2}\, dz\right| \le \int_{0}^{b} |e^{-(a^2 - y^2 + 2iay)}i|\, dy$$

$$= e^{-a^2}\int_{0}^{b} e^{y^2}\, dy$$

$$\le e^{-a^2}\int_{0}^{b} e^{b^2}\, dy \qquad \text{(since } 0 \le y \le b\text{)}$$

$$= \frac{be^{b^2}}{e^{a^2}} \to 0 \text{ as } a \to \infty \text{ and } b \text{ is fixed.}$$

Therefore, the value of $\int_{\Gamma_2} e^{-z^2}\, dz \to 0$ as $a \to \infty$.

Using a similar argument, the fourth integral can be shown to be zero as $a \to \infty$. Thus, by taking the limit $a \to \infty$ but keeping b fixed, the contour integral around Γ is found to be equal to the sum of the first and third integrals

$$\lim_{a \to \infty} \oint_{\Gamma} e^{-z^2}\, dz = \lim_{a \to \infty}\left(\int_{-a}^{a} e^{-x^2} dx - e^{b^2}\int_{-a}^{a} e^{-x^2} \cos 2bx\, dx\right)$$

$$+ i\lim_{a \to \infty}\left(e^{b^2}\int_{-a}^{a} e^{-x^2} \sin 2bx\, dx\right) = 0.$$

Using the known value of the following integral (see Problem 4.13)

$$\int_{-\infty}^{\infty} e^{-x^2}\, dx = \sqrt{\pi},$$

we obtain

$$\int_{-\infty}^{\infty} e^{-x^2} \cos 2bx \, dx - i \int_{-\infty}^{\infty} e^{-x^2} \sin 2bx \, dx = e^{-b^2} \int_{-\infty}^{\infty} e^{-x^2} dx = e^{-b^2} \sqrt{\pi}.$$

By equating the imaginary parts of the above equation, we observe

$$\int_{-\infty}^{\infty} e^{-x^2} \sin 2bx \, dx = 0.$$

Finally, we obtain

$$\int_{-\infty}^{\infty} e^{-x^2} e^{\pm 2ibx} \, dx = \int_{-\infty}^{\infty} e^{-x^2} \cos 2bx \, dx = e^{-b^2} \sqrt{\pi}.$$

Example 4.2.5 Let $W(z, \bar{z})$ be a function with continuous first-order partial derivatives in a domain $\mathcal{D}$ and on its boundary C, where $z = x + iy$ and $\bar{z} = x - iy$. In terms of the pair of conjugate complex variables z and $\bar{z}$, show that the complex variable formulation of Green's theorem is

$$\oint_C W(z, \bar{z}) \, dz = 2i \iint_{\mathcal{D}} \frac{\partial W}{\partial \bar{z}} \, dxdy.$$

Supposing a closed contour C encloses a region of area A, show that

$$A = \frac{1}{2i} \oint_C \bar{z} \, dz = \frac{1}{4i} \left[\oint_C \bar{z} \, dz - \oint_C z \, d\bar{z} \right] = \frac{1}{2} \operatorname{Im} \oint_C \bar{z} \, dz.$$

Use the formula to find the area enclosed by the ellipse with parametric representation: $x = 5 \cos t$, $y = 4 \sin t$, $0 \le t < 2\pi$.

Solution Let $W(z, \bar{z}) = P(x, y) + iQ(x, y)$. The contour integral can be written as

$$\oint_C W(z, \bar{z}) \, dz = \oint_C (P + iQ)(dx + idy)$$

$$= \oint_C P \, dx - Q \, dy + i \oint_C Q \, dx + P \, dy.$$

The above two real integrals can be transformed into double integrals by Green's theorem as follows:

$$\oint_C W(z, \bar{z}) \, dz = -\iint_{\mathcal{D}} \left(\frac{\partial Q}{\partial x} + \frac{\partial P}{\partial y} \right) dxdy + i \iint_{\mathcal{D}} \left(\frac{\partial P}{\partial x} - \frac{\partial Q}{\partial y} \right) dxdy$$

$$= i \iint_{\mathcal{D}} \left(\frac{\partial}{\partial x} + i \frac{\partial}{\partial y} \right) (P + iQ) \, dxdy$$

$$= 2i \iint_{\mathcal{D}} \frac{\partial W}{\partial \bar{z}} \, dxdy, \quad \text{since} \quad 2 \frac{\partial}{\partial \bar{z}} = \frac{\partial}{\partial x} + i \frac{\partial}{\partial y}.$$

By setting $W(z, \bar{z}) = \bar{z}$ in the above formula and observing that

$$\frac{\partial W}{\partial \bar{z}} = 1,$$

we obtain

$$\oint_C \bar{z}\, dz = 2i \iint_D dx dy = 2i\, A.$$

Taking the complex conjugate on both sides of the above equation, we have

$$\oint_C z\, d\bar{z} = -2i\, A.$$

The above results can be combined into the form

$$A = \frac{1}{2i} \oint_C \bar{z}\, dz = \frac{1}{4i} \left(\oint_C \bar{z}\, dz - \oint_C z\, d\bar{z} \right).$$

Since A is real, the value of $\oint_C \bar{z}\, dz$ must be purely imaginary. We then have

$$A = \frac{1}{2i} \oint_C \bar{z}\, dz = \frac{1}{2} \operatorname{Im} \oint_C \bar{z}\, dz.$$

Using the complex representation, the ellipse can be parametrized as

$$z(t) = 5 \cos t + 4i \sin t, \quad 0 \le t < 2\pi.$$

The area of the ellipse is given by

$$A = \frac{1}{2i} \int_0^{2\pi} (5 \cos t - 4i \sin t)(-5 \sin t + 4i \cos t)\, dt$$

$$= \frac{1}{2i} \left(\int_0^{2\pi} -9 \cos t\, \sin t\, dt + i \int_0^{2\pi} 20\, dt \right) = 20\pi.$$

4.3 Cauchy integral formula and its consequences

The Cauchy integral formula is considered to be one of the most important and useful results in the theory of analytic functions. At the elementary level, it provides a useful tool for the evaluation of a wide variety of complex integrals. More importantly, the Cauchy integral formula plays a major role in the development of more advanced topics in analytic function theory.

Theorem 4.3.1 (Cauchy integral formula) *Let the function $f(z)$ be analytic on and inside a positively oriented simple closed contour C and z be any point*

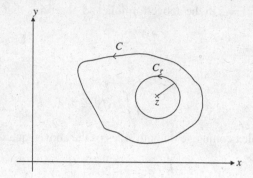

Figure 4.6. A circle C_r centered at the point z is drawn which lies completely inside the closed contour C.

inside C. Then

$$f(z) = \frac{1}{2\pi i} \oint_C \frac{f(\zeta)}{\zeta - z} \, d\zeta. \qquad (4.3.1)$$

Proof We draw a circle C_r, with radius r, centered at the point z, small enough to be completely inside C (see Figure 4.6). Since $\frac{f(\zeta)}{\zeta - z}$ is analytic in the region lying between C_r and C, by Corollary 3 of the Cauchy–Goursat theorem, we have

$$\frac{1}{2\pi i} \oint_C \frac{f(\zeta)}{\zeta - z} \, d\zeta = \frac{1}{2\pi i} \oint_{C_r} \frac{f(\zeta)}{\zeta - z} \, d\zeta$$

$$= \frac{1}{2\pi i} \oint_{C_r} \frac{f(\zeta) - f(z)}{\zeta - z} \, d\zeta + \frac{f(z)}{2\pi i} \oint_{C_r} \frac{1}{\zeta - z} \, d\zeta.$$

By virtue of the result obtained in Example 4.1.2, the last integral is seen to be equal to $f(z)$. To complete the proof, it suffices to show that the first integral has the value zero.

Since f is continuous at z, for each $\epsilon > 0$, there exists $\delta > 0$ such that

$$|f(\zeta) - f(z)| < \epsilon \quad \text{whenever} \quad |\zeta - z| < \delta.$$

Now, suppose we choose $r < \delta$ and thus we ensure that C_r lies completely inside the contour C. The modulus of the first integral is bounded by

$$\left| \frac{1}{2\pi i} \oint_{C_r} \frac{f(\zeta) - f(z)}{\zeta - z} \, d\zeta \right| \leq \frac{1}{\pi} \oint_{C_r} \frac{|f(\zeta) - f(z)|}{|\zeta - z|} \, |d\zeta|$$

$$= \frac{1}{2\pi r} \oint_{C_r} |f(\zeta) - f(z)| \, |d\zeta|$$

$$< \frac{\epsilon}{2\pi r} \oint_{C_r} |d\zeta| = \frac{\epsilon}{2\pi r} 2\pi r = \epsilon.$$

Since the modulus of the above integral is less than any positive number ϵ, however small, the value of that integral must be zero.

The Cauchy integral formula is a remarkable result. The value of $f(z)$ at any point inside the closed contour C is determined by the values of the function along the bounding contour C.

4.3.1 Derivatives of contour integrals

Suppose we differentiate both sides of the Cauchy integral formula in eq. (4.3.1) formally with respect to z (holding ζ fixed). Assuming that differentiation under the integral sign is legitimate, we obtain

$$f'(z) = \frac{1}{2\pi i} \oint_C \frac{d}{dz} \frac{f(\zeta)}{\zeta - z} \, d\zeta = \frac{1}{2\pi i} \oint_C \frac{f(\zeta)}{(\zeta - z)^2} \, d\zeta. \qquad (4.3.2)$$

Can we establish the legitimacy of direct differentiation of the integrand in the Cauchy integral formula? Assuming that z and $z + h$ both lie inside C, we consider the expression

$$\frac{f(z + h) - f(z)}{h} - \frac{1}{2\pi i} \oint_C \frac{f(\zeta)}{(\zeta - z)^2} \, d\zeta$$

$$= \frac{1}{h} \left\{ \frac{1}{2\pi i} \oint_C \left[\frac{f(\zeta)}{\zeta - z - h} - \frac{f(\zeta)}{\zeta - z} - h \frac{f(\zeta)}{(\zeta - z)^2} \right] d\zeta \right\}$$

$$= \frac{h}{2\pi i} \oint_C \frac{f(\zeta)}{(\zeta - z - h)(\zeta - z)^2} \, d\zeta.$$

To show the validity of eq. (4.3.2), it suffices to show that the value of the last integral goes to zero as $h \to 0$. To estimate the value of the last integral, we draw the circle C_{2d}: $|\zeta - z| = 2d$ that lies completely inside the domain bounded by C and choose h such that $0 < |h| < d$. Every point ζ on the curve C is then outside the circle C_{2d} so that

$$|\zeta - z| > d \quad \text{and} \quad |\zeta - z - h| > d.$$

Let M be the upper bound of $|f(z)|$ on C and L be the total arc length of C. Using the modulus inequality (4.1.6) and together with the above inequalities, we obtain

$$\left| \frac{h}{2\pi i} \oint_C \frac{f(\zeta)}{(\zeta - z - h)(\zeta - z)^2} \, d\zeta \right| \leq \frac{|h|}{2\pi} \frac{ML}{d^3}.$$

In the limit $h \to 0$, we observe that

$$\lim_{h \to 0} \left| \frac{h}{2\pi i} \oint_C \frac{f(\zeta)}{(\zeta - z - h)(\zeta - z)^2} \, d\zeta \right| \leq \lim_{h \to 0} \frac{|h|}{2\pi} \frac{ML}{d^3} = 0;$$

therefore,

$$f'(z) = \lim_{h \to 0} \frac{f(z + h) - f(z)}{h} = \frac{1}{2\pi i} \oint_C \frac{f(\zeta)}{(\zeta - z)^2} \, d\zeta.$$

By induction, the *generalized Cauchy integral formula* can be established as follows:

$$f^{(k)}(z) = \frac{k!}{2\pi i} \oint_C \frac{f(\zeta)}{(\zeta - z)^{k+1}} \, d\zeta, \qquad k = 1, 2, 3, \ldots. \qquad (4.3.3)$$

An immediate consequence of the generalized Cauchy integral formula (4.3.3) is that the mere assumption of analyticity of f at a point is sufficient to guarantee the existence of the derivatives of f of all orders at the same point. The precise statement of the result is summarized in the following theorem.

Theorem 4.3.2 *If a function $f(z)$ is analytic at a point, then its derivatives of all orders are also analytic at the same point.*

Proof Suppose f is analytic at a point z. Then there exists a neighborhood $|\zeta - z| < \epsilon$ around z such that f is analytic throughout the neighborhood. Let $C_{\epsilon'}$ denote the positively oriented circle $|\zeta - z| < \epsilon'$, where $\epsilon' < \epsilon$, so that f is analytic inside and on $C_{\epsilon'}$. According to eq. (4.3.3), the derivative f'' exists at each point inside $C_{\epsilon'}$, so f' is analytic at the point z. Repeating the argument for the function f' that is analytic at z, we can conclude that f'' is also analytic at z, and so forth for all higher-order derivatives of f.

Remarks

(i) The above theorem is limited to complex functions only. In fact, no similar statement can be made on real differentiable functions. It is easy to find examples of real-valued functions $f(x)$ such that $f'(x)$ exists at a point but $f''(x)$ does not exist at that point.

(ii) Suppose we express an analytic function inside a domain $\mathcal{D}$ as $f(z) = u(x, y) + iv(x, y)$, $z = x + iy$. Since the derivatives of f of all orders are analytic functions it then follows that the partial derivatives of $u(x, y)$ and $v(x, y)$ of all orders exist and are continuous. This result is consistent with the earlier claim on continuity of higher-order derivatives when we discuss the theory of harmonic functions in Section 2.6: the mere assumption of analyticity of f at a point would guarantee the continuity of all second-order derivatives of the real part and imaginary part of f.

The Cauchy integral formula can be extended to the case where the simple closed contour C can be replaced by the oriented boundary of a multiply connected domain as described in Corollary 3 of the Cauchy–Goursat theorem. In fact, suppose C, C_1, C_2, ..., C_n and $f(z)$ obey the same conditions as in Corollary 3; then for any point $z \in C \cup \text{ int } C \setminus \text{int } C_1 \setminus \text{int } C_2 \setminus \cdots \setminus \text{int } C_n$, we have

$$f(z) = \frac{1}{2\pi i} \oint_C \frac{f(\zeta)}{\zeta - z} \, d\zeta - \sum_{k=1}^{n} \frac{1}{2\pi i} \oint_{C_k} \frac{f(\zeta)}{\zeta - z} \, d\zeta. \tag{4.3.4}$$

The proof follows the same approach of introducing cuts that join the exterior contour with the interior contours (see Figure 4.3).

Cauchy inequality

Suppose $f(z)$ is analytic on and inside the disk $|z - z_0| = r$, $0 < r < \infty$, and let

$$M(r) = \max_{|z-z_0|=r} |f(z)|.$$

Then

$$\frac{|f^{(k)}(z)|}{k!} \leq \frac{M(r)}{r^k}, \quad k = 0, 1, 2, \ldots. \tag{4.3.5}$$

This inequality follows from the generalized Cauchy integral formula (4.3.3).

Example 4.3.1 Suppose $f(z)$ is defined by the integral

$$f(z) = \oint_{|\zeta|=3} \frac{3\zeta^2 + 7\zeta + 1}{\zeta - z} \, d\zeta.$$

Find $f'(1 + i)$.

Solution   By setting $k = 1$ in the generalized Cauchy integral formula (4.3.3), we have

$$f'(z) = \oint_{|\zeta|=3} \frac{3\zeta^2 + 7\zeta + 1}{(\zeta - z)^2} \, d\zeta$$

$$= \oint_{|\zeta|=3} \frac{3(\zeta - z)^2 + (6z + 7)(\zeta - z) + 3z^2 + 7z + 1}{(\zeta - z)^2} \, d\zeta$$

$$= \oint_{|\zeta|=3} 3 \, d\zeta + (6z + 7) \oint_{|\zeta|=3} \frac{1}{\zeta - z} \, d\zeta$$

$$+ (3z^2 + 7z + 1) \oint_{|\zeta|=3} \frac{1}{(\zeta - z)^2} \, d\zeta.$$

The first integral equals zero since the integrand is entire (as it is a constant function). For the second integral, we observe that

$$\oint_{|\zeta|=3} \frac{1}{\zeta - z} \, d\zeta = \begin{cases} 0 & \text{if } |z| > 3 \\ 2\pi i & \text{if } |z| < 3 \end{cases}.$$

Furthermore, we deduce that the third integral is zero since

$$\oint_{|\zeta|=3} \frac{1}{(\zeta - z)^2} \, d\zeta = \frac{d}{dz} \left[\oint_{|\zeta|=3} \frac{1}{\zeta - z} \, d\zeta \right] = 0.$$

Combining the results, we have

$$f'(z) = \begin{cases} (2\pi i)(6z + 7) & \text{if } |z| < 3 \\ 0 & \text{if } |z| > 3 \end{cases}.$$

We observe that $1 + i$ is inside $|z| < 3$ since $|1 + i| = \sqrt{2} < 3$. Therefore, we obtain

$$f'(1 + i) = 2\pi i \, [6(1 + i) + 7] = -12\pi + 26\pi i.$$

Example 4.3.2   Suppose $f(z)$ is analytic inside the unit circle $|z| = 1$ and

$$|f(z)| \le \frac{1}{1 - |z|}.$$

Show that

$$|f^{(n)}(0)| \le (n + 1)! \left(1 + \frac{1}{n} \right)^n.$$

Solution   We integrate $\frac{f(\zeta)}{\zeta^{n+1}}$ around the circle $|\zeta| = \frac{n}{n+1}$, where $f(\zeta)$ is analytic on and inside the circle. Using the generalized Cauchy integral formula, we have

$$f^{(n)}(0) = \frac{n!}{2\pi i} \oint_{|\zeta| - \frac{n}{n+1}} \frac{f(\zeta)}{\zeta^{n+1}} \, d\zeta$$

$$= \frac{n!}{2\pi i} \int_0^{2\pi} \frac{f\left(\frac{n}{n+1} e^{i\theta}\right)}{\left(\frac{n}{n+1}\right)^{n+1} e^{i(n+1)\theta}} \left(\frac{n}{n+1}\right) e^{i\theta} \, i \, d\theta$$

$$= \left(1 + \frac{1}{n}\right)^n \frac{n!}{2\pi} \int_0^{2\pi} f\left(\frac{n}{n+1} e^{i\theta}\right) e^{-in\theta} \, d\theta.$$

The modulus $|f^{(n)}(0)|$ is bounded by

$$|f^{(n)}(0)| \le \left(1 + \frac{1}{n}\right)^n \frac{n!}{2\pi} \int_0^{2\pi} \left| f\left(\frac{n}{n+1} e^{i\theta}\right) \right| d\theta$$

$$\le \left(1 + \frac{1}{n}\right)^n \frac{n!}{2\pi} \int_0^{2\pi} \frac{1}{1 - \frac{n}{n+1}} \, d\theta$$

$$= \left(1 + \frac{1}{n}\right)^n \frac{n!}{2\pi} [2\pi(n+1)]$$

$$= (n+1)! \left(1 + \frac{1}{n}\right)^n.$$

4.3.2 Morera's theorem

The converse of the Cauchy–Goursat theorem is Morera's theorem.

Theorem 4.3.3 (Morera's theorem) *Suppose $f(z)$ is continuous inside a simply connected domain $\mathcal{D}$ and*

$$\oint_C f(z) \, dz = 0, \qquad (4.3.6)$$

for any closed contour C lying inside $\mathcal{D}$. Then $f(z)$ is analytic throughout $\mathcal{D}$.

Proof The continuity property of $f(z)$ and the property defined in eq. (4.3.6) induce the following primitive function of $f(z)$:

$$F(z) = \int_{z_0}^z f(\zeta) \, d\zeta, \quad \text{for all } z \in \mathcal{D}, \qquad (4.3.7a)$$

where z_0 is a fixed point inside $\mathcal{D}$. The function $F(z)$ is single-valued; otherwise, the property in eq. (4.3.6) would be violated. Further, it can be shown that (see

Problem 4.9)

$$F'(z) = f(z), \quad \text{for all } z \in \mathcal{D}, \tag{4.3.7b}$$

and so $F(z)$ is analytic throughout $\mathcal{D}$. By virtue of Theorem 4.3.2, $F'(z)$ is also analytic in the same domain. Hence, $f(z)$ is analytic throughout $\mathcal{D}$.

Remark   This proof and the argument presented in Corollary 2 of the Cauchy–Goursat theorem look quite similar. However, readers are reminded that the assumption of the analyticity of $f(z)$ is not needed to establish eqs. (4.3.7a,b).

4.3.3 Consequences of the Cauchy integral formula

The Cauchy integral formula leads to a wide variety of important theorems, many of which find applications in the further development of complex analysis and in other fields of mathematics. We discuss a few of these theorems that are considered to be both important and interesting.

Theorem 4.3.4 (Gauss' mean value theorem)   *If $f(z)$ is analytic on and inside the disk $C_r : |z - z_0| = r$, then*

$$f(z_0) = \frac{1}{2\pi} \int_0^{2\pi} f(z_0 + re^{i\theta}) \, d\theta. \tag{4.3.8}$$

Proof   From the Cauchy integral formula, we have

$$\begin{aligned}
f(z_0) &= \frac{1}{2\pi i} \oint_{C_r} \frac{f(z)}{z - z_0} \, dz \\
&= \frac{1}{2\pi i} \int_0^{2\pi} \frac{f(z_0 + re^{i\theta}) i re^{i\theta}}{re^{i\theta}} \, d\theta \\
&= \frac{1}{2\pi} \int_0^{2\pi} f(z_0 + re^{i\theta}) \, d\theta.
\end{aligned}$$

Example 4.3.3   Find the mean value of $x^2 - y^2 + x$ on the circle $|z - i| = 2$.

Solution   First, we observe that $x^2 - y^2 + x = \text{Re}(z^2 + z)$. The mean value of $x^2 - y^2 + x$ on $|z - i| = 2$ is defined by

$$\frac{1}{2\pi} \int_0^{2\pi} f(i + 2e^{i\theta}) \, d\theta,$$

where $f(z) = \text{Re}(z^2 + z)$. By Gauss' mean value theorem, we have

$$\frac{1}{2\pi} \int_0^{2\pi} f(1 +)e^{i\theta})d\theta \quad \text{Re}(\tau^2 + \tau)\Big|_{\tau=1} = \text{Re}(-1 + i) = -1.$$

Theorem 4.3.5 (Liouville's theorem) *The only bounded entire functions are constant functions. Equivalently, suppose $f(z)$ is entire and there exists a constant $B \in \mathbb{R}$ such that $|f(z)| \leq B$, for all $z \in \mathbb{C}$; then $f(z) = K$ for some constant $K \in \mathbb{C}$.*

Proof It suffices to show that $f'(z) = 0$, for all $z \in \mathbb{C}$. We integrate $\frac{f(\zeta)}{(\zeta-z)^2}$ around the circle $C_R : |\zeta - z| = R$. By virtue of the generalized Cauchy integral formula, we have

$$f'(z) = \frac{1}{2\pi i} \oint_{C_R} \frac{f(\zeta)}{(\zeta - z)^2} \, d\zeta.$$

The above result remains valid for any sufficiently large R since $f(z)$ is an entire function. Using the modulus inequality (4.1.6), the bound on the modulus of the integral can be estimated as

$$|f'(z)| = \left|\frac{1}{2\pi i}\right| \left|\oint_{C_R} \frac{f(\zeta)}{(\zeta - z)^2} \, d\zeta\right| \leq \frac{1}{2\pi} \frac{B}{R^2} 2\pi R = \frac{B}{R}.$$

By letting $R \to \infty$, we have

$$|f'(z)| = 0.$$

This implies that $f(z) = K$ for some $K \in \mathbb{C}$.

An interesting deduction from the Liouville theorem is that non-constant entire functions must be unbounded. Actually we have already seen that entire functions like $\sin z$ and $\cos z$ are unbounded, unlike their real counterparts.

Theorem 4.3.6 (Maximum modulus theorem) *Let $\mathcal{D}$ be a bounded domain that is enclosed by the closed contour C. If $f(z)$ is analytic on the domain $\mathcal{D}$ and continuous on the bounding contour C, then the maximum value of $|f(z)|$ occurs on C, unless $f(z)$ is a constant function.*

Proof Given that the function $f(z)$ is continuous on and inside C, the modulus function $|f(z)|$ is also continuous on and inside C and the maximum value of $|f(z)|$ on or inside C always exists. To prove the theorem, it suffices to establish the following claim. Supposing $|f(z)|$ attains its maximum value at some interior point $\alpha \in \mathcal{D}$, then $f(z)$ is constant for all $z \in \mathcal{D}$.

First, we take a small neighborhood $N(\alpha; r)$ around α that lies completely inside $\mathcal{D}$. We then apply modulus inequality (4.1.3d) to the formula of Gauss' mean value theorem [see eq. (4.3.8)] and obtain

$$|f(\alpha)| \leq \frac{1}{2\pi} \int_0^{2\pi} |f(\alpha + re^{i\theta})|\, d\theta. \tag{4.3.9a}$$

Suppose $|f(\alpha)|$ were a maximum; then

$$|f(\alpha + re^{i\theta})| \leq |f(\alpha)|, \quad \text{for all } \theta, \tag{4.3.9b}$$

which implies that

$$\frac{1}{2\pi} \int_0^{2\pi} |f(\alpha + re^{i\theta})|\, d\theta \leq \frac{1}{2\pi} \int_0^{2\pi} |f(\alpha)|\, d\theta = |f(\alpha)|. \tag{4.3.9c}$$

Combining the results in inequalities (4.3.9a,c), we deduce that

$$\int_0^{2\pi} [|f(\alpha)| - |f(\alpha + re^{i\theta})|]\, d\theta = 0.$$

In addition, inequality (4.3.9b) dictates that the above integrand is non-negative. We then argue that $|f(\alpha)| = |f(\alpha + re^{i\theta})|$ for all points on the circle, in view that it is not possible to have $|f(\alpha + re^{i\theta})|$ less than $|f(\alpha)|$ even at some isolated points. Assume the contrary, suppose $|f(\alpha + re^{i\theta})|$ is less than $|f(\alpha)|$ at a single point; by continuity of $f(z)$, $|f(\alpha + re^{i\theta})|$ will be less than $|f(\alpha)|$ for a finite arc on the circle. This leads to

$$\frac{1}{2\pi} \int_0^{2\pi} |f(\alpha + re^{i\theta})|\, d\theta < |f(\alpha)|, \tag{4.3.9d}$$

a contradiction to inequality (4.3.9a). Furthermore, since r can be any value, it then follows that $|f(z)|$ is constant in any neighborhood of α lying inside $\mathcal{D}$.

Next, we would like to show that $|f(z)|$ is constant not only in every neighborhood of α, but also at every point in $\mathcal{D}$. We take any point $z \in \mathcal{D}$, and join α to z by a curve γ lying completely inside $\mathcal{D}$. A sequence of points

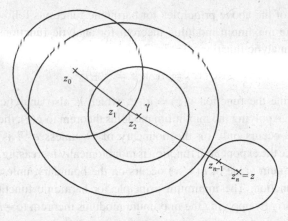

Figure 4.7. The point z_0 is joined to the point z via a curve γ in $\mathcal{D}$. The points $z_0, z_1, \ldots, z_n$ are points lying on γ and each of them is the center of a disk (plus its boundary) lying completely inside $\mathcal{D}$ and containing the preceding point.

$z_0 = \alpha, z_1, \ldots, z_n = z$ on γ are chosen with the property that each of these points is the center of a disk (plus its boundary) lying inside $\mathcal{D}$ and containing the preceding point (see Figure 4.7). Since z_1 is contained in the disk centered at α, from the above result, we have $|f(z_1)| = |f(\alpha)|$. Applying the same argument to $|f(z)|$ in the disk centered at z_1, we conclude that $|f(z)|$ must be constant within this disc. In particular, $|f(z_2)| = |f(z_1)| = |f(\alpha)|$. Repeating the argument, we finally have $|f(z)| = |f(z_{n-1})| = \cdots = |f(z_1)| = |f(\alpha)|$, valid for any point z in $\mathcal{D}$.

Now, we have established that $|f(z)|$ is constant for every point in $\mathcal{D}$ if $|f(z)|$ attains its maximum value at some interior point $\alpha \in \mathcal{D}$. Recall that $f(z)$ is constant in $\mathcal{D}$ if $|f(z)|$ is constant in $\mathcal{D}$ (see Problem 2.26). Therefore, for a non-constant $f(z)$, the maximum value of $|f(z)|$ cannot occur inside $\mathcal{D}$. Equivalently, the maximum value of $|f(z)|$ occurs only on the boundary curve C.

Corollary Suppose $\mathcal{D}$ is a bounded domain and $u(x, y)$ is harmonic inside $\mathcal{D}$ and non-constant; then $u(x, y)$ cannot attain its maximum or minimum value inside $\mathcal{D}$. In other words, if there exists $(x_0, y_0) \in \mathcal{D}$ such that

$$u(x_0, y_0) = \sup_{\mathcal{D}} u(x, y) \quad \text{or} \quad u(x_0, y_0) = \inf_{\mathcal{D}} u(x, y),$$

then $u(x, y)$ is constant inside $\mathcal{D}$. These results are called the *maximum and minimum principles for harmonic functions*.

The proof of the above principles for harmonic functions follows immediately from the maximum modulus theorem for analytic functions. First, we construct an analytic function $f(z)$ in $\mathcal{D}$ such that

$$\text{Re} f(z) = u(z), \; z = x + iy.$$

Next, we define the function $g(z) = e^{f(z)}$, which is also analytic in $\mathcal{D}$, and $|g(z)| = e^{u(z)}$. Applying the maximum modulus theorem to $g(z)$, the maximum value of $e^{u(z)}$ occurs only on the boundary of $\mathcal{D}$, unless $e^{u(z)}$ is a constant function. Since the exponential function is monotonically increasing, we deduce that the maximum value of $u(x, y)$ occurs on the boundary, unless $u(x, y)$ is a constant function. The minimum principle for harmonic functions can be proved similarly by applying the maximum modulus theorem to $-u(z)$.

Example 4.3.4 At what point inside the region $|z - z_0| \leq 1$ does $|e^z|$ attain its maximum value? Find this maximum value.

Solution By virtue of the maximum modulus principle, since e^z is an entire function, $|e^z|$ attains its maximum value inside the given region only along the boundary $|z - z_0| = 1$. On te circle $|z - z_0| = 1$, $z = z_0 + e^{i\theta}$, the value of $|e^z|$ along the boundary is given by

$$|e^z| = |e^{z_0 + e^{i\theta}}| = |e^{\text{Re } z_0 + \cos\theta + i(\text{Im } z_0 + \sin\theta)}| = e^{\text{Re } z_0} e^{\cos\theta}.$$

Note that $e^{\cos\theta}$ attains its maximum value at $\theta = 0$. Therefore, $|e^z|$ attains its maximum value $e^{\text{Re } z_0 + 1}$ at the boundary point $z = z_0 + 1$.

4.4 Potential functions of conservative fields

In this section, we examine the application of analytic function theory to the study of several prototype conservative fields in physics, like fluid flow, electrostatics and gravitation. The potential functions derived from these conservative fields are shown to be harmonic, thus explaining why complex variables techniques are commonly applied to solve physical problems in fluid flow, electrostatics and gravitation.

4.4.1 Velocity potential and stream function of fluid flows

The potential theory of fluid flows rests on the assumption of the incompressible, inviscid and irrotational properties of the fluid motion. When the Mach number (which is the ratio of the fluid velocity to the speed of sound) of the fluid flow is below 0.3, the assumption of incompressibility is acceptable. For

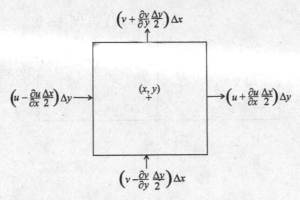

Figure 4.8. Volume fluxes across the four surfaces of a differential control volume in a two-dimensional flow field.

common fluids, like air and water, viscosity effects are negligible in regions distant from boundaries with solids. First, we derive the continuity equation based on the property of incompressibility of the fluid. We then examine the rotational property of fluid motion, characterized by the angular velocity of a fluid element. A fluid motion is *irrotational* if the angular velocity of each element throughout the whole flow field is zero. The condition of irrotationality allows the definition of a scalar quantity called the *velocity potential*, whose gradient gives the velocity vector. An irrotational, inviscid and incompressible flow is termed a *potential flow*.

Continuity equation

Let u and v denote the horizontal and vertical components of the fluid velocity vector. Consider a control volume of infinitesimal widths Δx and Δy in a two-dimensional flow field (see Figure 4.8). Let (x, y) be the coordinates of the center of the control volume. The horizontal velocities at the left and right faces are $u - \frac{\partial u}{\partial x} \frac{\Delta x}{2}$ and $u + \frac{\partial u}{\partial x} \frac{\Delta x}{2}$, respectively. Likewise, the vertical velocities at the bottom and top faces are $v - \frac{\partial v}{\partial y} \frac{\Delta y}{2}$ and $v + \frac{\partial v}{\partial y} \frac{\Delta y}{2}$, respectively. The volume flux of fluid *flowing into* the control volume (assuming unit depth of the control volume normal to the plane) equals $\left(u - \frac{\partial u}{\partial x} \frac{\Delta x}{2}\right) \Delta y + \left(v - \frac{\partial v}{\partial y} \frac{\Delta y}{2}\right) \Delta x$, while the volume flux of fluid *flowing out* of the control volume equals $\left(u + \frac{\partial u}{\partial x} \frac{\Delta x}{2}\right) \Delta y + \left(v + \frac{\partial v}{\partial y} \frac{\Delta y}{2}\right) \Delta x$. The net rate of accumulation of fluid in the control volume is equal to the volume flux inflow minus the volume flux outflow. Thus, within the differential control volume,

$$\text{net rate of accumulation} = -\left(\frac{\partial u}{\partial x} + \frac{\partial v}{\partial y}\right) \Delta x \Delta y.$$

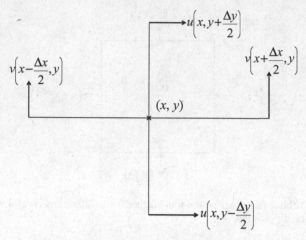

Figure 4.9. The local rotational motion of the horizontal and vertical differential line segments through the point (x, y).

Since the fluid is incompressible, the net rate of accumulation in any control volume in the flow field is zero. Therefore, the condition of incompressibility is given by

$$\frac{\partial u}{\partial x} + \frac{\partial v}{\partial y} = 0. \qquad (4.4.1)$$

The above equation is called the *continuity equation*.

Irrotationality condition

To derive the condition of irrotationality, let us consider the local rotational motion of the horizontal and vertical differential line segments through (x, y) as shown in Figure 4.9. The horizontal line segment of width Δx through (x, y) is subject to angular velocity (in the anti-clockwise sense) as given by

$$\frac{v\left(x + \frac{\Delta x}{2}, y\right) - v\left(x - \frac{\Delta x}{2}, y\right)}{\Delta x} \approx \frac{\partial v}{\partial x}(x, y).$$

Similarly, the vertical line segment of width Δy through the center is subject to angular velocity (in the anti-clockwise sense) as given by

$$-\frac{u\left(x, y + \frac{\Delta y}{2}\right) - u\left(x, y - \frac{\Delta y}{2}\right)}{\Delta y} \approx -\frac{\partial u}{\partial y}(x, y).$$

The average local angular velocity of the differential fluid element of widths Δx and Δy centered at the point (x, y) is then given by $\frac{1}{2}\left(\frac{\partial v}{\partial x} - \frac{\partial u}{\partial y}\right)$.

Let $\mathbf{v}(\dot{x}, y)$ denote the velocity vector of the fluid motion. We define the *vorticity* ω to be the curl of the velocity vector, that is,

$$\omega = \|\ \|\ \|\ \|$$ (4.4.2)

In a two-dimensional flow field, the vorticity has only one component that is normal to the x-y plane. Its magnitude is given by

$$|\omega| = \frac{\partial v}{\partial x} - \frac{\partial u}{\partial y}.$$ (4.4.3)

The local angular velocity of a fluid element is seen to be $\frac{|\omega|}{2}$. A fluid motion is said to be *irrotational* when the angular velocity at each point throughout the flow field is zero. Therefore, the condition of irrotationality is given by

$$\frac{\partial v}{\partial x} - \frac{\partial u}{\partial y} = 0.$$ (4.4.4)

To understand the physical interpretation of an irrotational fluid motion, we consider an infinitesimal straw immersed in the flow field. If the fluid is truly irrotational, the tiny straw always moves parallel to itself since the fluid has zero angular velocity everywhere. Consider the following interesting example: the flow field due to a vortex such as a tornado can be irrotational even though the global motion of the fluid is circulating around in concentric circular patterns.

In the above discussion, the continuity equation and irrotationality condition have been formulated using the differential approach based on the physics of the underlying processes. On the other hand, the integral formulations presented below exhibit better linkage with the theory of conservative fields, and lead naturally to the definitions of two scalar functions via a theorem in vector calculus. These two scalar functions are termed the velocity potential and the stream function, and they play fundamental roles in the description of potential fluid flows.

Flux and circulation

Let $\mathbf{r}(s) = x(s)\mathbf{i} + y(s)\mathbf{j}$ be a smooth curve parametrized by its arc length s. The unit tangent and normal vectors are, respectively, given by

$$\mathbf{t} = \frac{d\mathbf{r}}{ds} = \frac{dx}{ds}\mathbf{i} + \frac{dy}{ds}\mathbf{j},$$

$$\mathbf{n} = \frac{dy}{ds}\mathbf{i} - \frac{dx}{ds}\mathbf{j}.$$

The flux $\mathcal{F}_\gamma$ across a curve γ is defined by

$$\mathcal{F}_\gamma = \int_\gamma \mathbf{v} \cdot \mathbf{n}\, ds = \int_\gamma u\, dy - v\, dx.$$ (4.4.5)

When γ is a closed curve, by the Gauss theorem in vector calculus, the above line integral can be converted into a double integral:

$$\mathcal{F}_\gamma = \oint_\gamma \mathbf{v} \cdot \mathbf{n} \, ds = \iint_{A_\gamma} \nabla \cdot \mathbf{v} \, dxdy = \iint_{A_\gamma} \left(\frac{\partial u}{\partial x} + \frac{\partial v}{\partial y} \right) dxdy, \quad (4.4.6)$$

where A_γ is the area bounded by the closed curve γ. Physically, if the fluid is incompressible, then the flux across any closed curve γ is zero. Mathematically, we deduce from eq. (4.4.6) that the integrand in the double integral vanishes at any point in the flow field. This result is precisely the continuity equation given in eq. (4.4.1).

The circulation C_γ along a curve γ is defined by

$$C_\gamma = \int_\gamma \mathbf{v} \cdot d\mathbf{r} = \int_\gamma u \, dx + v \, dy. \quad (4.4.7)$$

When γ is a closed curve, by the Stokes theorem in vector calculus, C_γ can be expressed as a double integral:

$$C_\gamma = \oint_\gamma \mathbf{v} \cdot d\mathbf{r} = \iint_{A_\gamma} \nabla \times \mathbf{v} \cdot d\mathbf{A} = \iint_{A_\gamma} |\omega| \, dxdy, \quad (4.4.8)$$

where $d\mathbf{A}$ is the differential area vector and ω is the vorticity. We deduce that the irrotationality condition [see eq. (4.4.4)] is equivalent to the vanishing of C_γ for any closed curve γ inside the flow field.

Velocity potential and stream function

A continuously differentiable vector function $\mathbf{F}$ is said to be *conservative* in a domain $\mathcal{D}$ if and only if

$$\oint_\gamma \mathbf{F} \cdot d\mathbf{r} = 0, \quad (4.4.9)$$

for any closed curve γ inside $\mathcal{D}$. By a well-known theorem in vector calculus, the above condition is equivalent to the existence of a scalar potential $\phi(x, y)$ such that

$$\mathbf{F} = \nabla\phi. \quad (4.4.10)$$

Supposing the circulation C_γ along any closed curve γ inside the domain $\mathcal{D}$ is zero, by virtue of eq. (4.4.10), we claim that there exists a scalar function $\phi(x, y)$ such that

$$\mathbf{v} = \nabla\phi \quad \text{or} \quad u = \frac{\partial \phi}{\partial x}, \quad v = \frac{\partial \phi}{\partial y}. \quad (4.4.11)$$

This scalar function ϕ is called the *velocity potential* of the flow field in $\mathcal{D}$. The level curves $\phi(x, y) = $ constant are called the *equipotential lines* of the flow field.

If we write $\mathbf{F} = -v\mathbf{i} + u\mathbf{j}$, then the incompressibility condition is given by

$$\mathcal{F}_\gamma = \oint_\gamma \mathbf{v} \cdot \mathbf{n} \, ds = \oint_\gamma \mathbf{F} \cdot d\mathbf{r} = 0, \tag{4.4.12}$$

for any closed curve γ inside the flow field. Applying eq. (4.4.10) again, we find that there exists a scalar function $\psi(x, y)$ such that

$$-v\mathbf{i} + u\mathbf{j} = \nabla\psi \quad \text{or} \quad u = \frac{\partial\psi}{\partial y}, \quad v = -\frac{\partial\psi}{\partial x}. \tag{4.4.13}$$

This scalar function ψ is called the *stream function* of the flow field in $\mathcal{D}$. The level curves $\psi(x, y) = $ constant are called the *streamlines* of the flow field. Note that

$$\nabla\phi \cdot \nabla\psi = (u\mathbf{i} + v\mathbf{j}) \cdot (-v\mathbf{i} + u\mathbf{j}) = 0, \tag{4.4.14}$$

so that the streamlines and equipotential lines are orthogonal to each other.

Physical properties of the stream function

(i) A flow particle moves along a streamline. To verify this statement, we consider

$$d\psi = \frac{\partial\psi}{\partial x}dx + \frac{\partial\psi}{\partial y}dy = -v \, dx + u \, dy.$$

Along a streamline, we have $d\psi = 0$ so that

$$0 = -v \, dx + u \, dy \quad \text{or} \quad \left(\frac{dy}{dx}\right)_\psi = \frac{v}{u}. \tag{4.4.15}$$

Therefore, the direction of the velocity vector at a point on a streamline coincides with the slope of the streamline.

(ii) The difference between the values of ψ along two different streamlines gives the volume of fluid flowing between these two streamlines. To prove the statement, we consider two streamlines $\psi = \psi_1$ and $\psi = \psi_2$ and place a curve AB joining the two streamlines as shown in Figure 4.10. The volume flow rate Q between ψ_1 and ψ_2 (assuming $\psi_2 > \psi_1$) is given by

$$Q = \int_A^B u \, dy - v \, dx = \int_A^B \frac{\partial\psi}{\partial x}dx + \frac{\partial\psi}{\partial y}dy$$

$$= \int_A^B d\psi = \psi_2 - \psi_1. \tag{4.4.16}$$

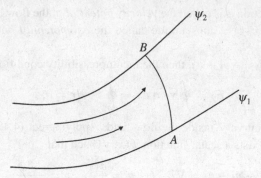

Figure 4.10. The two streamlines behave like impenetrable walls. The amount of fluid flowing within the streamlines ψ_1 and ψ_2 (assuming $\psi_2 > \psi_1$) is given by $\psi_2 - \psi_1$.

Complex potential

The incompressibility and irrotationality conditions are expressed by

$$\frac{\partial u}{\partial x} + \frac{\partial v}{\partial y} = 0 \quad \text{and} \quad \frac{\partial v}{\partial x} - \frac{\partial u}{\partial y} = 0,$$

respectively. The scalar functions $\phi(x, y)$ and $\psi(x, y)$ are derived based on irrotationality and incompressibility, respectively. Suppose we substitute $u = \frac{\partial \phi}{\partial x}$ and $v = \frac{\partial \phi}{\partial y}$ into the continuity equation (4.4.1). We obtain

$$\frac{\partial^2 \phi}{\partial x^2} + \frac{\partial^2 \phi}{\partial y^2} = 0, \tag{4.4.17}$$

where $\phi(x, y)$ is assumed to be twice differentiable. Hence, ϕ satisfies the Laplace equation, so ϕ is a harmonic function.

Similarly, if we substitute $u = \frac{\partial \psi}{\partial y}$ and $v = -\frac{\partial \psi}{\partial x}$ into the irrotationality condition [see eq. (4.4.4)], we obtain

$$\frac{\partial^2 \psi}{\partial x^2} + \frac{\partial^2 \psi}{\partial y^2} = 0, \tag{4.4.18}$$

so ψ is also harmonic.

Note that the velocity potential and stream function are related to the velocity components by

$$u = \frac{\partial \phi}{\partial x} = \frac{\partial \psi}{\partial y} \quad \text{and} \quad v = \frac{\partial \phi}{\partial y} = -\frac{\partial \psi}{\partial x}. \tag{4.4.19}$$

The above relations are recognized to be the Cauchy-Riemann relations for the harmonic functions $\phi(x, y)$ and $\psi(x, y)$. If we define the complex potential $f(z)$ of a potential flow field by

$$f(z) = \phi(x, y) + i\psi(x, y), \quad z = x + iy, \tag{4.4.20}$$

and assume that velocity components are continuous functions, then $f(z)$ is an analytic function by virtue of the relations in eq. (4.4.19) and the continuity properties of the velocity functions. Conversely, for any analytic function, its real and imaginary parts can be considered as a feasible velocity potential and stream function of a potential flow field, respectively.

Since the partial derivatives of ϕ and ψ are related to the velocity components, the derivative of the complex potential f is expected to have a similar physical interpretation. Indeed, we observe that

$$\frac{df}{dz} = \frac{\partial \phi}{\partial x} + i \frac{\partial \psi}{\partial x} = u - iv, \qquad (4.4.21)$$

and the quantity $u - iv$ is called the *complex velocity*.

For example, the complex velocity for a uniform flow with speed V_0 and at an angle α to the positive x-axis is given by $\overline{V_0 e^{i\alpha}}$. Using eq. (4.4.21) and performing the integration, the complex potential of this uniform flow is found to be

$$f(z) = V_0 e^{-i\alpha} z.$$

The complex potential is unique up to an additive constant.

Let u_r and u_θ denote the radial and tangential components of velocity. The relations between the velocity components in the rectangular and polar coordinates are

$$u = u_r \cos \theta - u_\theta \sin \theta \quad \text{and} \quad v = u_r \sin \theta + u_\theta \cos \theta,$$

so that the complex velocity in polar coordinates is given by

$$\begin{aligned} u - iv &= (u_r \cos \theta - u_\theta \sin \theta) - i(u_r \sin \theta + u_\theta \cos \theta) \\ &= (u_r - iu_\theta)e^{-i\theta}. \end{aligned} \qquad (4.4.22)$$

Example 4.4.1 Discuss the two-dimensional flow field as represented by the complex potential

$$f(z) = cz^{1/2}, \quad c \text{ is real.}$$

Solution It is more convenient to express the complex potential in polar coordinates $z = re^{i\theta}$. The given complex potential can be expressed as

$$f(r) = cr^{1/2} e^{i\theta/2} = cr^{1/2} \cos \frac{\theta}{2} + i cr^{1/2} \sin \frac{\theta}{2}.$$

The velocity potential and stream function are, respectively,

$$\phi(r, \theta) = cr^{1/2} \cos \frac{\theta}{2} \quad \text{and} \quad \psi(r, \theta) = cr^{1/2} \sin \frac{\theta}{2}, \quad -\pi < \theta \leq \pi.$$

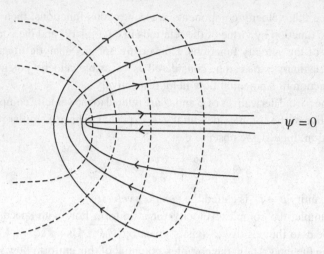

Figure 4.11. Pattern of the streamlines (solid) and equipotential lines (dashed) of the flow field around a flat plate that is placed along the whole positive x-axis.

The line $\theta = 0$ is the streamline corresponding to $\psi = 0$. The other streamlines are defined by $r^{1/2} \sin \frac{\theta}{2} = $ constant. Some of the streamlines are drawn in Figure 4.11. The equipotential lines are given by $r^{1/2} \cos \frac{\theta}{2} = $ constant. The complex potential represents the potential flow around a flat plate placed along the whole positive x-axis (the plate corresponds to the streamline $\psi = 0$).

The complex velocity is given by

$$\frac{df}{dz} = \frac{c}{2} z^{-1/2} = \frac{c}{2r^{1/2}} \left(\cos \frac{\theta}{2} + i \sin \frac{\theta}{2} \right) e^{-i\theta}.$$

Using eq. (4.4.22), the radial and tangential components of velocity are seen to be

$$u_r = \frac{c}{2r^{1/2}} \cos \frac{\theta}{2} \quad \text{and} \quad u_\theta = -\frac{c}{2r^{1/2}} \sin \frac{\theta}{2}.$$

At the corner of the plate where $r = 0$, the velocity components become infinite. The velocity is singular of order $r^{-1/2}$.

Complex potentials of basic fluid elements

The fluid source and fluid vortex are considered as the basic fluid elements. Here, we would like to derive the complex potential of each of them. Also, we give the physical interpretation of some flow fields obtained by superposition of these basic elements.

Source

The velocity function of a fluid source placed at the origin was derived in Subsection 2.1.1. The complex velocity of the source is given by

$$f'(z) = \overline{v(z)} = \frac{k}{z}, \quad k \text{ is real.}$$

Upon integration, the complex potential is found to be

$$f(z) = k \operatorname{Log} z.$$

Separating the complex potential into its real and imaginary parts, and writing $z = re^{i\theta}$, the velocity potential and stream function of the flow field due to the source are

$$\phi(r, \theta) = k \ln r \quad \text{and} \quad \psi(r, \theta) = k\theta.$$

The streamlines are the radial lines $\theta = \theta_0$, $-\pi < \theta_0 \leq \pi$. The equipotential lines are orthogonal to the streamlines, and they are concentric circles centered at the origin: $r = r_0$, $r_0 > 0$.

The flow pattern exhibits radial symmetry at the origin. The volume flow rate m (usually called the source strength) of the flow through the circle $r = r_0$ is found to be

$$m = \int_0^{2\pi} \frac{k}{r_0} r_0 \, d\theta = 2\pi k.$$

As expected, m is independent of r_0 since there is no accumulation of fluid within any circle. This is the volume of fluid per unit time flowing out from the source. The direction of the flow is radially outward and the velocity at the source is infinite. In terms of the source strength m, the complex potential of the flow field due to the source is given by

$$f(z) = \frac{m}{2\pi} \operatorname{Log} z. \tag{4.4.23}$$

The flux across any closed curve γ enclosing the origin in the flow field due to the above source is given by

$$\mathcal{F}_\gamma = \oint_\gamma u \, dy - v \, dx$$

$$= \operatorname{Im} \oint_\gamma f'(z) \, dz, \quad dz = dx + i \, dy,$$

$$= \operatorname{Im} \oint_\gamma \frac{m}{2\pi} \frac{1}{z} \, dz = m, \tag{4.4.24}$$

by virtue of the Cauchy integral formula. Again, as there is no accumulation of fluid inside γ, the flux across any closed curve γ enclosing the source is equal to the source strength. However, if γ does not encircle the source, the flux becomes zero. On the other hand, it can be shown that the circulation across any closed curve γ in the flow field due to a source is always zero.

In general, suppose there are n sources located at $z_1, z_2, \ldots, z_n$ in the flow field, whose respective source strengths are $m_1, m_2, \ldots, m_n$. The flux across a closed curve γ that encircles all the sources is given by

$$
\begin{aligned}
F_\gamma &= \text{Im} \oint_\gamma \left(\frac{m_1}{2\pi} \frac{1}{z - z_1} + \frac{m_2}{2\pi} \frac{1}{z - z_2} + \cdots + \frac{m_n}{2\pi} \frac{1}{z - z_n} \right) dz \\
&= m_1 + m_2 + \cdots + m_n,
\end{aligned}
\tag{4.4.25}
$$

which equals the sum of the source strengths.

Vortex

Another basic fluid element is the vortex, of which the common whirlpool or tornado is a close approximation. The streamlines of the flow pattern due to a vortex are concentric circles centered at the vortex. Supposing the vortex is placed at the origin, the stream function may take the form

$$
\psi(r, \theta) = -k \ln r, \quad k \text{ is real.}
$$

Based on this assumed form of ψ, the derived velocity components agree with the required physics, namely, the radial velocity is zero and the tangential velocity has $O\left(\frac{1}{r}\right)$ dependence (see Problem 2.5). Since ψ is the harmonic conjugate of the velocity potential ϕ, we deduce that

$$
\phi(r, \theta) = k\theta,
$$

so that the complex potential of the flow field due to the vortex becomes

$$
f(z) = k\theta - ik \ln r = -ik \, \text{Log} \, z, \quad z = re^{i\theta}.
\tag{4.4.26}
$$

The above complex potential resembles that of a source except that the multiplicative constant in front of Log z becomes an imaginary number.

The circulation around any closed curve γ enclosing the origin in the flow field due to the above vortex placed at the origin is given by

$$
\begin{aligned}
C_\gamma &= \oint_\gamma u \, dx + v \, dy \\
&= \text{Re} \oint_\gamma f'(z) \, dz, \quad dz = dx + i \, dy \\
&= \text{Re} \oint_\gamma \frac{-ik}{z} \, dz = 2\pi k.
\end{aligned}
\tag{4.4.27}
$$

We define the strength of a vortex to be the magnitude of the circulation around a closed curve enclosing the vortex. Let Γ denote the strength of the vortex; we then have $\Gamma = 2\pi k$, and so the complex potential of the fluid flow due to the vortex with strength Γ can be expressed as

$$f(z) = \frac{\Gamma}{2\pi i} \operatorname{Log} z. \qquad (4.4.28)$$

From eq. (4.4.27), it is seen that the circulation around any closed contour not enclosing the vortex is zero. By following a similar procedure to that in eq. (4.4.24), one can show that the flux around any closed contour in the flow field due to a vortex is always zero.

Doublet

A doublet is formed by the coalescence of a source and a sink (a negative source) of equal strength placed closed to each other (see Example 2.1.1). Consider a source of strength m placed at $z = -\epsilon$ and a sink of the same strength m placed at $z = \epsilon$, where ϵ is infinitesimal. The complex potential for this flow configuration is given by

$$\begin{aligned}
f(z) &= \frac{m}{2\pi} \operatorname{Log}(z + \epsilon) - \frac{m}{2\pi} \operatorname{Log}(z - \epsilon) \\
&= \frac{m}{2\pi} \operatorname{Log}\left(\left(1 + \frac{\epsilon}{z}\right)\left[1 + \frac{\epsilon}{z} + O\left(\frac{\epsilon^2}{z^2}\right)\right]\right), \quad \epsilon \ll |z| \\
&= \frac{m}{2\pi} \operatorname{Log}\left(1 + 2\frac{\epsilon}{z} + O\left(\frac{\epsilon^2}{z^2}\right)\right).
\end{aligned}$$

In the limits $\epsilon \to 0$ and $m \to \infty$ while keeping $m\epsilon = \pi\mu$, where μ is finite, the complex potential due to a doublet then becomes

$$f(z) = \frac{\mu}{z}. \qquad (4.4.29)$$

The stream function of the flow field due to the doublet at the origin is found to be

$$\psi(x, y) = -\mu \frac{y}{x^2 + y^2}.$$

For the streamline $\psi(x, y) = \psi_0$, the corresponding equation is

$$x^2 + \left(y + \frac{\mu}{2\psi_0}\right)^2 = \left(\frac{\mu}{2\psi_0}\right)^2. \qquad (4.4.30)$$

The pattern of the streamlines forms a coaxial system of circles along the y-axis (see Figure 4.12).

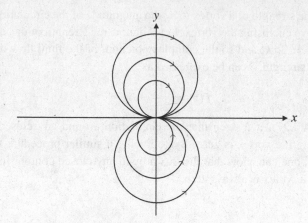

Figure 4.12. Pattern of the streamlines of the flow field due to a doublet at the origin.

The complex velocity of the doublet flow is given by

$$f'(z) = -\mu \frac{1}{z^2} = -\mu \frac{\cos\theta - i\sin\theta}{r^2} e^{-i\theta}. \qquad (4.4.31)$$

Therefore, the radial and tangential velocity components are, respectively,

$$u_r = -\mu \frac{\cos\theta}{r^2} \quad \text{and} \quad u_\theta = -\mu \frac{\sin\theta}{r^2}. \qquad (4.4.32)$$

Example 4.4.2 Discuss the potential flow field formed by the superposition of a uniform flow of speed U along the positive x-axis and a doublet of strength μ placed at the origin. Investigate the streamline pattern of the flow field.

Solution By the superposition principle, the complex potential of the resulting flow field is given by

$$f(z) = Uz + \frac{\mu}{z}, \quad z = re^{i\theta}.$$

The stream function of the flow is found to be

$$\psi(r, \theta) = \text{Im } f(z) = \left(Ur - \frac{\mu}{r} \right) \sin\theta.$$

We see that ψ becomes zero when $r = \sqrt{\frac{\mu}{U}}$, implying that the circle $|z| = \sqrt{\frac{\mu}{U}}$ is a streamline in the flow field. Apparently, we may place a solid circle of the same radius in the flow field where the body surface coincides with the streamline. The flow field is not disturbed since fluid particles are flowing past a streamline as if flowing tangentially past a solid surface.

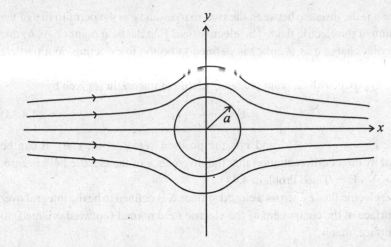

Figure 4.13. Streamline pattern of uniform potential flow past a circular obstacle of radius a placed at the origin.

The present flow field is exactly the same as that of a uniform flow past a circle of radius $a = \sqrt{\frac{\mu}{U}}$ placed at the origin. In other words, the complex potential of the uniform flow with speed U in the positive x-direction that streams past the circle $|z| = a$ is given by

$$f(z) = U\left(z + \frac{a^2}{z}\right).$$

The stream function can be expressed as

$$\psi(r, \theta) = U\left(r - \frac{a^2}{r}\right)\sin\theta.$$

The streamlines $\psi = $ constant of the above flow configuration are the family of cubic curves $y(x^2 + y^2 - a^2) = c(x^2 + y^2)$, where c is constant. The pattern of the streamlines of the flow field is plotted in Figure 4.13.

4.4.2 Electrostatic fields

An electrostatic field is a vector field defined by the forces caused by stationary charged bodies. The Coulomb law in electrostatic theory states that the magnitude of the force $\mathbf{F}$ between two point charges q_1 and q_2 is given by

$$|\mathbf{F}| = \frac{1}{4\pi\epsilon}\frac{q_1 q_2}{r^2}, \tag{4.4.33}$$

where r is the distance between the two charges and ϵ is the permittivity of the medium of the electric field. The electric field $\mathbf{E}$ at the field point (x, y, h) due to a point charge q at (ξ, η, ζ) is defined to be the force acting on a particle of positive unit charge at the point (x, y, h). Therefore, the electric field $\mathbf{E}$ at (x, y, h) produced by a point charge q placed at the origin is given by

$$\mathbf{E} = \frac{q}{4\pi \epsilon r^3} \mathbf{r}, \qquad (4.4.34)$$

where $r^2 = x^2 + y^2 + h^2$ and $\mathbf{r}$ is the position vector of (x, y, h). It can be shown by direct differentiation that the divergence of the electric field is zero, that is, $\nabla \cdot \mathbf{E} = 0$ (see Problem 4.44).

The electric flux F_S across a closed surface S is defined to be the integral over the surface of the component of the electric field normal (outward oriented) to the surface, that is,

$$F_S = \iint_S \mathbf{E} \cdot d\mathbf{S}, \qquad (4.4.35)$$

where $d\mathbf{S}$ is the differential area vector. We use the Gauss theorem in vector calculus to transform the above area integral into a volume integral and obtain

$$F_S = \iint_S \mathbf{E} \cdot d\mathbf{S} = \iiint_V \nabla \cdot \mathbf{E} \, dV,$$

where V is the volume bounded by the closed surface S.

We now compute the flux across a closed surface S due to the point charge q placed at the origin. Suppose the surface does not contain the origin (that is, not enclosing the point charge); since the divergence of the electric field is zero, we then have

$$F_S = \iint_S \frac{q}{4\pi \epsilon r^3} \mathbf{r} \cdot d\mathbf{S} = \iiint_V \nabla \cdot \frac{q}{4\pi \epsilon r^3} \mathbf{r} \, dV = 0.$$

However, when V does contain the origin, the above volume integral over V has the singularity $r = 0$. We delete a small ball B_d of radius d around the origin from the volume V. Inside the remaining volume $V \backslash B_d$, the electric field has zero divergence. By applying the Gauss theorem over the deleted volume $V \backslash B_d$, we obtain

$$0 = \iiint_{V \backslash B_d} \nabla \cdot \frac{q}{4\pi \epsilon r^3} \mathbf{r} \, dV$$

$$= \iint_S \frac{q}{4\pi \epsilon r^3} \mathbf{r} \cdot d\mathbf{S} - \iint_{S_d} \frac{q}{4\pi \epsilon r^3} \mathbf{r} \cdot d\mathbf{S},$$

where S_d is the surface of the ball B_d. The result indicates that the electric fluxes across S and S_d have the same value. The electric flux across S_d can be evaluated in a simple manner. On the spherical surface S_d, we have

$$\frac{\mathbf{r} \cdot d\mathbf{S}}{r^3} = \frac{d S}{d^2}.$$

and so

$$F_S = F_{S_d} = \iint_{S_d} \frac{q}{4\pi \epsilon r^3} \mathbf{r} \cdot d\mathbf{S} = \frac{q}{4\pi \epsilon d^2} \iint_{S_d} d S = \frac{q}{\epsilon}.$$

In summary, the electric flux across the closed surface S is given by

$$F_S = \iint_V \mathbf{E} \cdot d\mathbf{S} = \begin{cases} \dfrac{q}{\epsilon} & \text{if } V \text{ contains the point charge } q \\ 0 & \text{otherwise} \end{cases}. \qquad (4.4.36)$$

In general, if we consider a distributed charge of volume charge density ρ inside a closed volume V, the electric flux is then given by

$$F_S = \iint_S \mathbf{E} \cdot d\mathbf{S} = \iiint_V \frac{\rho}{\epsilon} \, dV,$$

where $\iiint \rho \, dV$ is the total charge enclosed inside V. By transforming the above surface integral into a volume integral, we obtain

$$\iiint_V \nabla \cdot \mathbf{E} \, dV = \iiint_V \frac{\rho}{\epsilon} \, dV.$$

Since the volume V is arbitrary, the integrand functions on both sides of the above equation must be equal. We then obtain

$$\nabla \cdot \mathbf{E} = \frac{\rho}{\epsilon}. \qquad (4.4.37)$$

Equation (4.4.37) is one of the *Maxwell equations* in electromagnetic theory.

The electric field $\mathbf{E}$ is known to be a conservative field, that is,

$$\oint_\gamma \mathbf{E} \cdot d\mathbf{r} = 0,$$

for any closed path γ. Therefore, there exists a scalar *electric potential* ϕ such that

$$\nabla \phi = -\mathbf{E}. \qquad (4.4.38)$$

By convention, the negative sign is included since the electric potential defines its value as the work required to move a positive unit test charge against the electric force from a point with a lower potential to another point with a higher potential. For example, using both eqs. (4.4.34) and (4.4.38), the electric potential at the field point (x, y, h) due to the point charge q placed at the origin is found to be

$$\phi(x, y, z) = \frac{q}{4\pi r}. \tag{4.4.39}$$

For $q > 0$, the electric potential value increases as the field point approaches the point charge.

Substituting eq. (4.4.38) into eq. (4.4.37), we then obtain the Poisson equation

$$\nabla^2 \phi = -\frac{\rho}{\epsilon}, \tag{4.4.40}$$

which is the governing equation for the electric potential. In particular, when $\rho = 0$, the Poisson equation reduces to the Laplace equation

$$\nabla^2 \phi = 0. \tag{4.4.41}$$

Suppose we confine ourselves to two-dimensional electrostatic problems in free space without generating charges; then the electric potential is a harmonic function. In this case, the complex variable techniques may provide useful tools for solving these problems. Given that ϕ is harmonic, there exists a harmonic conjugate ψ to ϕ such that

$$\Omega(z) = \phi(x, y) + i\psi(x, y), \quad z = x + iy, \tag{4.4.42}$$

is an analytic function. The harmonic conjugate ψ is called the *flux function* and Ω is called the *complex potential*. The orthogonal families of curves in the two-dimensional plane defined by

$$\phi(x, y) = \alpha \quad \text{and} \quad \psi(x, y) = \beta$$

are called the equipotential lines and flux lines of the electric field, respectively. The derivative of Ω is seen to be

$$\frac{d\Omega}{dz} = \frac{\partial \phi}{\partial x} + i\frac{\partial \psi}{\partial x} = \frac{\partial \phi}{\partial x} - i\frac{\partial \phi}{\partial y} = -(E_x - iE_y), \tag{4.4.43}$$

where E_x and E_y are the respective x- and y-components of the electric field vector **E**.

Example 4.4.3   Find the complex potential due to an infinitely long line charge with uniform linear charge density ρ.

Solution   Assume the line charge to be aligned with the vertical ζ-axis. The electric field vector due to the line charge is expected to have circular symmetry. By eq. (4.4.34), its value at the field point (x, y, h) is given by

$$\mathbf{E} = \int_{-\infty}^{\infty} \frac{\rho}{4\pi\epsilon} \frac{\mathbf{R}}{R^3} \, d\zeta,$$

where $\rho \, d\zeta$ gives the amount of charge over the differential segment $d\zeta$ along the line charge, $\mathbf{R} = x\mathbf{i} + y\mathbf{j} + (h - \zeta)\mathbf{k}$, and $R^2 = x^2 + y^2 + (h - \zeta)^2$. We see that

$$\int_{-\infty}^{\infty} \frac{1}{R^3} \, d\zeta = \frac{2}{x^2 + y^2} \quad \text{and} \quad \int_{-\infty}^{\infty} \frac{h - \zeta}{R^3} \, d\zeta = 0,$$

and so

$$\mathbf{E} = \frac{\rho}{2\pi\epsilon} \frac{x\mathbf{i} + y\mathbf{j}}{x^2 + y^2} = \frac{\rho}{2\pi\epsilon} \frac{\mathbf{r}}{r^2},$$

where $\mathbf{r} = x\mathbf{i} + y\mathbf{j}$ and $r^2 = x^2 + y^2$. Using eq. (4.4.38), the electric potential ϕ can be found by integrating

$$\nabla\phi = -\frac{\rho}{2\pi\epsilon r^2}\mathbf{r}$$

to give

$$\phi = -\frac{\rho}{2\pi\epsilon} \ln r.$$

The corresponding complex potential is then found to be

$$\Omega(z) = -\frac{\rho}{2\pi\epsilon} \operatorname{Log} z, \quad z = re^{i\theta}.$$

The equipotential lines are concentric circles around the line charge and the flux lines are rays emanating from the line charge. The line charge configuration closely resembles the fluid source in a potential flow field.

4.4.3 Gravitational fields

The gravitational vector field at the field point (x, y, h) due to a particle of mass m placed at (ξ, η, ζ) is defined to be the gravitational attractive force acting on a particle of unit mass at (x, y, h). Let

$$\mathbf{r} = (x - \xi)\mathbf{i} + (y - \eta)\mathbf{j} + (h - \zeta)\mathbf{k}$$

denote the vector from (ξ, η, ζ) to (x, y, h). By Newton's law of universal gravitation, the gravitational vector field is given by

$$\mathbf{F} = -G\frac{m}{r^3}\mathbf{r}, \tag{4.4.44}$$

where $r^2 = (x - \xi)^2 + (y - \eta)^2 + (h - \zeta)^2$ and G is the universal gravitational constant. The negative sign in the above equation reflects the attractive nature of the gravitational force. The gravitational vector field $\mathbf{F}$ is known to be conservative, which is reminiscent of the property that the work done in traversing around a closed loop in the field is zero. Therefore, there exists a scalar gravitational potential function ϕ such that

$$\nabla\phi = \mathbf{F}.$$

For example, the gravitational potential due to a particle of mass m placed at (ξ, η, ζ) can be found by integrating

$$\nabla\phi = -G\frac{m}{r^3}\mathbf{r}$$

to give

$$\phi = G\frac{m}{r}. \tag{4.4.45}$$

In general, suppose the mass in a body is distributed continuously with density $\rho(\xi, \eta, \zeta)$ throughout the volume V. The gravitational potential due to the body at the field point (x, y, h) is given by

$$\phi(x, y, h) = G \iiint_V \frac{\rho(\xi, \eta, \zeta)}{r} \, d\xi d\eta d\zeta. \tag{4.4.46}$$

Note that there is a singularity $r = 0$ of the integrand in the above integral when the field point is inside or on the surface of the body.

Following the usual convention, the vertical h-axis and ζ-axis are taken to be positive downward. The vertical gravity g_h due to a three-dimensional body is defined by

$$g_h = \frac{\partial \phi}{\partial h}(x, y, h) = G \iiint_V \rho(\xi, \eta, \zeta)\frac{\zeta - h}{r^3} \, d\xi d\eta d\zeta. \tag{4.4.47}$$

For example, consider a horizontal cylinder with uniform density ρ whose longitudinal axis is parallel to the η-axis. Suppose the longitudinal extent of the cylinder ranges from η_1 to η_2 (assuming $\eta_1 < \eta_2$); the vertical gravity due to the cylinder can be obtained by computing the integral

$$g_h = G\rho \iint_S \left(\int_{\eta_1}^{\eta_2} \frac{\zeta - h}{r^3} d\eta \right) d\xi d\zeta$$

$$= G\rho \iint_S \frac{\zeta - h}{(\xi - x)^2 + (\zeta - h)^2} \left[\frac{\eta_2 - y}{\sqrt{(\xi - x)^2 + (\eta_2 - y)^2 + (\zeta - h)^2}} \right.$$

$$\left. - \frac{\eta_1 - y}{\sqrt{(\xi - x)^2 + (\eta_1 - y)^2 + (\zeta - h)^2}} \right] d\xi d\zeta,$$

where S is the cross-section of the horizontal cylinder projected onto the ξ-ζ plane. Suppose we take the limits $\eta_1 \to -\infty$ and $\eta_2 \to \infty$ but keep y finite in the above integral. The resulting integral for g_h of a horizontal cylinder with an infinite extent becomes

$$
g_h = 2G\rho \iint_S \frac{\zeta - h}{(\zeta - x)^2 + (\zeta - h)^2} \, d\xi \, d\zeta \tag{4.4.48}
$$

Conjugate complex variables formulation

The evaluation of the above integral for g_h can be greatly facilitated by complex integration. We illustrate how to transform the above double integral into a contour integral by Green's theorem using the conjugate complex variables formulation. Let $\omega = \xi + i\zeta$ and $\overline{\omega} = \xi - i\zeta$ be a pair of conjugate complex variables. Given that the function $F(\omega, \overline{\omega})$ is continuous up to its first-order partial derivatives in a simply connected domain S and its boundary C, we show in Example 4.2.5 that Green's theorem can be formulated as

$$
\oint_C F(\omega, \overline{\omega}) \, d\omega = 2i \iint_S \frac{\partial F}{\partial \overline{\omega}} \, d\xi \, d\zeta. \tag{4.4.49}
$$

In the present gravitational problem, we further define another pair of conjugate complex variables: $s = x + ih$ and $\overline{s} = x - ih$. In terms of the two pairs of conjugate complex variables, we observe that eq. (4.4.49) can be expressed as

$$
g_h = -2G\rho \iint_S \operatorname{Im} \frac{1}{\omega - s} \, d\xi \, d\zeta. \tag{4.4.50}
$$

The function $\frac{1}{\omega - s}$ is analytic throughout S if the field point s is outside the boundary curve C. In this case, Green's theorem as formulated in eq. (4.4.49) can be applied. When the point s lies on or inside C, the above form of Green's theorem has to be modified accordingly.

We set

$$
\frac{\partial F}{\partial \overline{\omega}}(\omega, \overline{\omega}) = \frac{1}{\omega - s}
$$

so that

$$
F(\omega, \overline{\omega}) = \frac{\overline{\omega} - \overline{s}}{\omega - s}.
$$

By virtue of eq. (4.4.50), the double integral for g_h then becomes

$$
g_h = -2G\rho \operatorname{Im} \iint_S \frac{1}{\omega - s} \, d\xi \, d\zeta = G\rho \operatorname{Re} \oint_C \frac{\overline{\omega} - \overline{s}}{\omega - s} \, d\omega, \tag{4.4.51}
$$

where $s = x + ih$ lies outside C.

Example 4.4.4 Find the vertical gravity g_h of a horizontal infinite cylinder, whose cross-section is an n-sided polygon with vertices $P_j(\xi_j, \zeta_j)$,

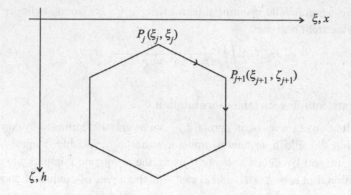

Figure 4.14. The cross-section of the horizontal infinite cylinder is an n-sided polygon with vertices $P_j(\xi_j, \zeta_j)$, $j = 1, 2, \ldots, n$. By convention, the vertical ζ-axis and h-axis are chosen to be positive downward, so correspondingly, the vertices are looped in the clockwise sense.

$j = 1, 2, \ldots, n$ looped in the positive sense (see Figure 4.14). The field point is assumed to be outside the bounding curve of the polygonal cross-section. For convenience of notation, it is taken to be at the origin, that is, $s = 0$.

Solution We write $\omega_j = \xi_j + i\zeta_j$, $j = 1, 2, \ldots, n$, and define $\Delta\omega_j = \omega_{j+1} - \omega_j$ (take ω_{n+1} to be ω_1). The equation of the polygonal side joining the two adjacent vertices P_j and P_{j+1} can be expressed as

$$\overline{\omega} = \alpha_j \omega + \beta_j,$$

where

$$\alpha_j = \frac{\overline{\Delta\omega_j}}{\Delta\omega_j} = e^{-2i\,\mathrm{Arg}\,\Delta\omega_j},$$

$$\beta_j = \overline{\omega_j} - \frac{\overline{\Delta\omega_j}}{\Delta\omega_j}\omega_j = \frac{2i(\xi_j\zeta_{j+1} - \xi_{j+1}\zeta_j)}{\Delta\omega_j}.$$

Substituting the above relations into the integral formula for g_h [see eq. (4.4.51)], the vertical gravity of the horizontal polygonal cylinder is given by

$$g_h = G\rho\,\mathrm{Re}\,\sum_{j=1}^{n}\int_{\omega_j}^{\omega_{j+1}}\frac{\alpha_j\omega + \beta_j}{\omega}\,d\omega$$

$$= G\rho\,\mathrm{Re}\,\sum_{j=1}^{n}\left(\alpha_j\Delta\omega_j + \beta_j\,\mathrm{Log}\,\frac{\omega_{j+1}}{\omega_j}\right).$$

Note that

$$\sum_{j=1}^{n} u_j \overline{\Delta w_j} - \sum_{j=1}^{n} \overline{w_{j+1} - w_j} = 0$$

and

$$\text{Log} \frac{w_{j+1}}{w_j} = \ln \frac{r_{j+1}}{r_j} + i(\theta_{j+1} - \theta_j),$$

where

$$r_j^2 = \xi_j^2 + \zeta_j^2 \quad \text{and} \quad \theta_j = \text{Arg} \, w_j.$$

Finally, we obtain

$$g_h = 2G\rho \sum_{j=1}^{n} \frac{\xi_j \zeta_{j+1} - \xi_{j+1} \zeta_j}{(\xi_{j+1} - \xi_j)^2 + (\zeta_{j+1} - \zeta_j)^2}$$

$$\times \left[(\zeta_{j+1} - \zeta_j) \ln \frac{r_{j+1}}{r_j} - (\xi_{j+1} - \xi_j)(\theta_{j+1} - \theta_j) \right]$$

$$= 2G\rho \sum_{j=1}^{n} \frac{\xi_j \zeta_{j+1} - \xi_{j+1} \zeta_j}{\sqrt{(\xi_{j+1} - \xi_j)^2 + (\zeta_{j+1} - \zeta_j)^2}}$$

$$\times \left[\sin(\text{Arg} \, \Delta w_j) \ln \frac{r_{j+1}}{r_j} - \cos(\text{Arg} \, \Delta w_j)(\theta_{j+1} - \theta_j) \right].$$

Remark If the field point is placed at (x, h) instead of the origin, then the corresponding formula for g_h can be obtained simply by changing ξ_j to $\xi_j - x$ and ζ_j to $\zeta_j - h$, $j = 1, 2, \ldots, n$.

4.5 Problems

4.1. For any curve Γ joining z_1 and z_2 in the complex plane, show that

$$\int_{\Gamma} z^2 \, dz = \frac{z_2^3 - z_1^3}{3}.$$

4.2. Evaluate the following integrals:

(a) $\int_0^{\pi+2i} \cos \frac{z}{2} \, dz;$ (b) $\int_1^i (2 + iz)^2 \, dz.$

In each case, the contour is the line segment joining the lower point to the upper point.

4.3. Evaluate the contour integral

$$\int_C \bar{z} \, dz,$$

where the contour C is

(a) the line segment joining $(0, 0)$ and $(1, 1)$;
(b) a union of two line segments: C_1 runs from $(0, 0)$ to $(1, 0)$ and C_2 runs from $(1, 0)$ to $(1, 1)$.

Do the two results agree? If not, explain why.

4.4. Show that

$$\left| \int_C dz \right| \le \int_C |dz|$$

and explain its geometric significance.

4.5. Show that

$$\int_C \overline{f(z)} \, \overline{dz} = \overline{\int_C f(z) \, dz}.$$

4.6. Estimate the upper bound of the modulus of the integral

$$I = \int_C \frac{\text{Log } z}{z - 4i} \, dz,$$

where C is the circle $|z| = 3$.

Hint: $\left| \dfrac{\text{Log } z}{z - 4i} \right| = \dfrac{\left| \ln |z| + i \text{ Arg } z \right|}{|z - 4i|} \le \dfrac{\left| \ln |z| \right| + |\text{Arg } z|}{\left| |z| - |4i| \right|}.$

4.7 Find an upper bound for

$$\left| \oint_{C_1(0)} e^{\frac{1}{z}} \, dz \right|,$$

where $C_1(0)$ is the positively oriented unit circle centered at the origin.

Hint: Along $C_1(0)$, we have

$$\left| e^{\frac{1}{z}} \right| \le e.$$

4.8. Let C be the union of three line segments, namely

$$C = \left\{ \frac{\pi}{2} + it : 0 \le t \le 1 \right\} \cup \left\{ i - t : -\frac{\pi}{2} \le t \le \frac{\pi}{2} \right\}$$
$$\cup \left\{ -\frac{\pi}{2} - it : -1 \le t \le 0 \right\}.$$

Estimate an upper bound of the modulus of each of the following integrals:

(a) $\int_C |z|^2 \, dz;$ (b) $\int_C |\sin z| \, dz;$ (c) $\int_C \operatorname{Im} e^{-z} \, dz.$

4.9. Suppose $f(z)$ is analytic throughout a simply connected domain $\mathcal{D}$. Take a fixed point $z_0 \in \mathcal{D}$ and define the function

$$F(z) = \int_{z_0}^z f(\zeta) \, d\zeta, \quad z \in \mathcal{D}.$$

Show that

$$F'(z) = f(z), \quad z \in \mathcal{D}.$$

Hint: Consider

$$F(z + \Delta z) - F(z) = \int_z^{z+\Delta z} f(\zeta) \, d\zeta.$$

By the Cauchy theorem, the last integral is independent of the path joining z and $z + \Delta z$ as long as the path is completely inside $\mathcal{D}$. Suppose we take the integration path from z to $z + \Delta z$ to be a line segment; then

$$\frac{F(z + \Delta z) - F(z)}{\Delta z} - f(z) = \frac{1}{\Delta z} \int_z^{z+\Delta z} [f(\zeta) - f(z)] \, d\zeta.$$

Using the continuity property of $f(z)$, show that the above right-hand integral tends to zero as $\Delta z \to 0$.

4.10. Suppose $0 < r < 1$; show that

$$\frac{1}{2\pi} \int_0^{2\pi} \left| \frac{re^{i\theta}}{(1 - re^{i\theta})^2} \right| \, d\theta = \frac{r}{1 - r^2}.$$

Hint: Consider

$$1 = \frac{1}{2\pi i} \oint_{|\zeta|=1} \left(\frac{1}{\zeta - z} - \frac{1}{\zeta - \frac{1}{z}} \right) d\zeta,$$

where $|z| = r < 1$ and $\zeta \bar{\zeta} = 1$.

4.11. Evaluate the following integrals:

(a) $\displaystyle\oint_{|z|=10} \frac{1}{1+z^3}\, dz$; (b) $\displaystyle\oint_{|z|=2} \frac{3z-1}{z(z-1)}\, dz$; (c) $\displaystyle\oint_{|z+1|=1} \frac{\sin\frac{\pi z}{4}}{z^2-1}\, dz$;

(d) $\displaystyle\oint_{|z+i|=1} \frac{e^z}{1+z^2}\, dz$; (e) $\displaystyle\oint_{x^2+y^2=2x} \frac{1}{1+z^4}\, dz$;

(f) $\displaystyle\oint_{|z|=r<1} \frac{1}{(z^2-1)(z^3-1)}\, dz$; (g) $\displaystyle\oint_{|z|=\frac{3}{2}} \frac{1}{(z^2+1)(z^2+4)}\, dz$;

(h) $\displaystyle\oint_{|z|=2} \frac{\sin z}{\left(z-\frac{\pi}{2}\right)^2}\, dz$; (i) $\displaystyle\oint_{|z|=1} \frac{e^z}{z^3}\, dz$; (j) $\displaystyle\oint_{|z|=1} |z-1|\,|dz|$.

4.12. Suppose the contour joining $-i$ and i lies in the complex plane excluding the origin and the negative real axis. Evaluate

$$\int_{-i}^{i} z^{\frac{1}{2}}\, dz$$

where the principal branch of $z^{\frac{1}{2}}(-\pi < \operatorname{Arg} z \le \pi)$ is taken.

4.13. By considering the contour integral

$$\oint_C \frac{e^{i\pi z^2 + 2\pi z}}{e^{2\pi z}+1}\, dz,$$

where $C = C_1 \cup C_2 \cup C_3 \cup C_4$ is the closed contour shown in the figure, show that

$$\int_0^\infty e^{-x^2}\, dx = \frac{\sqrt{\pi}}{2}.$$

The four line segments are

$$C_1 = \{x + ix : -R \le x \le R\},$$
$$C_2 = \{R + i(R+y) : 0 \le y \le 1\},$$
$$C_3 = \{x + i(x+i) : -R \le x \le R\},$$
$$C_4 = \{-R + i(y-R) : 0 \le y \le 1\}.$$

The integrand has a simple pole at $z = \frac{i}{2}$. Show that the line integral along the vertical line segment C_2 or C_4 vanishes as $R \to \infty$.

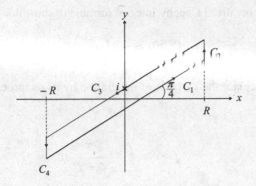

4.14. By evaluating the contour integral

$$\oint_{|z|=1} \left(z + \frac{1}{z} \right)^{2n} \frac{dz}{z},$$

show that

$$\int_0^{2\pi} \cos^{2n} \theta \, d\theta = 2\pi \frac{1 \cdot 3 \cdot 5 \cdots (2n-1)}{2 \cdot 4 \cdot 6 \cdots (2n)}.$$

4.15. Apply the Cauchy integral formula to the integral

$$\oint_{|z|=1} \frac{e^{kz}}{z} \, dz, \quad k \text{ is a real constant,}$$

to show that

$$\int_0^\pi e^{k \cos \theta} \cos(k \sin \theta) \, d\theta = \pi.$$

4.16. Suppose $f(z)$ is analytic on and inside the unit circle $|z| = 1$, and $f(0) = 1$. Show that

$$\frac{2}{\pi} \int_0^{2\pi} f(e^{i\theta}) \cos^2 \frac{\theta}{2} \, d\theta = 2 + f'(0)$$

and

$$\frac{2}{\pi} \int_0^{2\pi} f(e^{i\theta}) \sin^2 \frac{\theta}{2} \, d\theta = 2 - f'(0).$$

Hint: Evaluate $\dfrac{1}{2\pi i} \oint_{|z|=1} \left(2 \pm z + \dfrac{1}{z} \right) \dfrac{f(z)}{z} \, dz.$

4.17. Consider the function

$$f(z) = \frac{z-a}{z+a}.$$

Use the generalized Cauchy integral formula to show that

$$f^{(n)}(0) = -2 \left(-\frac{1}{a} \right)^n n!.$$

Hint: Replace the variable of integration by its reciprocal.

4.18. Let

$$f(z) = \oint_{|\zeta|=2} \frac{e^{\frac{\pi \zeta}{3}}}{\zeta - z} \, d\zeta.$$

Find the values $f(i)$ and $f(-i)$. Moreover, evaluate $f(z)$ when $|z| > 2$.

4.19 Suppose C is any contour that has the starting point $z = 0$ and ending point $z = 1$ but does not go through $z = i$ and $z = -i$. Show that the contour integral

$$\int_C \frac{1}{1 + z^2} \, dz$$

has value $\frac{\pi}{4} + k\pi$, where k is some integer.

4.20. Let

$$g(z) = \oint_{|\zeta|=2} \frac{2\zeta^2 - \zeta + 1}{\zeta - z} \, d\zeta.$$

Compute (a) $g(1)$; (b) $g(z_0)$, $|z_0| > 2$. Can we evaluate $g(2)$?

4.21. If C is any closed contour around the point $z = -1$, show that

$$\frac{1}{2\pi i} \oint_C \frac{z e^{zt}}{(z+1)^3} \, dz = \left(t - \frac{t^2}{2} \right) e^{-t}, \qquad t > 0.$$

4.22. If $f(z)$ is an nth-degree polynomial with nonzero leading coefficient,

$$f(z) = a_0 z^n + a_1 z^{n-1} + \cdots + a_n,$$

and C is a simple closed contour enclosing all the zeros of $f(z)$, show that

(a) $\dfrac{1}{2\pi i} \oint_C \dfrac{z f'(z)}{f(z)} \, dz = -\dfrac{a_1}{a_0};$

(b) $\dfrac{1}{2\pi i} \oint_C \dfrac{z^2 f'(z)}{f(z)} \, dz = \dfrac{a_1^2 - 2a_0 a_2}{a_0^2}.$

4.23. If $f(z)$ is analytic on and inside the unit circle $|z| = 1$, show that

$$(1 - |z|^2) f(z) = \frac{1}{2\pi i} \oint_{|\zeta|=1} f(\zeta) \frac{1 - \bar{z}\zeta}{\zeta - z} \, d\zeta, \qquad |z| < 1,$$

and deduce that

$$(1 - |a|^2)\,|f(z)| < \frac{1}{|z|} \int_0^{2\pi} |f(e^{i\theta})|\,d\theta, \quad |z| < 1.$$

Hint: $|1 - \bar{z}\zeta| = |\zeta - z|$ when $|\zeta| = 1$.

4.24. Supposing $f(z)$ is analytic on and inside the unit circle $|z| = 1$, and that $\operatorname{Re} f(z) > 0$, $f(0) = \alpha > 0$, show that

$$\left|\frac{f(z) - \alpha}{f(z) + \alpha}\right| \leq |z| \quad \text{and} \quad |f'(0)| \leq 2\alpha.$$

4.25. Suppose $f(z)$ is analytic and $|f(z)| \leq M$ inside the circle $|z| = R$, and $f(0) = 0$; show that

$$|f(z)| \leq \frac{M}{R}|z| \quad \text{and} \quad |f'(0)| \leq \frac{M}{R},$$

and equality holds only if $f(z) = \frac{M}{R} e^{i\theta} z$, where θ is real.

4.26. Suppose $f(z)$ is analytic inside the domain $0 < |z| < 1$, and

$$\oint_{|z|=r} f(z)\,dz = 0,$$

for all values of $r \in (0, 1)$. Is $f(z)$ analytic at $z = 0$? If not, give a counter example.

4.27. Suppose $f(z)$ is analytic and nonzero inside the domain $\mathcal{D}$. Explain why

$$\oint_\gamma \frac{f'(z)}{f(z)}\,dz = 0,$$

where γ is any simple closed contour inside $\mathcal{D}$.

4.28. Let f be entire and suppose $\operatorname{Im} f(z) \leq M$ for all z. Show that f must be a constant function.

4.29. Let f be analytic on and inside a simple closed curve C and $|f(z) - 1| < 1$ for all z on C. Use the maximum modulus theorem to show that f is nonzero inside C.

4.30. Find the maximum value of $|z^2 + 3z - 1|$ in the disk $|z| \leq 1$.

4.31. Let R denote the rectangular region:

$$0 \leq x \leq \pi, \quad 0 \leq y \leq 1.$$

By the maximum modulus theorem, the modulus of the entire function

$$f(z) = \sin z$$

has a maximum value in R that occurs on the boundary. Find the point on the boundary that gives the maximum value.

4.32. Suppose $f(z)$ is analytic in the domain $|z| < R$ and

$$M(r) = \max_{|z|=r} |f(z)|, \quad r < R.$$

Show that $M(r)$ is an increasing function of r.

4.33. Use the mean value theorem to show that

$$\int_0^\pi \ln(1 - 2r\cos\theta + r^2)\, d\theta = 0, \quad -1 < r < 1.$$

4.34. Suppose $u(z)$ and $v(z)$ are harmonic in a domain that contains the unit disk $|z| \le 1$; prove the following claims:

(a) If $u(z) > v(z)$ for $|z| = 1$, then $u(z) > v(z)$ for all z inside $|z| = 1$;
(b) If $u(z) = v(z)$ for $|z| = 1$, then $u(z) = v(z)$ for all z inside $|z| = 1$;
(c) If $u(0) \ge u(z)$ for $|z| = 1$, then $u(z)$ is constant for $|z| \le 1$.

If $u(z)$ and $v(z)$ are treated as the temperature functions in a steady state temperature field, do the above results agree with your physical intuition about temperature fields?

4.35. Show that the integral of the complex velocity around any simple closed contour in a flow field is equal to $\Gamma + im$, where Γ is the net strength of vortices inside the contour, and m is the net strength of sources and sinks inside the contour.

4.36. Interpret the flow field with the complex potential

$$f(z) = \frac{K_1 - iK_2}{2\pi} \operatorname{Log} \frac{z - \alpha}{z - \beta},$$

where K_1 and K_2 are real constants. Find the velocity potential and stream function, and sketch the pattern of streamlines and equipotential lines of the flow field.

4.37. Suppose the velocity potential and stream function of a potential flow are defined by

$$x + iy = c\cos(\phi + i\psi).$$

Show that

$$\frac{x^2}{c^2 \cosh^2 \psi} + \frac{y^2}{c^2 \sinh^2 \psi} = 1.$$

Explain why the streamlines are confocal ellipses, and show that the circulation around any one of these ellipses is 2π.

4.38. Suppose the complex potential $f = \phi + i\psi$ of a potential flow in the complex plane is defined by

$$z = \cosh f$$

Show that the streamlines are confocal hyperbolas and explain why the pattern might represent the flow through an aperture.

4.39. Suppose a source of strength m is located at $z = a$, $a > 0$, and a wall is placed along the vertical line $x = 0$. Explain why the complex potential of this flow configuration in the right half-plane, Re $z > 0$, is given by

$$f(z) = \frac{m}{2\pi}[\text{Log}(z - a) + \text{Log}(z + a)].$$

Note that the vertical wall $x = 0$ must be a streamline. Sketch some other streamlines of the flow field. This solution approach is called the *method of images*. The image source is placed at $z = -a$, which is the mirror image of $z = a$ with respect to the vertical wall $x = 0$.

4.40. A uniform flow of velocity $U_\infty > 0$ is disturbed by the vertical line segment joining $z = -i$ and $z = i$. Find the complex potential and the equation representing the family of streamlines. Sketch some of the streamlines of the flow field.

Hint: Note that the imaginary part of the function $\sqrt{z^2 + 1}$ becomes zero when z assumes values along the line segment joining $z = -i$ and $z = i$. Also, see Example 8.1.5 for a more sophisticated approach to solving this problem.

4.41. Consider an n-sided polygonal vortex patch with uniform vorticity K placed in a two-dimensional flow field. Let the vertices of the polygon be denoted by $P_j(\xi_j, \eta_j)$, $j = 1, 2, \ldots, n$, and $P(x, y)$ be an arbitrary field point outside the vortex patch. Show that the velocity components at the field point (x, y) induced by the n-sided polygonal vortex patch are given by

$$u(x, y) = -\frac{K}{2\pi} \iint_S \frac{y - \eta}{(x - \xi)^2 + (y - \eta)^2}\, d\xi d\eta,$$

$$v(x, y) = \frac{K}{2\pi} \iint_S \frac{x - \xi}{(x - \xi)^2 + (y - \eta)^2}\, d\xi d\eta,$$

where S is the region covered by the vortex patch. Find the explicit formulas for the velocity components by evaluating the above integrals.

4.42. Suppose a potential flow field is described by the complex potential $w = f(z)$ which is free from singularities inside the circle $|z| = a$. A stationary circular cylinder of radius a is now placed at the origin to modify the flow. Show that the complex potential of the modified flow configuration is given by

$$f(z) + \overline{f}\left(\frac{a^2}{\overline{z}}\right).$$

This is called the *circle theorem* in potential flow theory.

4.43. Consider the uniform potential flow of speed U in the positive x-direction which streams past a circle of radius a, whose center coincides with the origin (see Example 4.4.2). Show that the flow speed V at an arbitrary point (r, θ), $r \geq a$, in the flow field is given by

$$V = U\sqrt{1 + \frac{a^4}{r^4} + \frac{2a^2}{r^2}(\sin^2\theta - \cos^2\theta)}.$$

Find the position in the flow field that has the greatest flow speed. Also, find the equation representing the family of equipotential lines.

4.44. Show that

$$\nabla \cdot \mathbf{r}f(r) = 3f(r) + r\frac{df(r)}{dr}.$$

By setting $f(r) = r^{-3}$, verify that

$$\nabla \cdot \frac{\mathbf{r}}{r^3} = 0, \quad r \neq 0.$$

Can you give the physical interpretation of the above result with reference to the electric field generated by a point charge placed at the origin?

4.45. Suppose a line charge with linear charge density ρ is placed at the point $z_1 = x_1 + iy_1$, $z_1 \neq 0$, in the two-dimensional plane. An infinite plane is placed along the real axis which is earthed to the potential zero. Find the complex potential of the resulting electrostatic field. Describe the pattern of the equipotential lines of this electrostatic field.

4.46. Consider the region bounded by two infinitely long concentric cylindrical conductors of radii a and b (assuming $b > a$) which are charged to uniform electric potentials ϕ_a and ϕ_b, respectively. Find the electric potential of the electrostatic field inside the region.

Hint: The solution of the electric potential takes the form

$$\phi(r) = A\ln r + B,$$

where A and B are arbitrary constants.

4.47. For the horizontal polygonal cylinder considered in Example 4.4.4, show that the x- and h-derivatives of the vertical gravity g_h are given by

$$\frac{\partial g_h}{\partial x} = -G\rho \sum_{j=1}^{n} \left[\cos(2\mathrm{Arg}\, \Delta\omega_j)\ln \frac{r_{j+1}}{r_j} + \sin(2\mathrm{Arg}\, \Delta\omega_j)(\theta_{j+1} - \theta_j) \right],$$

$$\frac{\partial g_h}{\partial h} = G\rho \sum_{j=1}^{n} \left[\cos(2\mathrm{Arg}\, \Delta\omega_j)(\theta_{j+1} - \theta_j) - \sin(2\mathrm{Arg}\, \Delta\omega_j)\ln \frac{r_{j+1}}{r_j} \right].$$

4.48. Consider the horizontal infinite cylinder whose cross-section is a circle of radius r and centered at (ξ_0, ζ_0). Find the corresponding contour integral for computing the vertical gravity g_h due to the body. Evaluate the integral to obtain an explicit formula for g_h, assuming that the field point $(0, 0)$ is outside the circular cross-section.

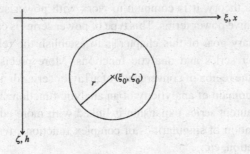

4.49. Suppose the density function of an infinite horizontal cylinder has linear variation of the form

$$\rho(\xi, \zeta) = a_0 + a_1\xi + a_2\zeta.$$

Show that the integral formula for the vertical gravity g_h [see eq. (4.4.51) with $s = 0$] becomes

$$g_h = G \left[a_0 \,\mathrm{Re} \oint_C \frac{\bar{\omega}}{\omega}\, d\omega - \frac{a_1}{4} \,\mathrm{Re} \oint_C \frac{\bar{\omega}^2}{2}\, d\omega \right.$$
$$\left. + a_2(\text{area of } S) - \frac{a_2}{4} \,\mathrm{Im} \oint_C \frac{\bar{\omega}^2}{\omega}\, d\omega \right].$$

5

Taylor and Laurent Series

A power series with non-negative power terms is called a *Taylor series*. In complex variable theory, it is common to work with power series with both positive and negative power terms. This type of power series is called a *Laurent series*. The primary goal of this chapter is to establish the relation between convergent power series and analytic functions. More precisely, we try to understand how the region of convergence of a Taylor series or a Laurent series is related to the domain of analyticity of an analytic function. The knowledge of Taylor and Laurent series expansion is linked with more advanced topics, like the classification of singularities of complex functions, residue calculus, analytic continuation, etc.

This chapter starts with the definitions of convergence of complex sequences and series. Many of the definitions and theorems for complex sequences and series are inferred from their counterparts in real variable calculus.

5.1 Complex sequences and series

An infinite sequence of complex numbers, denoted by $\{z_n\}$, can be considered as a function defined on a set of positive integers into the unextended complex plane. In other words, the sequence of complex numbers z_1, z_2, $z_3, \ldots$ is arranged sequentially and defined by some specific rule.

5.1.1 Convergence of complex sequences

Given a complex sequence $\{z_n\}$, if for each positive quantity ϵ, there exists a positive integer N such that

$$|z_n - z| < \epsilon \qquad \text{whenever } n > N, \qquad (5.1.1)$$

then the sequence is said to *converge* to the limit z. In general, the choice of N depends on ϵ. The definition implies that every ϵ-neighborhood of z contains all except a finite number of members of the sequence. Symbolically, we write

$$\lim_{n\to\infty} z_n = z. \qquad (5.1.2)$$

The limit of a convergent sequence is unique. If the sequence fails to converge, it is said to be *divergent*.

Suppose we separate z_n, $n = 1, 2, \ldots$, and z into their real and imaginary parts and write $z_n = x_n + i y_n$ and $z = x + iy$. Then

$$|x_n - x| \le |z_n - z| \le |x_n - x| + |y_n - y|,$$
$$|y_n - y| \le |z_n - z| \le |x_n - x| + |y_n - y|.$$

Using the above inequalities, it is easy to show that

$$\lim_{n\to\infty} z_n = z \iff \lim_{n\to\infty} x_n = x \quad \text{and} \quad \lim_{n\to\infty} y_n = y. \qquad (5.1.3)$$

Therefore, the study of the convergence of a complex sequence is equivalent to the consideration of the convergence of two real sequences. Therefore most theorems concerning the convergence of real series can be generalized and extended to complex series. For example, the Cauchy criterion for convergence of a real sequence can also be extended to a complex sequence.

Cauchy criterion

A complex sequence $\{z_n\}$ converges if and only if for each positive ϵ, there exists $N(\epsilon)$ such that

$$|z_m - z_n| < \epsilon \quad \text{for} \quad n, m > N. \qquad (5.1.4)$$

This can be proved easily by referring the convergence of $\{z_n\}$ to the convergence of the real sequences $\{x_n\}$ and $\{y_n\}$ and then applying the Cauchy criterion for the two real sequences.

As expected, theorems on the sum, difference, product and quotient of two complex sequences are the same as those for two real sequences. If the two complex sequences $\{a_n\}$ and $\{b_n\}$ converge to their respective limits A and B, then

$$\lim_{n\to\infty} (a_n \pm b_n) = A \pm B,$$

$$\lim_{n\to\infty} a_n b_n = AB,$$

$$\lim_{n\to\infty} \frac{a_n}{b_n} = \frac{A}{B}, \qquad \text{provided } B \ne 0.$$

Limit superior

The root test discussed in Section 5.1.3 requires the concept of the limit superior of a real sequence, denoted by the symbol $\overline{\lim}$. Consider a sequence $\{x_n\}$ of real numbers, and let S denote the set of all of its limit points. The *limit superior* of the sequence $\{x_n\}$ is defined to be the supremum (least upper bound) of S. For example, consider the real sequence $x_n = 3 + (-1)^n, n = 1, 2, \ldots$. The limit points of this real sequence are 2 and 4, so the limit superior is $\max\{2, 4\} = 4$.

5.1.2 Infinite series of complex numbers

An infinite series of complex numbers $z_1,\ z_2,\ z_3,\ \ldots$ is the infinite sum of the sequence $\{z_n\}$ given by

$$z_1 + z_2 + z_3 + \cdots = \lim_{n \to \infty} \left(\sum_{k=1}^{n} z_k \right). \tag{5.1.5}$$

To study the properties of an infinite series, we define the sequence of *partial sums* $\{S_n\}$ by

$$S_n = \sum_{k=1}^{n} z_k. \tag{5.1.6}$$

If the limit of the sequence $\{S_n\}$ converges to S, then the series is said to be convergent and S is its sum; otherwise, the series is divergent. The consideration of an infinite series is relegated to that of an infinite sequence of partial sums.

Given that an infinite series converges, we define the remainder after n terms by

$$R_n = S - S_n, \tag{5.1.7}$$

and obviously

$$\lim_{n \to \infty} R_n = 0. \tag{5.1.8}$$

Conversely, it can be shown that if eq. (5.1.8) holds, then the infinite series is convergent.

A necessary condition for the infinite series in eq. (5.1.5) to converge is that

$$\lim_{n \to \infty} z_n = 0. \tag{5.1.9}$$

However, the above condition is not sufficient for convergence. This can be revealed by the harmonic sequence defined by $z_n = \frac{1}{n}$. Though $\frac{1}{n} \to 0$ as $n \to \infty$, the harmonic series $\sum_{n=1}^{\infty} \frac{1}{n}$ is known to be divergent.

Absolute convergence

All infinite series $\sum z_n$ is said to be *absolutely convergent* if the associated series of absolute values $\sum |z_n|$ converges. It is relatively straightforward to show that an absolutely convergent series is always convergent (see Problem 5.1). If $\sum z_n$ converges but $\sum |z_n|$ does not, the series $\sum z_n$ is said to be *conditionally convergent*. For example, the complex series $\sum_{n=1}^{\infty} \frac{e^{in\theta}}{n}$, $\theta \neq 0$, can be shown to be conditionally convergent (see Problem 5.13).

5.1.3 Convergence tests of complex series

The following tests are most commonly used for examining the convergence of series.

Comparison test

If $|z_n| \leq \alpha_n$ and $\sum \alpha_n$ converges, then the series $\sum z_n$ is absolutely convergent.

Ratio test

Suppose $\lim\limits_{n \to \infty} \left| \dfrac{z_{n+1}}{z_n} \right|$ converges to L; then $\sum z_n$ is absolutely convergent if $L < 1$ and divergent if $L > 1$. When $L = 1$, the ratio test fails.

Root test

Suppose the limit superior of the real sequence $\{|z_n|^{\frac{1}{n}}\}$ equals L. The series $\sum z_n$ converges absolutely if $L < 1$ and diverges if $L > 1$. The root test fails when $L = 1$.

Gauss' test

When the ratio test fails, this method may be useful. If the ratio has the asymptotic expansion

$$\left| \frac{z_{n+1}}{z_n} \right| = 1 - \frac{k}{n} + \frac{\alpha_n}{n^2} + \cdots ,$$

where $|\alpha_n|$ is bounded for all $n > N$ for sufficiently large N, then the series $\sum z_n$ converges absolutely if $k > 1$, and diverges or converges conditionally if $k \leq 1$.

Remark

   (i) A bounded monotonically increasing or decreasing sequence is conver-
      gent.
  (ii) The removal or addition of a finite number of terms from or to an infinite
      series does not change the convergence or divergence of the series.

Example 5.1.1 Check the convergence of the following series:

(a) $\sum_{n=1}^{\infty} \frac{1}{n^2} + in^{\frac{1}{n}}$, (b) $\sum_{n=1}^{\infty} \frac{e^{in\theta}}{n^2}$.

Solution

(a) First, we observe that $\lim_{n\to\infty} n^{1/n} = 1$. The series $\sum_{n=1}^{\infty} n^{1/n}$ is divergent
    since $\lim_{n\to\infty} n^{1/n} \neq 0$. Therefore, $\sum_{n=1}^{\infty} \frac{1}{n^2} + in^{1/n}$ is also divergent.
(b) The associated series of absolute values of the terms in the given complex
    series is seen to be $\sum_{n=1}^{\infty} \frac{1}{n^2}$, which is known to be convergent. By the
    comparison test, the given series $\sum_{n=1}^{\infty} \frac{e^{in\theta}}{n^2}$ is absolutely convergent, and
    so the series is convergent.

Example 5.1.2 Consider the complex sequence $\{z_n\}$ defined by

$$z_n = \frac{q^n}{1 + q^{2n}}.$$

Give a judicious guess for the limit

$$\alpha = \lim_{n\to\infty} z_n$$

for various values of q, then prove the claims.

Solution We distinguish the following three cases:

   (i) $|q| > 1$

$$|z_n| = \left| \frac{q^n}{1 + q^{2n}} \right| \leq \frac{|q|^n}{|q|^{2n} - 1} \sim \frac{|q|^n}{|q|^{2n}} \to 0 \text{ as } n \to \infty;$$

  (ii) $|q| < 1$

$$|z_n| = \left| \frac{q^n}{1 + q^{2n}} \right| \leq \frac{|q|^n}{1 - |q|^{2n}} \sim |q|^n \to 0 \text{ as } n \to \infty;$$

(iii) $|q| = 1$, that is, $q = e^{i\theta}$ for some real value of θ

$$z_n = \frac{e^{in\theta}}{1 + e^{2in\theta}} = \frac{1}{2\cos n\theta},$$

and obviously $\lim\limits_{n \to \infty} z_n$ does not exist.

The proofs of the first two claims are presented below.

(i) $|q| > 1$

Given any $\epsilon > 0$, we want to find N such that for all $n > N$,

$$|z_n| = \left| \frac{q^n}{1 + q^{2n}} \right| \le \frac{|q|^n}{|q|^{2n} - 1} < \epsilon.$$

The above inequality is equivalent to either

$$|q|^n > \frac{1 + \sqrt{1 + 4\epsilon^2}}{2\epsilon} \quad \text{or} \quad |q|^n < \frac{1 - \sqrt{1 + 4\epsilon^2}}{2\epsilon}.$$

The second inequality is discarded since we have assumed $|q| > 1$. The first inequality can be expressed as

$$n > \frac{\ln\left(\frac{1 + \sqrt{1 + 4\epsilon^2}}{2\epsilon}\right)}{\ln |q|}.$$

Suppose we choose N to be the largest integer smaller than or equal to the right-hand term in the above inequality. Then for all $n > N$, the above inequality holds. Reversing the above procedure by rearranging the terms, we can deduce

$$|z_n| < \epsilon \text{ for any } \epsilon > 0 \text{ whenever } n > N.$$

This establishes the proof that

$$\lim_{n \to \infty} z_n = 0.$$

(ii) $|q| < 1$

Given any $\epsilon > 0$, we consider

$$\left| \frac{q^n}{1 + q^{2n}} \right| \le \frac{|q|^n}{1 - |q|^{2n}} = \frac{\frac{1}{|q|^n}}{\frac{1}{|q|^{2n}} - 1} < \epsilon,$$

which takes a similar form as in case (i) if we replace $|q|$ by $\frac{1}{|q|}$. Correspondingly, we choose N to be the largest integer smaller than or equal to

$$\frac{\ln\left(\frac{1 + \sqrt{1 + 4\epsilon^2}}{2\epsilon}\right)}{\ln\left(\frac{1}{|q|}\right)}.$$

For any $\epsilon > 0$ and all $n > N$, we then have

$$|z_n| = \left| \frac{q^n}{1 + q^{2n}} \right| \leq \frac{\frac{1}{|q|^n}}{\frac{1}{|q|^{2n}} - 1} < \epsilon,$$

thus establishing

$$\lim_{n \to \infty} z_n = 0.$$

In summary, we have

$$\lim_{n \to \infty} \frac{q^n}{1 + q^{2n}} = 0 \quad \text{when } |q| \neq 1;$$

and the limit of the sequence $\left\{ \frac{q^n}{1+q^{2n}} \right\}$ does not exist when $|q| = 1$.

5.2 Sequences and series of complex functions

In this section, we discuss the convergence of sequences and series of complex functions. Let $f_1(z),\ f_2(z), \ldots, f_n(z), \ldots$, denoted by $\{f_n(z)\}$, be a sequence of complex functions of z that are defined and single-valued in a point set $\mathcal{R}$ in the complex plane. For some point $z_0 \in \mathcal{R}$, $\{f_n(z_0)\}$ becomes a sequence of complex numbers. Supposing the sequence $\{f_n(z_0)\}$ converges, the limit is unique. The value of the limit depends on z_0, and we write

$$f(z_0) = \lim_{n \to \infty} f_n(z_0). \tag{5.2.1}$$

If this holds for every $z \in \mathcal{R}$, the sequence of complex functions $\{f_n(z)\}$ defines a complex function $f(z)$ in the point set $\mathcal{R}$. Symbolically, we write

$$f(z) = \lim_{n \to \infty} f_n(z). \tag{5.2.2}$$

Definition 5.2.1   A sequence of complex functions $\{f_n(z)\}$ defined in a set $\mathcal{R}$ is said to *converge* to a complex function $f(z)$ defined in the same set if and only if, for any given small positive quantity ϵ, we can find a positive integer $N(\epsilon; z)$ [in general, $N(\epsilon; z)$ depends on ϵ and z] such that $|f(z) - f_n(z)| < \epsilon$ for all $n > N(\epsilon; z)$. The set $\mathcal{R}$ is called the *region of convergence*[†] of the sequence of complex functions.

In general, we may not be able to find a single $N(\epsilon)$ that works for all points in $\mathcal{R}$. In this case, the convergence is said to be *pointwise*. Otherwise, $\{f_n(z)\}$ is said to *converge uniformly* to $f(z)$ in $\mathcal{R}$ when N depends on ϵ only.

[†] The conventional use of "region" in characterizing "region of convergence" is often but not always in the same sense of "region" that has been defined in Section 1.4.

5.2.1 Convergence of series of complex functions

An infinite series of complex functions

$$f_1(z) + f_2(z) + f_3(z) + \cdots = \sum_{k=1}^{\infty} f_k(z) \tag{5.2.3}$$

is related to the sequence of partial sums $\{S_n(z)\}$, the relation being defined by

$$S_n(z) = \sum_{k=1}^{n} f_k(z). \tag{5.2.4}$$

Similarly, the discussion of the convergence of the infinite series in eq. (5.2.3) can be relegated to that of the sequence of partial sums defined in eq. (5.2.4). The infinite series in eq. (5.2.3) is said to be *convergent* if

$$\lim_{n \to \infty} S_n(z) = S(z), \tag{5.2.5}$$

where $S(z)$ is called the *sum*; otherwise the series is *divergent*.

Many of the properties related to convergence in sequences and series of complex functions can be extended from their counterparts in sequences and series of complex numbers. For example, a necessary but not sufficient condition for the infinite series in eq. (5.2.3) to converge is that

$$\lim_{k \to \infty} f_k(z) = 0. \tag{5.2.6}$$

Example 5.2.1 Consider the following infinite series of complex functions

$$\sum_{k=1}^{\infty} \frac{\sin kz}{k^2}.$$

Show that it is absolutely convergent when z is real, but that when z is non-real, it becomes divergent.

Solution

(i) When z is real, we have

$$\left| \frac{\sin kz}{k^2} \right| \le \frac{1}{k^2}, \qquad \text{for all positive integer values of } k.$$

Since $\sum_{k=1}^{\infty} \frac{1}{k^2}$ is known to be convergent, $\sum_{k=1}^{\infty} \frac{\sin kz}{k^2}$ is absolutely convergent.

(ii) When z is non-real, we let $z = x + iy$, $y \ne 0$. From the relation

$$\frac{\sin kz}{k^2} = \frac{e^{-ky}e^{ikx} - e^{ky}e^{-ikx}}{2k^2i},$$

we deduce that

$$\left|\frac{\sin kz}{k^2}\right| \geq \frac{e^{k|y|} - e^{-k|y|}}{2k^2} \to \infty \text{ as } k \to \infty.$$

Since $\left|\frac{\sin kz}{k^2}\right|$ is unbounded as $k \to \infty$, $\sum_{k=1}^{\infty} \frac{\sin kz}{k^2}$ is divergent.

Uniform convergence

Suppose we let

$$R_n(z) = S(z) - \sum_{k=1}^{n} f_k(z) = S(z) - S_n(z) \qquad (5.2.7)$$

be the remainder after n terms of the infinite complex series in eq. (5.2.3). The infinite series is said to converge *uniformly* to $S(z)$ in some set $\mathcal{R}$ if and only if for any positive quantity ϵ, we can find a sufficiently large positive integer N, *independent of z*, such that for all $z \in \mathcal{R}$,

$$|R_n(z)| < \epsilon \qquad \text{whenever } n > N. \qquad (5.2.8)$$

Uniform convergence is a strong property of an infinite series of complex functions. Properties such as continuity or analyticity of the constituent functions $f_k(z)$ are carried over to the sum $S(z)$. To be precise, suppose $\sum f_k(z)$ converges uniformly to $S(z)$ in some set $\mathcal{R}$ and let $f_k(z)$ be continuous in $\mathcal{R}$. Then the sum $S(z)$ is also continuous in the set $\mathcal{R}$.

Moreover, a uniformly convergent series of continuous functions on a contour C can be integrated term by term; that is,

$$\int_C S(z)\,dz = \int_C \sum_{k=1}^{\infty} f_k(z)\,dz = \sum_{k=1}^{\infty} \int_C f_k(z)\,dz, \qquad (5.2.9)$$

where C is any contour on which the constituent functions $f_k(z)$ are continuous.

A similar statement can be made for term by term differentiation. Given that each constituent function $f_k(z)$ is analytic in a simply connected domain $\mathcal{D}$ and the infinite series $\sum_{k=1}^{\infty} f_k(z)$ converges uniformly to $S(z)$ in any compact subset of $\mathcal{D}$, then $S(z)$ is also analytic in $\mathcal{D}$ and

$$S'(z) = \frac{d}{dz} \sum_{k=1}^{\infty} f_k(z) = \sum_{k=1}^{\infty} f'_k(z). \qquad (5.2.10)$$

Weierstrass M test

The *Weierstrass M test* provides a simple tool for testing the uniform convergence of an infinite series of complex functions. It states that if $|f_k(z)| \leq M_k$, where M_k is independent of z in the set $\mathcal{R}$ and the infinite series $\sum_{k=1}^{\infty} M_k$ converges, then the infinite series $\sum_{k=1}^{\infty} f_k(z)$ is absolutely convergent and uniformly convergent in $\mathcal{R}$.

We use the comparison test and the property on uniform convergence shown in eq. (5.2.8) to establish the M test. Given that $|f_k(z)| \leq M_k$ in $\mathcal{R}$ and $\sum_{k=1}^{\infty} M_k$ converges, the absolute convergence of $\sum_{k=1}^{\infty} f_k(z)$ in $\mathcal{R}$ follows from the comparison test. To prove uniform convergence, we consider the bound on the remainder after n terms

$$|R_n(z)| \leq \left| \sum_{k=n+1}^{\infty} f_k(z) \right| \leq \sum_{k=n+1}^{\infty} |f_k(z)| \leq \sum_{k=n+1}^{\infty} M_k.$$

Given that $\sum_{k=1}^{\infty} M_k$ converges, the corresponding remainder $\sum_{k=n+1}^{\infty} M_k$ can be made less than any ϵ by choosing $n > N$ for some N. Clearly, N is independent of z. We have

$$|R_n(z)| < \epsilon \quad \text{for} \quad n > N,$$

so uniform convergence of $\sum_{k=1}^{\infty} f_k(z)$ in $\mathcal{R}$ follows.

Example 5.2.2 Consider the following infinite series of complex functions

$$z(1-z) + z^2(1-z) + \cdots + z^k(1-z) + \cdots = \sum_{k=1}^{\infty} z^k(1-z).$$

Show that the series converges for $|z| < 1$ and find its sum. Examine the uniform convergence of the series for (i) $|z| \leq r_0 < 1$, a closed disk that lies completely inside the unit circle; (ii) $|z| \leq 1$, the closed unit disk.

Solution Define the partial sum

$$\begin{aligned} S_n(z) &= \sum_{k=1}^{\infty} z^k(1-z) \\ &= (z - z^2) + (z^2 - z^3) + \cdots + (z^n - z^{n+1}) \\ &= z - z^{n+1}. \end{aligned}$$

Supposing $|z| < 1$, a judicious guess of the limit of the partial sum as $n \to \infty$ is z. For $|z| < 1$, given any $\epsilon > 0$, we consider

$$|S_n(z) - z| = |z - z^{n+1} - z| = |z|^{n+1} < \epsilon.$$

The above inequality is valid provided that we choose n such that

$$(n+1)\ln|z| < \ln\epsilon \quad \text{for} \quad z \neq 0;$$

or

$$n > \frac{\ln \epsilon}{\ln |z|} - 1 \quad \text{for} \quad z \neq 0.$$

We set

$$N(\epsilon; z) = fl\left(\frac{\ln \epsilon}{\ln |z|} - 1\right),$$

where $fl(x)$ denotes the largest integer less than or equal to x. We then have

$$|S_n(z) - z| < \epsilon \quad \text{whenever } n > N(\epsilon; z), \quad z \neq 0.$$

Here, $N(\epsilon; z)$ has dependence on both ϵ and z. Lastly, when $z = 0$, we observe

$$|S_n(0) - 0| = 0 < \epsilon \quad \text{for all } n.$$

Hence, the series $\sum_{k=1}^{\infty} z^k(1 - z)$ converges to its sum z for $|z| < 1$.
Next, we examine uniform convergence of the series.

(i) Given that $|z| \leq r_0 < 1$, we may choose N to be $fl\left(\frac{\ln \epsilon}{\ln |r_0|} - 1\right)$, which has dependence on ϵ only. Hence, the series $\sum_{k=1}^{\infty} z^k(1 - z)$ converges uniformly for $|z| \leq r_0 < 1$.

(ii) However, the above choice of N cannot be applied to the case when $r_0 = 1$ since $\ln |r_0|$ becomes infinite. Therefore, we cannot establish uniform convergence of the series when $|z| \leq 1$.

Example 5.2.3 Using the Weierstrass M test, establish the uniform convergence of the following series in their respective regions of convergence.

(a) $\displaystyle\sum_{n=1}^{\infty} \frac{z^n}{n\sqrt{n + 1}}, \quad |z| \leq 1;$

(b) $\displaystyle\sum_{n=1}^{\infty} \frac{1}{n^2 + z^2}, \quad 1 < |z| < 2.$

Solution

(a) Given that $|z| \leq 1$, we observe

$$\left|\frac{z^n}{n\sqrt{n + 1}}\right| \leq \frac{1}{n^{3/2}}.$$

Accordingly, we choose $M_n = \frac{1}{n^{3/2}}$. Since $\sum_{n=1}^{\infty} \frac{1}{n^{3/2}}$ is known to be convergent,

$$\sum_{n=1}^{\infty} \frac{z^n}{n\sqrt{n+1}} \text{ converges uniformly for } |z| \leq 1$$

(b) For $n \geq 3$ and $1 < |z| < 2$, we observe

$$|n^2 + z^2| \geq |n^2| - |z|^2 \geq n^2 - 4 \geq \frac{n^2}{2}$$

so that

$$\left| \frac{1}{n^2 + z^2} \right| \leq \frac{2}{n^2}.$$

We take $M_n = \frac{2}{n^2}$ and note that $\sum_{n=3}^{\infty} \frac{2}{n^2}$ converges. Hence, $\sum_{n=3}^{\infty} \frac{1}{n^2 + z^2}$ converges uniformly for $1 < |z| < 2$. Adding two extra terms does not affect uniform convergence, so $\sum_{n=1}^{\infty} \frac{1}{n^2 + z^2}$ shares the same property of uniform convergence.

Example 5.2.4 Consider the Riemann zeta function defined by

$$\zeta(z) = \sum_{n=1}^{\infty} n^{-z}.$$

Show that the function is analytic inside the region $\{z : \operatorname{Re} z > 1\}$. Also, find its derivative.

Solution The constituent functions $n^{-z} = e^{-(\ln n)z}$, $n = 1, 2, \ldots$, are seen to be analytic in $\operatorname{Re} z > 1$. We would like to establish the uniform convergence of $\sum_{n=1}^{\infty} n^{-z}$ for $\operatorname{Re} z \geq \gamma > 1$. Noting that for $\operatorname{Re} z \geq \gamma > 1$,

$$|n^{-z}| = |n^{-\operatorname{Re} z}| \leq n^{-\gamma},$$

we choose $M_n = n^{-\gamma}$, $\gamma > 1$, in the M test. It is well known that $\sum_{n=1}^{\infty} \frac{1}{n^{\gamma}}$ converges for $\gamma > 1$. By the Weierstrass M test, $\sum_{n=1}^{\infty} n^{-z}$ converges uniformly for $\operatorname{Re} z \geq \gamma > 1$. The infinite series $\sum_{n=1}^{\infty} n^{-z}$ is seen to be uniformly convergent in any compact subset of $\operatorname{Re} z > 1$. By the termwise differentiation property of uniformly convergent series and eq. (5.2.10), we deduce that $\zeta(z)$ is analytic in $\operatorname{Re} z > 1$ and its derivative is given by

$$\zeta'(z) = \sum_{n=1}^{\infty} \frac{d}{dz} n^{-z} = \sum_{n=1}^{\infty} (\ln n) n^{-z},$$

valid for $\operatorname{Re} z > 1$.

5.2.2 Power series

The choice of the constituent functions

$$f_n(z) = a_n(z - z_0)^n, \quad n = 0, 1, 2, \ldots,$$

in an infinite complex series leads to a *power series* expanded at the point $z = z_0$. A power series defines a function $f(z)$ for those points z at which it converges. The properties of absolute convergence of infinite power series are stated in Theorem 5.2.1.

Theorem 5.2.1 (Absolute convergence of power series) *If the infinite power series $\sum_{n=0}^{\infty} a_n(z - z_0)^n$ converges at $z = z_1$, where z_1 is some point other than z_0, then it is absolutely convergent at each point z inside the open disk $|z - z_0| < R_1$, where $R_1 = |z_1 - z_0|$.*

Proof Since $\sum_{n=0}^{\infty} a_n(z_1 - z_0)^n$ converges for $z_1 \neq z_0$, $a_n(z_1 - z_0)^n$ tends to 0 as $n \to \infty$. Therefore, there exists N such that when $n > N$ we have $|a_n(z_1 - z_0)^n| < 1$. Suppose z lies in the open disk $|z - z_0| < |z_1 - z_0|$; we derive the following bound on the remainder after N terms of the associated series of absolute terms

$$\sum_{n=N+1}^{\infty} |a_n(z - z_0)^n| = \sum_{n=N+1}^{\infty} |a_n| \, |z - z_0|^n \leq \sum_{n=N+1}^{\infty} \frac{|z - z_0|^n}{|z_1 - z_0|^n}.$$

The ratio of the successive terms in the last infinite series is less than 1 since z satisfies $|z - z_0| < |z_1 - z_0|$. By the ratio test, the infinite series

$$\sum_{n=N+1}^{\infty} \frac{|z - z_0|^n}{|z_1 - z_0|},$$

converges for $|z - z_0| < |z_1 - z_0|$. By the comparison test, the infinite power series $\sum_{n=0}^{\infty} a_n(z - z_0)^n$ converges absolutely inside the circle $|z - z_0| < |z_1 - z_0|$.

The above theorem states that the infinite power series converges absolutely at all points inside some circle centered at z_0 provided that it converges at some point other than z_0. One then deduces that there is a largest circle centered at z_0 such that the series converges absolutely for all points inside that circle. This circle is termed the *circle of convergence* of the series. The series fails to converge at any point outside the circle of convergence. Otherwise, the series converges at all points inside a new circle that is centered at z_0 and passes through an outside point. This violates the assumption that the earlier circle is the defined circle of convergence since there exists a larger circle with absolute convergence of the series at all points inside it.

In summary, given an infinite power series $\sum_{n=0}^{\infty} a_n(z - z_0)^n$, there exists a non-negative real number R, where R can be zero or infinity, such that the series converges absolutely for $|z - z_0| < R$ and diverges for $|z - z_0| > R$. Here, R is called the *radius of convergence* and the circle $|z - z_0| = R$ is called the *circle of convergence*. The power series may or may not converge at a point on the circle of convergence. Convergence of the power series must be determined for each point on the circle of convergence (see Example 5.2.7).

The radius of convergence of a power series can be found by the following formulas:

(i) $R = \lim\limits_{n\to\infty} \left| \dfrac{a_n}{a_{n+1}} \right|$ if the limit exists.

(ii) $R = \dfrac{1}{\lim_{n\to\infty} \sqrt[n]{|a_n|}}$ if the limit exists; in general, we always have $R = \dfrac{1}{\overline{\lim}_{n\to\infty} \sqrt[n]{|a_n|}}$.

These formulas are derived directly from the ratio test and root test, respectively. Applications of the above formulas in finding the radius of convergence of a given infinite power series are illustrated in Examples 5.2.5 and 5.2.6.

Example 5.2.5 Suppose the radii of convergence of $\sum_{k=0}^{\infty} a_k z^k$ and $\sum_{k=0}^{\infty} b_k z^k$ are R_a and R_b, respectively. Find the corresponding radii of convergence of the following series:

(a) $\sum_{k=0}^{\infty} (a_k + b_k) z^k$, (b) $\sum_{k=0}^{\infty} a_k b_k z^k$.

Solution

(a) Let $R = \min(R_a, R_b)$. Inside the domain $|z| < R$, the two limits

$$\lim_{n\to\infty} \sum_{k=0}^{n} a_k z^k \quad \text{and} \quad \lim_{n\to\infty} \sum_{k=0}^{n} b_k z^k$$

exist; so the summed infinite series

$$\lim_{n\to\infty} \sum_{k=0}^{n} (a_k + b_k) z^k$$

also exists and the radius of convergence of the summed series is at least R. The following example shows that the radius of convergence of a summed series can be greater than R. Consider

$$a_k = \frac{1}{2^k} \quad \text{and} \quad b_k = \frac{1}{3^k} - \frac{1}{2^k}.$$

The series $\sum_{k=0}^{\infty} a_k z^k$ and $\sum_{k=0}^{\infty} b_k z^k$ have the same radius of convergence, namely 2. However, their sum

$$\sum_{k=0}^{\infty}(a_k + b_k)z^k = \sum_{k=0}^{\infty}\left[\frac{1}{2^k} + \left(\frac{1}{3^k} - \frac{1}{2^k}\right)\right]z^k = \sum_{k=0}^{\infty}\frac{z^k}{3^k}$$

has the radius of convergence 3. This is greater than the minimum of the two radii of convergence of the constituent series.

(b) Let R be the radius of convergence of $\sum_{k=0}^{\infty} a_k b_k z^k$. Then

$$\frac{1}{R} = \overline{\lim_{k \to \infty}} \sqrt[k]{|a_k b_k|} \le \left(\overline{\lim_{k \to \infty}} \sqrt[k]{|a_k|}\right)\left(\overline{\lim_{k \to \infty}} \sqrt[k]{|b_k|}\right) = \frac{1}{R_a} \cdot \frac{1}{R_b},$$

so $R \ge R_a R_b$.

Example 5.2.6 Find the circle of convergence of each of the following power series:

(a) $\displaystyle\sum_{k=1}^{\infty} \frac{1}{k}(z - i)^k$, (b) $\displaystyle\sum_{k=1}^{\infty} k^{\ln k}(z - 2)^k$, (c) $\displaystyle\sum_{k=1}^{\infty} \left(\frac{z}{k}\right)^k$,

(d) $\displaystyle\sum_{k=1}^{\infty} \left(1 + \frac{1}{k}\right)^{k^2} z^k$, (e) $\displaystyle\sum_{k=1}^{\infty}(-1)^k z^{2^k}$.

Solution

(a) By the ratio test, we have

$$R = \lim_{k \to \infty} \frac{\frac{1}{k}}{\frac{1}{k+1}} = 1;$$

so the circle of convergence is $|z - i| = 1$.

(b) Using the root test, we have

$$\frac{1}{R} = \lim_{k \to \infty} \sqrt[k]{|a_k|} = \lim_{k \to \infty} \sqrt[k]{k^{\ln k}}.$$

To evaluate the limit, we consider the logarithm

$$\ln \lim_{k \to \infty} \sqrt[k]{k^{\ln k}} = \lim_{k \to \infty} \ln \sqrt[k]{k^{\ln k}} = \lim_{k \to \infty} \frac{(\ln k)^2}{k} = 0;$$

so

$$\frac{1}{R} = \lim_{k \to \infty} \sqrt[k]{k^{\ln k}} = e^0 = 1.$$

The circle of convergence is $|z| = 1$.

(c) By the ratio test, we have

$$\frac{1}{R} = \lim_{k \to \infty} \frac{\left(\frac{1}{k+1}\right)^{k+1}}{\left(\frac{1}{k}\right)^k}.$$

$$= \lim_{k \to \infty} \frac{1}{k+1} \lim_{k \to \infty} \frac{1}{\left(1+\frac{1}{k}\right)^k}$$

$$= 0 \cdot \frac{1}{e} = 0;$$

so the circle of convergence is the whole complex plane.

(d) By the root test, we have

$$\frac{1}{R} = \lim_{k \to \infty} \sqrt[k]{\left(1+\frac{1}{k}\right)^{k^2}} = \lim_{k \to \infty} \left(1+\frac{1}{k}\right)^k = e,$$

so the circle of convergence is $|z| = \frac{1}{e}$.

(e) The coefficients are of the form

$$a_m = \begin{cases} 0 & \text{if } m \neq 2^k \\ (-1)^k & \text{if } m = 2^k \end{cases};$$

so

$$\frac{1}{R} = \overline{\lim_{m \to \infty}} \sqrt[m]{|a_m|} = 1.$$

The circle of convergence is then found to be $|z| = 1$.

Example 5.2.7   Consider the infinite power series

$$\sum_{n=0}^{\infty} \frac{z^n}{n^p} = 1 + \frac{z}{1^p} + \frac{z^2}{2^p} + \cdots, \quad p \geq 0;$$

find its circle of convergence. Show that

(a) when $p > 1$, the series converges for all points on the circle of convergence;

(b) when $p = 0$, the series diverges for all points on the circle of convergence;

(c) when $p = 1$, the series converges for all points on the circle of convergence except one point.

Solution   We apply the root test and consider

$$|a_n|^{1/n} = \left(\frac{1}{n^p}\right)^{1/n} = \left(\frac{1}{e^{p \ln n}}\right)^{1/n} = e^{-p \frac{\ln n}{n}},$$

and as

$$\lim_{n \to \infty} \frac{\ln n}{n} = 0$$

we have

$$\lim_{n \to \infty} |a_n|^{1/n} = 1.$$

The circle of convergence is $|z| = 1$, for all values of p.

(a) When $p > 1$, at any point on the circle of convergence, we have

$$\left|\frac{z^n}{n^p}\right| = \frac{1}{n^p}.$$

Now, the infinite power series is dominated by the convergent series $1 + \sum_{n=1}^{\infty} \frac{1}{n^p}$. Hence, the infinite series $\sum_{n=0}^{\infty} \frac{z^n}{n^p}$ converges absolutely at all points on the circle of convergence.

(b) When $p = 0$, the infinite power series becomes the geometric series $\sum_{n=0}^{\infty} z^n$. At any point on the circle of convergence, $|z|^n = 1 \neq 0$, so the series diverges.

(c) When $p = 1$, the infinite power series becomes

$$1 + \sum_{n=1}^{\infty} \frac{z^n}{n}.$$

At any point z on the circle of convergence, we may write $z = e^{i\theta}$. It can be shown that (see Problem 5.13)

$$\sum_{n=1}^{\infty} \frac{\cos n\theta}{n} = -\ln\left(2 \sin \frac{\theta}{2}\right), \quad \theta \neq 0;$$

$$\sum_{n=1}^{\infty} \frac{\sin n\theta}{n} = \frac{\pi - \theta}{2}, \quad 0 < \theta < 2\pi.$$

At $z = 1$, the infinite series becomes the harmonic series $1 + \sum_{n=1}^{\infty} \frac{1}{n}$, which is known to be divergent. Therefore, the series converges at all points on the circle of convergence except at the point $z = 1$.

Uniform convergence of infinite power series

We have observed the absolute convergence property of an infinite power series $\sum_{n=0}^{\infty} a_n(z - z_0)^n$ at all points inside its circle of convergence $|z - z_0| < R$. Does an infinite power series also observe uniform convergence at all points inside the circle of convergence? It turns out that at best we can only establish uniform convergence in any closed disk $|z - z_0| \leq r_0 < R$ that lies inside the circle of convergence (see Theorem 5.2.2).

To motivate the concept of uniform convergence of infinite power series, let us consider uniform convergence of the infinite geometric series $\sum_{n=0}^{\infty} z^n$ for any closed disk $|z| \leq r_0 < 1$ inside its circle of convergence $|z| < 1$. To apply the M test, it suffices to find an appropriate choice of M_n such that $|z^n|$ is bounded by M_n. Since z lies inside the closed disk $|z| \leq r_0 < 1$, we have

$$|z|^n \leq r_0^n < 1;$$

so we choose $M_n = r_0^n$. The associated series

$$\sum_{n=0}^{\infty} M_n = \sum_{n=0}^{\infty} r_0^n$$

is known to be convergent for $r_0 < 1$. Hence, uniform convergence of the infinite geometric series is established for $|z| \leq r_0 < 1$.

Interestingly, though we can establish absolute convergence in the whole open disk $|z| < 1$, uniform convergence fails in the same open disk. Let us consider the remainder after n terms, which is found to be

$$R_n(z) = \sum_{k=n+1}^{\infty} z^k = \frac{z^{n+1}}{1 - z}.$$

To establish absolute convergence of the infinite geometric series in the open disk $|z| < 1$, we follow a similar procedure to Example 5.2.2. In order that

$$|R_n(z)| = \left| \frac{z^{n+1}}{1 - z} \right| < \epsilon, \quad z \neq 0,$$

we have to choose $n > N$, where

$$N(\epsilon; z) = fl \left(\frac{\ln \epsilon}{\ln |z|} + \frac{\ln |1 - z|}{\ln |z|} - 1 \right).$$

Note that $N(\epsilon; z)$ has dependence on z and it becomes infinite when $|z| \to 1$, so uniform convergence fails.

Alternatively, we may demonstrate the failure of uniform convergence in the open disk $|z| < 1$ from another perspective. Take any point $z = re^{i\theta}$ inside the open disk $|z| < 1$, where $r < 1$. By virtue of the relation

$$|1 - z| \leq 1 + |z| = 1 + r$$

we have

$$\frac{1}{|1 - z|} \geq \frac{1}{1 + r};$$

we then obtain

$$|R_n(z)| = \frac{|z|^{n+1}}{|1 - z|} \geq \frac{r^{n+1}}{1 + r}, \quad r < 1.$$

As r tends to 1 from below, $\frac{r^{n+1}}{1+r}$ tends to $\frac{1}{2}$. Uniform convergence requires that the partial sum $S_n(z) = \sum_{k=0}^{\infty} z^k$ converges to the sum $S(z) = \frac{1}{1-z}$ for all z in the open disk $|z| < 1$ simultaneously. This is equivalent to requiring

$$\max_{|z|<1} |S(z) - S_n(z)| = \max_{|z|<1} |R_n(z)| \to 0$$

for all z in $|z| < 1$. Since this condition fails, uniform convergence of the infinite geometric series is not observed in the open disk $|z| < 1$.

Theorem 5.2.2 *Consider the infinite power series $\sum_{n=0}^{\infty} a_n(z - z_0)^n$, whose circle of convergence is $|z - z_0| < R$, $0 < R < \infty$. The series is uniformly convergent to its pointwise sum function in the closed disk $|z - z_0| \leq r_0$, where $0 < r_0 < R$.*

Proof For $|z - z_0| \leq r_0$, we have

$$|a_n(z - z_0)^n| \leq |a_n| r_0^n, \quad \text{for all } n.$$

By Theorem 5.2.1, the infinite power series converges absolutely so $\sum_{n=0}^{\infty} |a_n| r_0^n$ converges. Applying the M test, we take $M_n = |a_n| r_0^n$. Since $\sum_{n=0}^{\infty} M_n$ converges, the power series converges uniformly in any closed disk $|z - z_0| \leq r_0$, where $0 < r_0 < R$.

As a consequence of the uniform convergence property of an infinite power series in any closed disk centered at z_0 that lies inside the circle of convergence, we can deduce the termwise integration property of an infinite power series inside its circle of convergence. For any contour C lying inside the circle of convergence, $\sum_{n=0}^{\infty} a_n(z - z_0)^n$ is a uniformly convergent series of continuous functions on C, so

$$\int_C \sum_{n=0}^{\infty} a_n(z - z_0)^n \, dz = \sum_{n=0}^{\infty} a_n \int_C (z - z_0)^n \, dz. \tag{5.2.11}$$

For example, consider the infinite geometric series

$$\frac{1}{1-z} = \sum_{n=0}^{\infty} z^n, \quad |z| < 1,$$

we consider the termwise integration along a contour C starting from 0 and ending at z inside the circle of convergence. Since $\frac{1}{1-z}$ and $z^n, n = 0, 1, \ldots,$ are analytic functions inside $|z| < 1$, the integrals of these functions along C are path independent. We then obtain

$$\int_C \frac{1}{1-\zeta} \, d\zeta = \sum_{n=0}^{\infty} \int_C \zeta^n \, d\zeta$$

$$\Leftrightarrow \int_0^z \frac{1}{1-\zeta} \, d\zeta = \sum_{n=0}^{\infty} \int_0^z \zeta^n \, d\zeta$$

$$\Leftrightarrow -\text{Log}(1-z) = \sum_{n=0}^{\infty} \frac{z^{n+1}}{n+1}, \quad |z| < 1, \tag{5.2.12}$$

where $\text{Log}\, 1 = 0$.

A similar statement can be made for termwise differentiation of an infinite power series inside its circle of convergence. Suppose that the infinite power series $\sum_{n=0}^{\infty} a_n(z-z_0)^n$ converges absolutely to $S(z)$ inside its circle of convergence, where $|z - z_0| < R$; then $\sum_{n=1}^{\infty} na_n(z-z_0)^{n-1}$ converges absolutely to $S'(z)$ inside the same circle of convergence. The validity of termwise differentiation stems from (i) analyticity of the power functions $a_n(z-z_0)^n, n = 0, 1, \ldots,$ inside the circle of convergence, and (ii) uniform convergence of $\sum_{n=0}^{\infty} a_n(z-z_0)^n$ in any closed disk centered at z_0 inside the circle of convergence. Let $S(z) = \sum_{n=0}^{\infty} a_n(z-z_0)^n$; then $S(z)$ is analytic inside $|z - z_0| < R$ and its derivative is given by [see eq. (5.2.10)]

$$S'(z) = \frac{d}{dz}\left(\sum_{n=0}^{\infty} a_n(z-z_0)^n\right) = \sum_{n=1}^{\infty} na_n(z-z_0)^{n-1}, \quad |z - z_0| < R.$$

$$\tag{5.2.13}$$

Hence, an infinite power series defines an analytic function inside its circle of convergence.

For example, consider the *Bessel function* of order n as defined by

$$J_n(z) = \sum_{k=0}^{\infty} \frac{(-1)^k}{k!(k+n)!} \left(\frac{z}{2}\right)^{2k+n}. \tag{5.2.14}$$

The radius of convergence R of the above infinite power series is found to be

$$\frac{1}{R} = \lim_{k \to \infty} \left| \frac{a_{k+1}}{a_k} \right|$$

$$= \lim_{k \to \infty} \left| \frac{2^{2k+n} k! (k+n)!}{2^{2(k+1)+n} (k+1)! (k+1+n)!} \right|$$

$$= \lim_{k \to \infty} \left| \frac{1}{2^2} \frac{1}{(k+1)(k+1+n)} \right| = 0.$$

Hence, the infinite power series converges absolutely for all z in the whole (unextended) complex plane. The power series defines an analytic function $J_n(z)$ for all z, so $J_n(z)$ is entire. The derivative of $J_n(z)$ is given by termwise differentiation of the infinite power series, where

$$J_n'(z) = \sum_{k=0}^{\infty} \frac{(-1)^k}{k!(k+n)!} \frac{d}{dz} \left(\frac{z}{2} \right)^{2k+n}$$

$$= \sum_{k=0}^{\infty} \frac{1}{2} \frac{(-1)^k (2k+n)}{k!(k+n)!} \left(\frac{z}{2} \right)^{2k+n-1}. \tag{5.2.15}$$

Power series in its Taylor series representation

Suppose a power series represents a function $f(z)$ inside the circle of convergence; that is,

$$f(z) = \sum_{n=0}^{\infty} a_n (z - z_0)^n. \tag{5.2.16}$$

Since it can be differentiated termwise, we have

$$f'(z) = \sum_{n=1}^{\infty} n a_n (z - z_0)^{n-1},$$

$$f''(z) = \sum_{n=2}^{\infty} n(n-1) a_n (z - z_0)^{n-2},$$

etc. By putting $z = z_0$ successively in the original series and in the above derivative series, we obtain the relations

$$a_n = \frac{f^{(n)}(z_0)}{n!}, \qquad n = 0, 1, 2, \ldots.$$

The corresponding series representation

$$f(z) = \sum_{n=0}^{\infty} \frac{f^{(n)}(z_0)}{n!} (z - z_0)^n \tag{5.2.17}$$

is called the *Taylor series* of $f(z)$. The special case $z_0 = 0$ is called the *Maclaurin series* of $f(z)$.

Suppose a function can be represented by a Taylor series; then implicitly it is differentiable of infinite order. This is not surprising since every Taylor series with a nonzero radius of convergence defines an analytic function and an analytic function is differentiable at all orders. If a Taylor series converges at every point inside the circle of convergence $\mathcal{R}$, then it converges to a function that is analytic at least in $\mathcal{R}$.

Cauchy product

If the two power series

$$f(z) = \sum_{n=0}^{\infty} a_n (z - z_0)^n$$

and

$$g(z) = \sum_{n=0}^{\infty} b_n (z - z_0)^n$$

are convergent for $|z - z_0| < R$, then the sum and difference,

$$f(z) \pm g(z) = \sum_{n=0}^{\infty} (a_n \pm b_n)(z - z_0)^n, \qquad (5.2.18)$$

and the product,

$$f(z)g(z) = \sum_{n=0}^{\infty} \alpha_n (z - z_0)^n, \qquad \text{where } \alpha_n = \sum_{k=0}^{n} a_k b_{n-k}, \quad (5.2.19)$$

are convergent inside the same circle of convergence. The new series $\sum_{n=0}^{\infty} \alpha_n (z - z_0)^n$ is called the *Cauchy product* of $f(z)$ and $g(z)$.

5.3 Taylor series

We have seen that an infinite power series defines an analytic function inside its circle of convergence. In this section, we show how an analytic function can

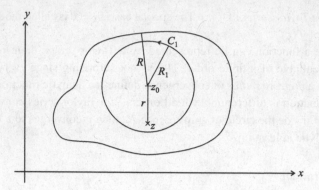

Figure 5.1. The circle C_1: $|z - z_0| = R_1$ lies completely inside the domain $\mathcal{D}$ and contains the point z. Also, R is the minimum distance from z_0 to the boundary of the domain.

be expanded in an infinite power series. Indeed, the class of analytic functions is closely related to the class of convergent power series.

Theorem 5.3.1 (Taylor series theorem) *Let $f(z)$ be analytic in a domain $\mathcal{D}$ with boundary $\partial \mathcal{D}$ and $z_0 \in \mathcal{D}$. We determine R such that*

$$R = \min \{|z - z_0|, z \in \partial \mathcal{D}\}.$$

Then there exists an infinite power series

$$\sum_{k=0}^{\infty} a_k (z - z_0)^k$$

which converges to $f(z)$ for $|z - z_0| < R$. The coefficients a_k are given by

$$a_k = \frac{1}{2\pi i} \oint_C \frac{f(\zeta)}{(\zeta - z_0)^{k+1}} \, d\zeta = \frac{f^{(k)}(z_0)}{k!}, \quad k = 0, 1, \ldots, \quad (5.3.1)$$

where C is any simple closed contour enclosing z_0 and lying completely inside $\mathcal{D}$. The power series is said to be the Taylor series of $f(z)$ expanded at z_0.

Proof We take a point z inside the circle $N(z_0; R)$ and denote $|z - z_0|$ by r so that $r < R$. A circle C_1 is drawn around z_0 with radius R_1, where $r < R_1 < R$ (see Figure 5.1). Since the point z also lies inside C_1, the Cauchy integral formula gives

$$f(z) = \frac{1}{2\pi i} \oint_{C_1} \frac{f(\zeta)}{\zeta - z} \, d\zeta .$$

By observing the relation $|z - z_0| < |\zeta - z_0|$, where ζ is any point on C_1, we may perform the following binomial expansion:

$$\frac{1}{\zeta - z} = \frac{1}{\zeta - z_0} \frac{1}{1 - \frac{z - z_0}{\zeta - z_0}}$$

$$= \frac{1}{\zeta - z_0} \left[1 + \frac{z - z_0}{\zeta - z_0} + \cdots + \left(\frac{z - z_0}{\zeta - z_0} \right)^n + \frac{\left(\frac{z - z_0}{\zeta - z_0} \right)^{n+1}}{1 - \frac{z - z_0}{\zeta - z_0}} \right].$$

Next, we multiply both sides of the above equation by $\frac{f(\zeta)}{2\pi i}$ and perform the contour integration along C_1. Using the property

$$\frac{1}{2\pi i} \oint_{C_1} \frac{f(\zeta)}{(\zeta - z_0)^{k+1}} \, d\zeta = \frac{f^{(k)}(z_0)}{k!} \tag{5.3.2}$$

and collecting the terms, we obtain

$$f(z) = \sum_{k=0}^{n} \frac{f^{(k)}(z_0)}{k!} (z - z_0)^k + R_n,$$

where the remainder R_n can be expressed as

$$R_n = \frac{1}{2\pi i} \oint_{C_1} \frac{f(\zeta)}{\zeta - z} \left(\frac{z - z_0}{\zeta - z_0} \right)^{n+1} d\zeta. \tag{5.3.3}$$

To complete the proof, it suffices to show that

$$\lim_{n \to \infty} R_n = 0.$$

We estimate $|R_n|$ using the modulus inequality from Section 4.1. The arc length along the integration path is given by $L = 2\pi R_1$. Since $f(z)$ is continuous inside $\mathcal{D}$, its modulus is bounded by some constant M on C_1, that is,

$$|f(\zeta)| \leq M, \quad \zeta \in C_1.$$

Note that

$$|z - z_0| = r \quad \text{and} \quad |\zeta - z_0| = R_1.$$

Moreover, we have

$$|\zeta - z| = |(\zeta - z_0) - (z - z_0)| \geq |\zeta - z_0| - |z - z_0| = R_1 - r,$$

so

$$\left| \frac{f(\zeta)}{\zeta - z} \right| \left| \frac{z - z_0}{\zeta - z_0} \right|^{n+1} \leq \frac{M}{R_1 - r} \left(\frac{r}{R_1} \right)^{n+1}.$$

These results are combined to give

$$|R_n| \leq \left| \frac{1}{2\pi i} \right| \frac{M}{R_1 - r} \left(\frac{r}{R_1} \right)^{n+1} 2\pi R_1 = \frac{MR_1}{R_1 - r} \left(\frac{r}{R_1} \right)^{n+1}.$$

It is then obvious that $|R_n|$ vanishes as $n \to \infty$ since $\frac{r}{R_1} < 1$.

From eq. (5.3.2), the Taylor coefficients a_k are given by

$$a_k = \frac{1}{2\pi i} \oint_{C_1} \frac{f(\zeta)}{(\zeta - z_0)^{k+1}} \, d\zeta, \quad k = 0, 1, \dots. \tag{5.3.4}$$

Since the integrand in the above integral is analytic in the domain $\mathcal{D}$ except at z_0, by virtue of the Cauchy–Goursat integral theorem, the integration path C_1 can be replaced by any simple closed contour C enclosing z_0 and lying entirely inside $\mathcal{D}$. This completes the proof.

Remark The Taylor theorem states that a complex function can be expanded in an infinite power series at a point around which the function is analytic. In fact, any infinite power series expansion of an analytic function $f(z)$ must be its Taylor series [see eq. (5.2.17)]. In other words, the expansion of $f(z)$ in a Taylor power series at a given point is *unique*.

Example 5.3.1 Let $f(z) = \frac{1}{(1+z)^2}$. Use the relation

$$f(z)(1 + z)^2 = 1$$

to compute the Taylor series of $f(z)$ at $z = 0$ and find the radius of convergence of the power series.

Solution If we assume the Taylor series expansion

$$f(z) = \sum_{k=0}^{\infty} a_k z^k,$$

then

$$(1 + z)^2 f(z) = a_0 + (2a_0 + a_1)z + \sum_{k=2}^{\infty} (a_{k-2} + 2a_{k-1} + a_k)z^k = 1.$$

By equating coefficients of like power terms on both sides, we deduce that

$$a_0 = 1, \quad a_1 = -2, \quad a_2 = 3, \quad a_3 = -4, \dots.$$

The Taylor series of $f(z)$ at $z = 0$ is then found to be

$$f(z) \cdot \sum_{k=0}^{\infty} (-1)^k z^k + \cdots z^k .$$

The radius of convergence R of the Taylor series can be found by the ratio test, and is shown to be

$$R = \lim_{k \to \infty} \left| \frac{(-1)^k (k + 1)}{(-1)^{k+1}(k + 2)} \right| = 1.$$

As expected, the radius of convergence equals the distance from $z = 0$ to the point $z = -1$, where the function is not analytic.

Remark The Taylor series expansion of the generalized power function $(1 + z)^\alpha$, where α is complex, is considered in Problem 5.17.

Example 5.3.2 Find the Maclaurin series up to z^3 for each of the following functions:

(a) $\sin^{-1} z$ (principal value), (b) e^{e^z}.

Solution

(a) Let $f(z) = \sin^{-1} z$; then $z = \sin f(z)$. Differentiating the equation with respect to z repeatedly, we obtain

$$1 = f'(z) \cos f(z),$$
$$0 = f''(z) \cos f(z) - [f'(z)]^2 \sin f(z),$$
$$0 = f'''(z) \cos f(z) - 3f''(z)f'(z) \sin f(z) - [f'(z)]^3 \cos f(z).$$

Putting $z = 0$ in the above relations, we have

$$f(0) = 0 \text{ (principal value)}, \quad f'(0) = 1, \quad f''(0) = 0, \quad f'''(0) = 1.$$

The first four terms of the Maclaurin series are found to be

$$\sin^{-1} z = f(0) + f'(0)z + \frac{f''(0)}{2!}z^2 + \frac{f'''(0)}{3!}z^3 + \cdots$$

$$= z + \frac{z^3}{3!} + \cdots .$$

(b) Writing $f(z) = e^{e^z}$, the successive higher-order derivatives of $f(z)$ are found to be

$$f'(z) = e^z \, e^{e^z},$$
$$f''(z) = e^z e^{e^z} + (e^z)^2 \, e^{e^z},$$
$$f'''(z) = e^z e^{e^z} + 3(e^z)^2 \, e^{e^z} + (e^z)^3 \, e^{e^z}, \quad \text{etc.}$$

Putting $z = 0$ into the above relations, we obtain

$$f(0) = e, \ f'(0) = e, \ f''(0) = 2e, \ f'''(0) = 5e.$$

Therefore, the Maclaurin series is found to be

$$e^{e^z} = f(0) + f'(0)z + \frac{f''(0)}{2!}z^2 + \frac{f'''(0)}{3!}z^3 + \cdots$$

$$= e + ez + ez^2 + \frac{5e}{6}z^3 + \cdots.$$

Example 5.3.3 By using an appropriate auxiliary analytic function, expand each of the following real functions:

(a) $y = e^x \cos x$ in a Taylor series of the real variable x,
(b) $y = e^{\cos x} \cos(\sin x)$ in a Fourier cosine series.

Solution

(a) Consider the entire function e^z and let $z = (1 + i)x$, x being real. We take the real part of the function and obtain

$$e^x \cos x = \mathrm{Re} \, e^z = \mathrm{Re} \, e^{(1+i)x} = \mathrm{Re} \sum_{n=0}^{\infty} \frac{(1+i)^n x^n}{n!}$$

$$= \sum_{n=0}^{\infty} 2^{n/2} \cos \frac{n\pi}{4} \frac{x^n}{n!}.$$

(b) First, we consider the expansion

$$e^{e^{i\theta}} = \sum_{n=0}^{\infty} \frac{e^{in\theta}}{n!} = \sum_{n=0}^{\infty} \frac{\cos n\theta}{n!} + i \sum_{n=0}^{\infty} \frac{\sin n\theta}{n!}.$$

By taking the real parts of both sides of the above equation, we obtain

$$\mathrm{Re} \, e^{e^{i\theta}} = \mathrm{Re} \, e^{\cos x + i \sin x} = e^{\cos x} \cos(\sin x) = \sum_{n=0}^{\infty} \frac{\cos n\theta}{n!}.$$

The expanded series is in the form of a Fourier cosine series.

Example 5.3.4 The generating function of the Bernoulli numbers B_n is given by

$$\frac{z}{e^z - 1} = \sum_{n=0}^{\infty} \frac{B_n}{n!} z^n \qquad \text{where} \qquad B_0 = 1, \ B_1 = -\frac{1}{2}, \ B_{2n+1} = 0, \ n \geq 1$$

Show that the Maclaurin series for $\tan z$ is given by

$$\tan z = \sum_{n=1}^{\infty} (-1)^{n-1} \frac{2^{2n}(2^{2n} - 1)}{(2n)!} B_{2n} z^{2n-1}.$$

What is the radius of convergence of the series?

Solution We consider

$$\tan z = \frac{\frac{1}{2i}(e^{iz} - e^{-iz})}{\frac{1}{2}(e^{iz} + e^{-iz})}$$

$$= \frac{2i}{e^{2iz} - 1} - \frac{4i}{e^{4iz} - 1} - i$$

$$= \left[\sum_{n=0}^{\infty} \frac{B_n}{n!} (2i)^n z^{n-1} \right] - \left[\sum_{n=0}^{\infty} \frac{B_n}{n!} (4i)^n z^{n-1} \right] - i$$

$$= [(-2i)B_1 - i] + \sum_{n=2}^{\infty} \frac{B_n}{n!} [(2i)^n - (4i)^n] z^{n-1}.$$

By observing that $B_1 = -\frac{1}{2}$ and $B_{2n+1} = 0$, $n \geq 1$, the above expression can be simplified to give

$$\tan z = \sum_{n=1}^{\infty} (-1)^{n-1} \frac{2^{2n}(2^{2n} - 1)}{(2n)!} B_{2n} z^{2n-1}.$$

The tangent function is not analytic at the zeros of $\cos z$, namely, $z = \pm\frac{\pi}{2}, \pm\frac{3\pi}{2}, \ldots$. The distance from $z = 0$ to the nearest singular point is $\frac{\pi}{2}$, so the radius of convergence of the above Maclaurin series is deduced to be $\frac{\pi}{2}$.

5.4 Laurent series

The Taylor series expansion represents an analytic function inside its circle of convergence. It is common to encounter functions which are analytic in some punctured domains. In these cases, the Taylor series representation is not the correct form for describing the infinite power series expansion of these types of complex function.

Let us consider an infinite series with negative powers of the form

$$b_1(z - z_0)^{-1} + b_2(z - z_0)^{-2} + \cdots = \sum_{n=1}^{\infty} b_n(z - z_0)^{-n}. \qquad (5.4.1)$$

To find the region of convergence of this series, we set $w = \frac{1}{z - z_0}$. The series in eq. (5.4.1) then becomes a Taylor series in w. By the ratio test, supposing

$$R' = \lim_{n \to \infty} \left| \frac{b_n}{b_{n+1}} \right|$$

exists, then the new series converges for all w such that $|w| < R'$. In terms of z, the region of convergence is the region outside the circle $|z - z_0| = \frac{1}{R'}$.

More generally, we consider an infinite series with positive and negative power terms of the form

$$\sum_{n=0}^{\infty} a_n(z - z_0)^n + \sum_{n=1}^{\infty} b_n(z - z_0)^{-n}.$$

This is called a *Laurent series* expanded at $z = z_0$. The second summation term with negative power terms is called the *principal part* of the Laurent series.

Suppose the limits

$$R = \lim_{n \to \infty} \left| \frac{a_n}{a_{n+1}} \right| \quad \text{and} \quad R' = \lim_{n \to \infty} \left| \frac{b_n}{b_{n+1}} \right|$$

exist, and $RR' > 1$; then inside the annular domain $\left\{ z : \frac{1}{R'} < |z - z_0| < R \right\}$, the Laurent series is convergent. The annulus may degenerate into a hollow plane if $R = \infty$, or a punctured disk if $R' = \infty$. When $RR' \leq 1$, such an annular region of convergence does not exist.

A Laurent series defines a function $f(z)$ in its annular region of convergence. Conversely, the Laurent series theorem states that a function defined and analytic in an annulus can be expanded in a Laurent series.

Theorem 5.4.1 (Laurent series theorem) *Let $f(z)$ be analytic in the annulus $A : R_1 < |z - z_0| < R_2$; then $f(z)$ can be represented by the Laurent series,*

$$f(z) = \sum_{k=-\infty}^{\infty} c_k(z - z_0)^k, \qquad (5.4.2)$$

which converges to $f(z)$ throughout the annulus. The Laurent coefficients are given by

$$c_k = \frac{1}{2\pi i} \oint_C \frac{f(\zeta)}{(\zeta - z_0)^{k+1}} \, d\zeta, \qquad k = 0, \pm 1, \pm 2, \dots, \qquad (5.4.3)$$

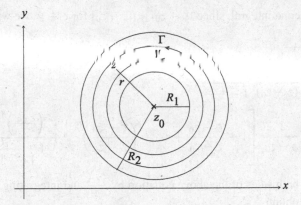

Figure 5.2. The function $f(z)$ is analytic inside the annulus $A: R_1 < |z - z_0| < R_2$. We choose two positively oriented circular paths γ and Γ, centered at z_0 and with radii r_1 and r_2, respectively. The point z lies within the two concentric circles γ and Γ.

where C is any simple closed contour lying completely inside the annulus and going around the point z_0.

Proof Take any point $z \in A$ and write $r = |z - z_0|$. Choose two positively oriented circles Γ and γ with radii r_2 and r_1, respectively, where $R_1 < r_1 < r < r_2 < R_2$ (see Figure 5.2). Since the annular region bounded by Γ and γ is doubly connected, the application of the Cauchy integral formula to this region gives

$$f(z) = \frac{1}{2\pi i} \oint_\Gamma \frac{f(\zeta)}{\zeta - z}\, d\zeta - \frac{1}{2\pi i} \oint_\gamma \frac{f(\zeta)}{\zeta - z}\, d\zeta \, .$$

For the first integral, we perform exactly the same computation as in the proof of the Taylor series theorem to obtain

$$\frac{1}{2\pi i} \oint_\Gamma \frac{f(\zeta)}{\zeta - z}\, d\zeta = \sum_{k=0}^{\infty} a_k (z - z_0)^k,$$

where

$$a_k = \frac{1}{2\pi i} \oint_\Gamma \frac{f(\zeta)}{(\zeta - z_0)^{k+1}}\, d\zeta, \qquad k = 0, 1, \dots . \tag{5.4.4}$$

For the second integral, since $|\zeta - z_0| < |z - z_0|$ for $\zeta \in \gamma$, we write

$$-\frac{1}{\zeta - z}$$

$$= \frac{1}{(z - z_0)} \frac{1}{1 - \frac{\zeta - z_0}{z - z_0}}$$

$$= \frac{1}{z - z_0} \left[1 + \frac{\zeta - z_0}{z - z_0} + \cdots + \left(\frac{\zeta - z_0}{z - z_0} \right)^{n-1} + \frac{\left(\frac{\zeta - z_0}{z - z_0} \right)^n}{1 - \frac{\zeta - z_0}{z - z_0}} \right].$$

Multiplying both sides of the above equation by $\frac{f(\zeta)}{2\pi i}$ and integrating along the circle γ, we obtain

$$-\frac{1}{2\pi i} \oint_\gamma \frac{f(\zeta)}{\zeta - z} \, d\zeta$$

$$= \left[\frac{1}{2\pi i} \oint_\gamma f(\zeta) \, d\zeta \right] \frac{1}{z - z_0}$$

$$+ \cdots + \left[\frac{1}{2\pi i} \oint_\gamma f(\zeta)(\zeta - z_0)^{n-1} \, d\zeta \right] \frac{1}{(z - z_0)^n} + \widehat{R}_n,$$

where

$$\widehat{R}_n = \frac{1}{2\pi i} \oint_\gamma \frac{f(\zeta)}{z - \zeta} \left(\frac{\zeta - z_0}{z - z_0} \right)^n \, d\zeta.$$

It remains to be shown that $\widehat{R}_n \to 0$ as $n \to \infty$. Let

$$M_\gamma = \max_{\zeta \in \gamma} |f(\zeta)|,$$

and note that for ζ lying on γ, we have

$$|z - \zeta| = |(z - z_0) - (\zeta - z_0)| \geq |z - z_0| - |\zeta - z_0| = r - r_1.$$

The modulus of the remainder $\widehat{R}_n$ is bounded by

$$|\widehat{R}_n| \leq \left| \frac{1}{2\pi i} \right| \frac{M_\gamma}{r - r_1} \left(\frac{r_1}{r} \right)^n 2\pi r_1 = M_\gamma \frac{r_1}{r - r_1} \left(\frac{r_1}{r} \right)^n,$$

which tends to zero as $n \to \infty$ since $\frac{r_1}{r} < 1$. Once the result

$$\lim_{n \to \infty} \widehat{R}_n = 0$$

is established, we then have

$$-\frac{1}{2\pi i} \oint_\gamma \frac{f(\zeta)}{\zeta - z} \, d\zeta = \sum_{k=1}^{\infty} b_k (z - z_0)^{-k},$$

where

$$b_k = \frac{1}{2\pi i} \oint_\gamma f(\zeta)(\zeta - z_0)^{k-1} \, d\zeta, \qquad k = 1, 2, \ldots . \qquad (5.4.5)$$

Since the integrand functions in eqs. (5.4.4) and (5.4.5) are analytic inside the annulus A, we may replace both circles Γ and γ by any simple closed contour C lying entirely inside A and enclosing the point z_0. The two results can be combined to give the Laurent series expansion as defined in eqs. (5.4.2) and (5.4.3).

Remark

(i) Suppose $f(z)$ is analytic in the full disk $|z - z_0| < R_2$ without the punctured hole (the same assumption on the domain of analyticity of $f(z)$ as in the Taylor series theorem); then the integrand in eq. (5.4.5) becomes analytic inside $|z - z_0| < R_2$. By the Cauchy–Goursat theorem, we then have $b_k = 0$, $k = 1, 2, \ldots$. This is expected since the Laurent series should reduce to a Taylor series.

(ii) As revealed by later examples, we normally do not find the Laurent series by computing the Laurent coefficients using eq. (5.4.3). The Laurent series can be found by any method, and by virtue of the uniqueness property of the Laurent series expansion, the Laurent series obtained using different methods would all agree.

(iii) Some of the Laurent series coefficients may be related to an integral whose value is desired. For example, when $k = -1$ in eq. (5.4.3), we have

$$2\pi i c_{-1} = \oint_C f(\zeta) \, d\zeta . \qquad (5.4.6)$$

Example 5.4.1 Find all the possible Laurent series of each of the following functions at the given point α:

(a) $\dfrac{1}{1 - z^2}$, $\alpha = -1$; (b) $\dfrac{1}{z^3}$, $\alpha = i$.

Solution

(a) The function $\frac{1}{1-z^2}$ is not analytic at $z = -1$ and $z = 1$. There are only two possible annular domains centered at $z = -1$ inside which the function is analytic throughout. These two annular domains are $0 < |z + 1| < 2$

and $|z + 1| > 2$. Recall that the binomial expansion

$$\frac{1}{1 - \zeta} = \sum_{k=0}^{\infty} \zeta^k$$

is valid provided that $|\zeta| < 1$. In the first annular domain, we have $\left|\frac{z+1}{2}\right| < 1$. This motivates us to express $\frac{1}{1-z^2}$ as $\frac{1}{2(z+1)\left(1-\frac{z+1}{2}\right)}$ and perform the binomial expansion of $\frac{1}{1-\frac{z+1}{2}}$ accordingly. We then have

$$\frac{1}{1 - z^2} = \frac{1}{2(z + 1)\left(1 - \frac{z+1}{2}\right)}$$

$$= \frac{1}{2(z + 1)} \sum_{k=0}^{\infty} \frac{(z + 1)^k}{2^k} = \sum_{k=0}^{\infty} \frac{(z + 1)^{k-1}}{2^{k+1}}.$$

Note that there is only one negative power term in the Laurent expansion.

In the second annular domain $|z + 1| > 2$, we have $\frac{2}{|z+1|} < 1$; so the Laurent expansion can be found as follows:

$$\frac{1}{1 - z^2} = -\frac{1}{(z + 1)^2 \left(1 - \frac{2}{z+1}\right)}$$

$$= -\frac{1}{(z + 1)^2} \sum_{k=0}^{\infty} \frac{2^k}{(z + 1)^k} = -\sum_{k=0}^{\infty} \frac{2^k}{(z + 1)^{k+2}}.$$

Here, the Laurent expansion has an infinite number of negative power terms but no positive power term.

(b) The function $\frac{1}{z^3}$ is not analytic at $z = 0$. We seek annular domains, centered at $z = i$, such that the function is analytic throughout. Two such domains are found to be $|z - i| < 1$ and $|z - i| > 1$. To expand the function $\frac{1}{z^3}$ in both domains, we make use of the binomial expansion

$$\frac{1}{(1 + \zeta)^3} = 1 + (-3)\zeta + \frac{(-3)(-4)}{2!}\zeta^2 + \frac{(-3)(-4)(-5)}{3!}\zeta^3 + \cdots$$

$$= \sum_{k=2}^{\infty} (-1)^k \frac{k(k - 1)}{2} \zeta^{k-2},$$

which holds for $|\zeta| < 1$. By setting $\zeta = \frac{z-i}{i}$ in the above binomial expansion, we obtain

$$\frac{1}{z^3} = \frac{1}{i^3 \left(1 + \frac{z-i}{i}\right)^3}$$

$$= \frac{1}{2i^3} \sum_{k=2}^{\infty} (-1)^k k(k-1) \frac{(z-i)^{k-2}}{i^{k-2}}$$

$$= i - 3(z-i) - 6i(z-i)^2 + 10(z-i)^3 + \cdots.$$

The series converges in the region $|z - i| < 1$, which is precisely the first domain. Since $|z - i| < 1$ is a solid disk, the power series expansion in this domain is actually a Taylor series with no negative power term.

In the second domain $|z - i| > 1$, the property $\left|\frac{i}{z-i}\right| < 1$ holds. We now set $\zeta = \frac{i}{z-i}$; the corresponding Laurent expansion inside $|z - i| > 1$ is found to be

$$\frac{1}{z^3} = \frac{1}{(z-i)^3 \left(1 + \frac{i}{z-i}\right)^3}$$

$$= \frac{1}{2(z-i)^3} \sum_{k=2}^{\infty} (-1)^k k(k-1) \frac{i^{k-2}}{(z-i)^{k-2}}$$

$$= \frac{1}{2} \sum_{k=2}^{\infty} (-1)^k k(k-1) \frac{i^{k-2}}{(z-i)^{k+1}}$$

$$= \frac{1}{(z-i)^3} - \frac{3i}{(z-i)^4} - \frac{6}{(z-i)^5} + \frac{10i}{(z-i)^6} + \cdots.$$

Example 5.4.2 Find the Laurent series for

$$f(z) = \frac{1}{e^z - e^{2z}}$$

expanded at $z = 0$ that is convergent in the annulus $0 < |z| < 2\pi$.

Solution The given function is not analytic at the zeros of $e^z - e^{2z}$. These zeros are given by $z = 2k\pi i$, where k is any integer. To perform the Laurent expansion of the function at $z = 0$, we choose an annular domain centered at $z = 0$ inside which the function is analytic. One such possible choice is the annulus $0 < |z| < 2\pi$. First, we decompose $f(z)$ into

$$f(z) = e^{-z} - \frac{1}{z}\left(\frac{z}{e^z - 1}\right).$$

The Taylor series expansions of e^{-z} and $\dfrac{z}{e^z - 1}$ are

$$e^{-z} = \sum_{n=0}^{\infty} (-1)^n \frac{z^n}{n!}, \quad \text{valid for } |z| < \infty,$$

$$\frac{z}{e^z - 1} = \sum_{n=0}^{\infty} \frac{B_n}{n!} z^n, \quad \text{valid for } |z| < 2\pi,$$

where B_n are the Bernoulli numbers (see Example 5.3.4). Combining the results, the Laurent series is found to be

$$f(z) = \sum_{n=0}^{\infty} \frac{(-1)^n}{n!} z^n - \sum_{n=0}^{\infty} \frac{B_n}{n!} z^{n-1}$$

$$= -\frac{1}{z} + \sum_{n=0}^{\infty} \left[\frac{(-1)^n}{n!} - \frac{B_{n+1}}{(n+1)!} \right] z^n,$$

where $B_0 = 1$. The Laurent series has only one negative power term and it is seen to be convergent in the annulus $0 < |z| < 2\pi$.

Example 5.4.3   Find the Laurent expansion of the function

$$f(z) = \frac{1}{z - k}$$

that is valid inside the domain $|z| > |k|$, where k is real and $|k| < 1$. Using the Laurent expansion, deduce that

$$\sum_{n=1}^{\infty} k^n \cos n\theta = \frac{k \cos \theta - k^2}{1 - 2k \cos \theta + k^2},$$

$$\sum_{n=1}^{\infty} k^n \sin n\theta = \frac{k \sin \theta}{1 - 2k \cos \theta + k^2}.$$

Solution   For $|z| > |k|$, by performing the appropriate binomial expansion, the Laurent expansion of $f(z)$ at $z = 0$ is found to be

$$\frac{1}{z - k} = \frac{1}{z \left(1 - \frac{k}{z}\right)} = \frac{1}{z} \left(1 + \frac{k}{z} + \frac{k^2}{z^2} + \cdots + \frac{k^n}{z^n} + \cdots \right) \quad \text{for } \left| \frac{k}{z} \right| < 1.$$

Since $|k| < 1$, the unit circle $|z| = 1$ lies inside the region of convergence of the above Laurent series. Therefore, the series obtained by substituting $z = e^{i\theta}$ into the above Laurent series is guaranteed to be convergent. We then have

$$\frac{1}{e^{i\theta} - k} = \frac{1}{e^{i\theta}} \left(1 + k e^{-i\theta} + k^2 e^{-2i\theta} + \cdots + k^n e^{-in\theta} + \cdots \right).$$

Rearranging the terms, we obtain

$$\frac{e^{i\theta}\,(e^{-i\theta} - k)}{(e^{i\theta} - k)(e^{-i\theta} - k)} - 1 = \frac{k\cos\theta - k^2 - ik\sin\theta}{1 - 2k\cos\theta + k^2}$$

$$= \sum_{n=1}^{\infty} k^n \cos n\theta - i \sum_{n=1}^{\infty} k^n \sin n\theta.$$

By equating the real and imaginary parts of the above expressions, we obtain the required results.

Example 5.4.4 The *Bessel functions* of integer order are defined by the Laurent series

$$e^{\frac{z}{2}(w - \frac{1}{w})} = \sum_{n=-\infty}^{\infty} J_n(z)w^n.$$

The function $e^{\frac{z}{2}(w - \frac{1}{w})}$ is called the generating function of the Bessel function. Show that

$$J_n(z) = \frac{1}{2\pi} \int_0^{2\pi} \cos(nt - z\sin t)\,dt = \sum_{k=0}^{\infty} \frac{(-1)^k}{(n+k)!k!}\left(\frac{z}{2}\right)^{n+2k},$$

$$n = 0, \pm 1, \pm 2, \ldots,$$

and deduce the relation

$$J_n(z) = (-1)^n J_{-n}(z).$$

Solution The generating function is analytic in the annulus $0 < |w| < \infty$. The Bessel functions are the Laurent coefficients of the generating function. Using eq. (5.4.3), we have

$$J_n(z) = \frac{1}{2\pi i} \oint_C e^{\frac{z}{2}(w - \frac{1}{w})} \frac{dw}{w^{n+1}}, \quad n = 0, \pm 1, \pm 2, \ldots,$$

where C is any simple closed contour lying completely inside the annulus of analyticity of $e^{\frac{z}{2}(w - \frac{1}{w})}$ and enclosing the origin. Here, we choose C to be the unit circle $|w| = 1$. We parametrize the unit circle by $w = e^{it}$, $0 \leq t < 2\pi$. The above integral for $J_n(z)$ then becomes

$$J_n(z) = \frac{1}{2\pi i} \int_0^{2\pi} i\, e^{iz\sin t} e^{-int}\,dt$$

$$= \frac{1}{2\pi} \int_0^{2\pi} \cos(nt - z\sin t)\,dt - \frac{i}{2\pi} \int_0^{2\pi} \sin(nt - z\sin t)\,dt,$$

$$n = 0, \pm 1, \pm 2, \ldots.$$

Setting $t = 2\pi - \tau$ in the second integral, we have

$$\int_0^{2\pi} \sin(nt - z\sin t)\,dt = -\int_0^{2\pi} \sin(n\tau - z\sin \tau)\,d\tau;$$

so the second integral is seen to have the value zero. Hence, the integral representation of $J_n(z)$ reduces to

$$J_n(z) = \frac{1}{2\pi}\int_0^{2\pi}\cos(nt - z\sin t)\,dt, \quad n = 0, \pm 1, \pm 2, \ldots.$$

To find the Laurent series expansion of $J_n(z)$, we consider the multiplication of the series for $e^{\frac{zw}{2}}$ and $e^{-\frac{z}{2w}}$ using the Cauchy product formula [see eq. (5.2.19)]. This gives

$$e^{\frac{z}{2}\left(w - \frac{1}{w}\right)} = \left[\sum_{n=0}^{\infty}\frac{1}{n!}\left(\frac{z}{2}\right)^n w^n\right]\left[\sum_{n=0}^{\infty}\frac{(-1)^n}{n!}\left(\frac{z}{2}\right)^n\frac{1}{w^n}\right]$$

$$= \sum_{n=-\infty}^{\infty}\left[\sum_{k=0}^{\infty}\frac{(-1)^k}{(n+k)!k!}\left(\frac{z}{2}\right)^{n+2k}\right]w^n.$$

By equating like power terms of w^n in the Laurent expansion, we obtain

$$J_n(z) = \sum_{k=0}^{\infty}\frac{(-1)^k}{(n+k)!k!}\left(\frac{z}{2}\right)^{n+2k}, \quad n = 0, \pm 1, \pm 2, \ldots.$$

Lastly, by swapping w with $-\frac{1}{w}$ in the defining relation for $J_n(z)$, we have

$$e^{\frac{z}{2}\left(w - \frac{1}{w}\right)} = \sum_{n=-\infty}^{\infty}J_n(z)\left(-\frac{1}{w}\right)^n = \sum_{n=-\infty}^{\infty}(-1)^n J_{-n}(z)w^n;$$

so

$$J_n(z) = (-1)^n J_{-n}(z).$$

5.4.1 Potential flow past an obstacle

Consider a potential flow with uniform upstream velocity U_∞ past an obstacle (see Figure 5.3). Assuming that the origin is inside the obstacle, the complex velocity $V(z)$ of the potential flow can be represented by the following Laurent series expansion in z:

$$V(z) = \overline{U_\infty} + \frac{c_{-1}}{z} + \frac{c_{-2}}{z^2} + \cdots + \frac{c_{-n}}{z^n} + \cdots, \tag{5.4.7}$$

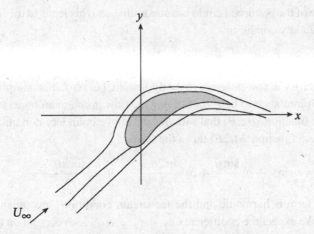

Figure 5.3. Plot of the streamlines that represent the potential flow with uniform upstream velocity U_∞ past an obstacle.

where $U_\infty, c_{-1}, c_{-2}, \ldots, c_{-n}$ are complex numbers. To substantiate the claim, we observe that $V(z)$ is analytic in the domain outside the obstacle, and at points far from the obstacle, $|z| \to \infty$, $\overline{V(z)}$ tends to the uniform velocity U_∞.

The coefficient c_{-1} has a special physical interpretation. The contour integral of the complex velocity around the body defined by the closed curve γ is equal to $\mathcal{C}_\gamma + i\mathcal{F}_\gamma$, where $\mathcal{C}_\gamma$ and $\mathcal{F}_\gamma$ are the circulation and flux around the body, respectively [see eqs. (4.4.8) and (4.4.12)]. On the other hand, from eq. (5.4.6), the Laurent coefficient c_{-1} and the contour integral around γ are related by

$$2\pi i c_{-1} = \oint_\gamma V(z) \, dz. \tag{5.4.8}$$

Combining the results, we then have

$$2\pi i c_{-1} = \mathcal{C}_\gamma + i\mathcal{F}_\gamma. \tag{5.4.9}$$

If the circulation and flux around the body are both zero, then $c_{-1} = 0$.

The determination of the other coefficients, $c_{-2}, \ldots, c_{-n}, \ldots$, depends on the configuration of the obstacle. As an illustrative example, we consider the potential flow past the perturbed circle $r = 1 - \epsilon \sin^2 \theta$, where ϵ is a small quantity. To find the flow past the perturbed circle, it is more convenient to formulate the problem in terms of the stream function $\psi(r, \theta)$. Let the uniform upstream velocity be parallel to the x-axis and have unit magnitude so that the far-stream conditions are

$$\frac{\partial \psi}{\partial y} \to 1 \quad \text{and} \quad \frac{\partial \psi}{\partial x} \to 0 \quad \text{as} \quad r \to \infty. \tag{5.4.10}$$

The surface of the perturbed circle is a streamline and this leads to the following surface boundary condition:

$$\psi(1 - \epsilon \sin^2 \theta, \theta) = 0. \qquad (5.4.11)$$

The flux across a non-penetrative body should be zero. For simplicity, we assume the circulation to be zero. Solving the flow problem amounts to finding a harmonic function $\psi(r, \theta)$ that satisfies the above boundary conditions.

The stream function $\psi(r, \theta)$ takes the form

$$\psi(r, \theta) = r \sin \theta + a_1 \frac{\sin \theta}{r} + a_2 \frac{\sin 2\theta}{r^2} + \cdots + a_n \frac{\sin n\theta}{r^n} + \cdots, \qquad (5.4.12)$$

since each term is harmonic and the far-stream conditions are satisfied automatically. We expect the coefficients $a_1, a_2, \ldots, a_n, \ldots$ to depend on the small parameter ϵ in the problem. When $\epsilon = 0$, the perturbed circle becomes the perfect unit circle. The stream function for the flow past the unit circle is known to be (see Example 4.4.2)

$$\psi(r, \theta; \epsilon = 0) = \left(r - \frac{1}{r} \right) \sin \theta. \qquad (5.4.13)$$

Alternatively, we may write the stream function as a *perturbation series* in powers of ϵ:

$$\psi(r, \theta) = \psi_0(r, \theta) + \epsilon \psi_1(r, \theta) + \epsilon^2 \psi_2(r, \theta) + \cdots + \epsilon^n \psi_n(r, \theta) + \cdots.$$

When ϵ is small, the first few terms of the perturbation series may already give a very good approximation to the true solution.

We now determine the successive order terms $\psi_0, \psi_1, \psi_2, \ldots$, sequentially. The zeroth-order solution corresponds to $\epsilon = 0$, and it has been given in eq. (5.4.13). To find the higher-order solutions, we apply the boundary condition (5.4.11). When expressed in perturbation expansion, it becomes

$$\psi_0(1 - \epsilon \sin^2 \theta, \theta) + \epsilon \psi_1(1 - \epsilon \sin^2 \theta, \theta)$$
$$+ \epsilon^2 \psi_2(1 - \epsilon \sin^2 \theta, \theta) + \cdots = 0. \qquad (5.4.14)$$

The perturbation parameter ϵ appears explicitly in front of the successive order terms and also implicitly in the arguments of the functions. Next, we expand the above series in powers of ϵ and equate like power terms of ϵ to generate the successive boundary conditions for $\psi_1, \psi_2, \ldots$, etc. To determine ψ_1, we keep only the linear terms in ϵ and obtain

$$\psi_0(1, \theta) - \epsilon \sin^2 \theta \frac{\partial \psi_0}{\partial r}(1, \theta) + \epsilon \psi_1(1, \theta) + \cdots = 0.$$

Hence $\psi_1(\psi, \theta)$ has to satisfy the boundary condition

$$\psi_1(1, \theta) = \sin^2 \theta \frac{\partial \psi_0}{\partial r}(1, \theta) = 2 \sin^2 \theta \sin \theta = \frac{3 \sin \theta - \sin 3\theta}{2}.$$

To satisfy the above boundary condition, the harmonic function $\psi_1(r, \theta)$ is found to be

$$\psi_1(r, \theta) = \frac{3}{2} \frac{\sin \theta}{r} - \frac{1}{2} \frac{\sin 3\theta}{r^3}. \tag{5.4.15}$$

By equating like power terms of $O(\epsilon^2)$ in the expansion of eq. (5.4.14), we obtain the following boundary condition for $\psi_2(r, \theta)$:

$$\frac{\sin^4 \theta}{2} \frac{\partial^2 \psi_0}{\partial r^2}(1, \theta) - \sin^2 \theta \frac{\partial \psi_1}{\partial r}(1, \theta) + \psi_2(1, \theta) = 0.$$

Simplifying the trigonometric terms and rearranging, the boundary condition can be expressed as

$$\psi_2(1, \theta) = -\frac{7}{8} \sin \theta + \frac{13}{16} \sin 3\theta - \frac{5}{16} \sin 5\theta. \tag{5.4.16}$$

The corresponding harmonic function $\psi_2(r, \theta)$ that satisfies the above boundary condition is found to be

$$\psi_2(r, \theta) = -\frac{7}{8} \frac{\sin \theta}{r} + \frac{13}{16} \frac{\sin 3\theta}{r^3} - \frac{5}{16} \frac{\sin 5\theta}{r^5}. \tag{5.4.17}$$

The solution procedure can be routinely applied to find the higher-order solutions, though the algebraic manipulations become more daunting. Fortunately, there exists an ingenious algorithm for computing the coefficients of the terms in the higher-order solutions, the details of which are shown in Problem 5.33.

5.5 Analytic continuation

Suppose we are given an infinite power series or an analytic formula which defines an analytic function f in a domain $\mathcal{D}$. It is natural to ask whether one can extend its domain of analyticity. More precisely, can we find a function $\mathcal{F}$ which is analytic in a larger domain and whose values agree with those of f for points in $\mathcal{D}$? Here, $\mathcal{F}$ is called an *analytic continuation* of f. For example, the complex exponential function e^z is the analytic continuation of the real exponential function e^x defined over the real interval $(-\infty, \infty)$. The complex function e^z is analytic in the finite complex plane and $e^z = e^x$ when $z = x$, x being real.

More generally, suppose f_1 is analytic in a domain $\mathcal{D}_1$ and f_2 is analytic in another domain $\mathcal{D}_2$. If $\mathcal{D}_1 \cap \mathcal{D}_2 \neq \emptyset$ and $f_1(z) = f_2(z)$ in the common intersection $\mathcal{D}_1 \cap \mathcal{D}_2$, then f_2 is said to be the analytic continuation of f_1 to $\mathcal{D}_2$ and

f_1 is the analytic continuation of f_2 to $\mathcal{D}_1$. Now, the function $\mathcal{F}$ defined by

$$\mathcal{F}(z) = \begin{cases} f_1(z) & \text{when } z \in \mathcal{D}_1 \\ f_2(z) & \text{when } z \in \mathcal{D}_2 \end{cases} \tag{5.5.1}$$

is analytic in the union $\mathcal{D}_1 \cup \mathcal{D}_2$. The function $\mathcal{F}$ is the analytic continuation to $\mathcal{D}_1 \cup \mathcal{D}_2$ of either f_1 or f_2; and f_1 and f_2 are called *elements* of $\mathcal{F}$.

Example 5.5.1   Show that the two functions

$$f(z) = \sum_{k=0}^{\infty} a^k z^k = 1 + az + a^2 z^2 + \cdots$$

and

$$g(z) = \sum_{k=0}^{\infty} \frac{(-1)^k (1-a)^k z^k}{(1-z)^{k+1}} = \frac{1}{1-z} - \frac{(1-a)z}{(1-z)^2} + \frac{(1-a)^2 z^2}{(1-z)^3} + \cdots$$

are analytic continuations of each other.

Solution   Consider the function

$$w(z) = \frac{1}{1 - az}$$

which is analytic everywhere except at $z = \frac{1}{a}$. The Taylor expansion of $w(z)$ about $z = 0$ in the domain $|z| < \frac{1}{|a|}$ is given by

$$\frac{1}{1 - az} = 1 + az + a^2 z^2 + \cdots = \sum_{k=0}^{\infty} a^k z^k.$$

The above Taylor series is precisely $f(z)$. Suppose we write $w(z)$ in the alternative form

$$\frac{1}{1 - az} = \frac{1}{1 - z} \frac{1}{1 + \frac{(1-a)z}{1-z}}.$$

Provided that

$$\left| \frac{(1-a)z}{1-z} \right| < 1,$$

$w(z)$ can be expressed in the form

$$\frac{1}{1 - az} = \frac{1}{1-z} - \frac{(1-a)z}{(1-z)^2} + \frac{(1-a)^2 z^2}{(1-z)^3} + \cdots,$$

which is precisely $g(z)$. Both $f(z)$ and $g(z)$ are elements of the analytic function $w(z)$. The respective regions of convergence of $f(z)$ and $g(z)$ are given by

$$|z| < \frac{1}{|a|} \quad \text{and} \quad |z - a| < |1 + a|,$$

which overlap with each other. Actually, it can be easily shown that both domains contain a circle centered around $z = 0$ with a finite radius. Therefore, $f(z)$ and $g(z)$ are analytic continuations of each other.

Example 5.5.2 The two power series

$$z + \frac{z^2}{2} + \frac{z^3}{3} + \cdots$$

and

$$i\pi - (z - 2) + \frac{1}{2}(z - 2)^2 - \frac{1}{3}(z - 2)^3 + \cdots$$

have no common region of convergence. However, they are analytic continuations of the same function. Find the function and explain why.

Solution Let the function $f_1(z)$ be defined by

$$f_1(z) = z + \frac{z^2}{2} + \frac{z^3}{3} + \cdots .$$

The infinite power series converges in the domain $|z| < 1$. In fact, it is the Maclaurin series expansion of the function

$$g(z) = -\mathrm{Log}(1 - z).$$

Similarly, the second series

$$f_2(z) = i\pi - (z - 2) + \frac{1}{2}(z - 2)^2 - \frac{1}{3}(z - 2)^3 + \cdots$$

is seen to be the Taylor series expansion of

$$h(z) = i\pi - \mathrm{Log}(z - 1)$$

at $z = 2$. The expansion is valid for $|z - 2| < 1$. Note that the two regions of convergence, $|z| < 1$ and $|z - 2| < 1$, are disjoint. However, when $\mathrm{Im}(z - 1) > 0$, $h(z) = g(z)$; also, both of the above regions overlap with the domain $\mathrm{Im}(z - 1) > 0$. Hence

$$-\mathrm{Log}(1 - z), \quad \{z : \mathrm{Im}(z - 1) > 0\},$$

is a direct continuation of both $f_1(z)$ and $f_2(z)$. Therefore, the two series are each an analytic continuation of the other.

Theorem 5.5.1   *If a function f is analytic in a domain $\mathcal{D}$ and $f(z) = 0$ at all points on an arc inside $\mathcal{D}$, then $f(z) = 0$ throughout $\mathcal{D}$.*

Proof   We take any point z_0 on the arc where $f(z) = 0$. Inside some circle C that is centered at z_0 and lying completely inside $\mathcal{D}$ (the circle can be extended at least to the boundary of $\mathcal{D}$), $f(z)$ can be expanded in a Taylor expansion in the form

$$f(z) = f(z_0) + f'(z_0)(z - z_0) + \frac{f''(z_0)}{2!}(z - z_0)^2 + \cdots.$$

Since $f(z) = 0$ for all points on the arc that lies inside the circle C, this would imply that

$$f(z_0) = f'(z_0) = f''(z_0) = \cdots = 0.$$

Therefore, $f(z) = 0$ for all points inside the circle C. Continuing the process with another arc that lies completely inside C and expanding $f(z)$ in a region bounded by another circle, $f(z)$ can be similarly shown to be zero for all points inside the new circle. Eventually, we can find a sufficient number of circles to cover the whole domain $\mathcal{D}$. Since $f(z) = 0$ inside all these circles, $f(z) = 0$ throughout $\mathcal{D}$.

Corollary   Let the two functions $f_1(z)$ and $f_2(z)$ be analytic in some domain $\mathcal{D}$. Suppose $f_1(z) = f_2(z)$ on an arc inside $\mathcal{D}$; then $f_1(z) = f_2(z)$ throughout the domain $\mathcal{D}$. The validity of the corollary can be revealed by writing $f(z) = f_1(z) - f_2(z)$ and applying the theorem. For an interesting application of this corollary, see Problem 5.38.

5.5.1 Reflection principle

Some functions like $z^2 + 1$ and $\cos z$ possess the property that $\overline{z^2 + 1} = \overline{z}^2 + 1$ and $\cos \overline{z} = \overline{\cos z}$. However, other functions, like $z^2 + i$ and $i \cos z$, do not observe the property $f(\overline{z}) = \overline{f(z)}$; that is, reflection of z with respect to the real axis does not correspond to reflection of $f(z)$ with respect to the real axis. The following theorem states precisely the condition under which $f(\overline{z}) = \overline{f(z)}$ holds.

Theorem 5.5.2 (Reflection principle) *Let the function f be analytic in some domain $\mathcal{D}$, which contains a segment of the real axis and is symmetric with respect to the real axis. For any point z in $\mathcal{D}$, the property*

$$f(\bar{z}) = \overline{f(z)} \tag{5.5.2}$$

holds if and only if $f(z)$ assumes real values on the real axis contained in $\mathcal{D}$.

Proof Writing $f(z) = u(x, y) + iv(x, y)$, $z = x + iy$, we have

$$\overline{f(\bar{z})} = u(x, -y) - iv(x, -y).$$

Note that $\overline{f(\bar{z})}$ is well defined for all $z \in \mathcal{D}$ since $\mathcal{D}$ is symmetric with respect to the real axis; so for any $z \in \mathcal{D}$, we have $\bar{z} \in \mathcal{D}$. Suppose $\overline{f(\bar{z})} = f(z)$; we then have $u(x, -y) - iv(x, -y) = u(x, y) + iv(x, y)$. When z is real, we put $y = 0$ in the above relation and obtain

$$u(x, 0) - iv(x, 0) = u(x, 0) + iv(x, 0), \tag{5.5.3}$$

which implies that $v(x, 0) = 0$. Hence, $f(z)$ is real when z is real.

Conversely, given that $f(z)$ assumes real values on the segment of the real axis within $\mathcal{D}$, from eq. (5.5.3) we deduce that $\overline{f(\bar{z})}$ and $f(z)$ have the same values along the segment of the real axis. To prove $\overline{f(\bar{z})} = f(z)$ for all $z \in \mathcal{D}$, it suffices to show that $\overline{f(\bar{z})}$ is analytic in $\mathcal{D}$. This is because once $f(z)$ and $\overline{f(\bar{z})}$ are known to be analytic in the same domain $\mathcal{D}$ and they share the same values on a segment in $\mathcal{D}$, then by the corollary to Theorem 5.5.1, we have $\overline{f(\bar{z})} = f(z)$ throughout $\mathcal{D}$.

The real and imaginary parts of $\overline{f(\bar{z})}$ are $u(x, -y)$ and $-v(x, -y)$, respectively. Since $f(z)$ is analytic in $\mathcal{D}$, both $u(x, y)$ and $v(x, y)$ have continuous first-order partial derivatives in $\mathcal{D}$; correspondingly, $u(x, -y)$ and $-v(x, -y)$ share the same continuity properties. The next step is to show that both $u(x, -y)$ and $-v(x, -y)$ satisfy the Cauchy–Riemann relations. From the analyticity of $f(z)$, we have

$$\frac{\partial u}{\partial x}(x, y) = \frac{\partial v}{\partial y}(x, y) \quad \text{and} \quad \frac{\partial u}{\partial y}(x, y) = -\frac{\partial v}{\partial x}(x, y),$$

from which we can deduce that

$$\frac{\partial u}{\partial x}(x, \tilde{y}) = \frac{\partial(-v)}{\partial \tilde{y}}(x, \tilde{y}) \quad \text{and} \quad \frac{\partial u}{\partial \tilde{y}}(x, \tilde{y}) = -\frac{\partial(-v)}{\partial x}(x, \tilde{y}),$$

where $\tilde{y} = -y$. These are precisely the Cauchy–Riemann relations for the real and imaginary parts of $\overline{f(\bar{z})}$. Hence, the analyticity of $\overline{f(\bar{z})}$ in $\mathcal{D}$ is established.

5.6 Problems

5.1. Let $z_n = x_n + iy_n$, $n = 1, 2, \ldots$.

(a) Given that $\sum_{n=1}^{\infty} |z_n|$ converges, show that $\sum_{n=1}^{\infty} z_n$ also converges. That is, an absolutely convergent series is always convergent.

(b) Show that $\sum_{n=1}^{\infty} z_n$ converges absolutely if and only if both $\sum_{n=1}^{\infty} x_n$ and $\sum_{n=1}^{\infty} y_n$ converge absolutely.

Hint: Use the comparison test to establish the convergence of the two real series

$$\sum_{n=1}^{\infty} |x_n| \quad \text{and} \quad \sum_{n=1}^{\infty} |y_n|.$$

5.2. Suppose $z_1, z_2, \ldots, z_n, \ldots$ all lie in the right half-plane Re $z \geq 0$, and both series $\sum_{n=1}^{\infty} z_n$ and $\sum_{n=1}^{\infty} z_n^2$ converge; show that $\sum_{n=1}^{\infty} |z_n|^2$ also converges.

5.3. Suppose $z_1, z_2, \ldots, z_n$, all lie within the sector $-\alpha \leq \text{Arg } z \leq \alpha$, $0 < \alpha < \frac{\pi}{2}$; show that the two series

$$\sum_{n=1}^{\infty} z_n \quad \text{and} \quad \sum_{n=1}^{\infty} |z_n|$$

both either converge or diverge.

Hint: Observe that $|z_n| \leq \frac{\text{Re } z_n}{\cos \alpha}$.

5.4. Examine the convergence of the following series:

(a) $\displaystyle\sum_{n=1}^{\infty} \frac{\cos in}{3^n}$;   (b) $\displaystyle\sum_{n=1}^{\infty} \frac{2n}{\sqrt{n} + in}$;   (c) $\displaystyle\sum_{n=1}^{\infty} \left[\frac{i(2n+i)}{5n} \right]^n$.

5.5. Discuss the convergence of

$$\sum_{n=0}^{\infty} (z^{n+1} - z^n).$$

5.6. Show that the series

$$\sum_{n=0}^{\infty} \frac{z^2}{(1 + z^2)^n}$$

does not converge uniformly when z assumes values along the real axis.

5.7. Let $R > 0$ be the radius of convergence of the series $\sum_{n=0}^{\infty} a_n z^n$. Show that the radius of convergence of the series $\sum_{n=0}^{\infty} (\operatorname{Re} a_n) z^n$ is at least R.

5.8. Suppose the radius of convergence of the series $\sum_{n=0}^{\infty} a_n z^n$ is R; find the corresponding radius of convergence of each of the following series:

(a) $\displaystyle\sum_{n=0}^{\infty} n^{10} a_n z^n$; (b) $\displaystyle\sum_{n=0}^{\infty} (2^n - 1) a_n z^n$; (c) $\displaystyle\sum_{n=0}^{\infty} \frac{a_n}{n!} z^n$; (d) $\displaystyle\sum_{n=0}^{\infty} a_n^k z^n$.

Hint: For part (c), consider the following examples: $a_n = \alpha^n n!$; $a_n = (n!)^2$; $a_n = \sqrt{n!}$.

5.9. Show that the series

$$\sum_{n=1}^{\infty} \frac{z^n}{1 + z^{2n}}$$

converges in both the domains $|z| < 1$ and $|z| > 1$. Discuss the convergence of the series on the unit circle $|z| = 1$.

5.10. Consider the series of complex functions

$$\sum_{k=1}^{\infty} \frac{z^2}{(1 + |z|^2)^k}.$$

The partial sum $S_n(z)$ is defined by

$$S_n(z) = \sum_{k=0}^{n} \frac{z^2}{(1 + |z|^2)^k}.$$

(a) Show that

$$S(z) = \lim_{n \to \infty} S_n(z) = \frac{z^2}{|z|^2} (1 + |z|^2),$$

valid for $|z| > 0$.

(b) Show that the convergence of $S_n(z)$ to $S(z)$ inside $|z| < 1$ as $n \to \infty$ is not uniform convergence.

5.11. Find the radii of convergence of the following series:

(a) $\displaystyle\sum_{n=0}^{\infty} n^2 q^{n^2} z^n$, $|q| < 1$; (b) $\displaystyle\sum_{n=0}^{\infty} z^{n!}$; (c) $\displaystyle\sum_{n=0}^{\infty} [3 + (-1)^n]^n z^n$;

(d) $\displaystyle\sum_{n=1}^{\infty} \frac{n!}{n^n} z^n$.

5.12 Suppose

$$L = \lim_{n \to \infty} \frac{a_{n+1}}{a_n}$$

exists; show that the following three power series have the same radius of convergence.

(a) $\displaystyle\sum_{n=0}^{\infty} a_n z^n$;   (b) $\displaystyle\sum_{n=0}^{\infty} \frac{a_n}{n+1} z^{n+1}$;   (c) $\displaystyle\sum_{n=0}^{\infty} n a_n z^{n-1}$.

5.13. Show that

$$\sum_{n=1}^{\infty} \frac{z^n}{n} = -\text{Log}(1-z) \quad \text{for } z \in \mathcal{D} = \{z : |z| < 1\}.$$

Explain why the series $\sum_{n=1}^{\infty} \frac{e^{in\theta}}{n}$, $\theta \neq 0$, is conditionally convergent. What happens when $\theta = 0$? Use the above series to show that

(a) $\displaystyle\sum_{n=1}^{\infty} \frac{\cos n\theta}{n} = -\ln\left(2 \sin \frac{\theta}{2}\right)$;

(b) $\displaystyle\sum_{n=1}^{\infty} \frac{\sin n\theta}{n} = \frac{\pi - \theta}{2}, \quad 0 < \theta < 2\pi$.

Hint: Note that $-\text{Log}(1-z)$ is analytic inside the domain $\mathcal{D}$ and

$$\frac{d}{dz}[-\text{Log}(1-z)] = \frac{1}{1-z}.$$

5.14. Expand both of the following functions in Maclaurin series up to the z^4 term:

(a) $e^{-z} \sin z$;   (b) $\text{Log}(1 + e^z)$.

5.15. Find the Maclaurin series of the following functions:

(a) $\dfrac{1-z}{1+z+z^2+z^3}$;   (b) $\dfrac{\sin z}{1+z^2}$;   (c) $\sinh(z^3)$;

(d) $\displaystyle\int_0^z \frac{1 - \cos \zeta}{\zeta} d\zeta$.

5.16. Suppose the Maclaurin series expansion of $\sec z$ is given by

$$\sec z = \sum_{k=0}^{\infty} \frac{(-1)^k E_{2k}}{(2k)!} z^{2k} = E_0 - \frac{E_2}{2!} z^2 + \frac{E_4}{4!} z^4 - \cdots + \cdots.$$

Find the values for E_0, E_2, E_4, E_6. These numbers are called the *Euler numbers*. Find the circle of convergence of the above series.

Consider

$$\sec z = \frac{1}{\cos z} = \frac{1}{1 - \frac{z^2}{2!} + \frac{z^4}{4!} - \cdots}$$

so that

$$1 = \left(E_0 - \frac{E_2}{2!} z^2 + \frac{E_4}{4!} z^4 - \cdots + \cdots \right) \left(1 - \frac{z^2}{2!} + \frac{z^4}{4!} - \cdots \right).$$

5.17. Consider the principal branch of the *generalized power function* $(1 + z)^\alpha$ for α complex; show that its Taylor series expansion at $z = 0$ is given by

$$(1 + z)^\alpha = 1 + \alpha z + \frac{\alpha(\alpha - 1)}{2!} z^2 + \frac{\alpha(\alpha - 1)(\alpha - 2)}{3!} z^3$$
$$+ \cdots + \frac{\alpha(\alpha - 1) \cdots (\alpha - n + 1)}{n!} z^n + \cdots.$$

Find the corresponding circle of convergence of the above series.

Hint: Suppose we write the function as

$$f(z) = e^{\alpha \mathrm{Log}(1+z)}, \quad f(0) = 1;$$

show that

$$(1 + z) f'(z) = \alpha f(z).$$

5.18. Show that the coefficient c_n of the Taylor series expansion

$$\frac{1}{1 - z - z^2} = \sum_{n=0}^{\infty} c_n z^n$$

satisfies the recurrence relation

$$c_n = c_{n-1} + c_{n-2}, \ n \geq 2.$$

Find the general form for c_n and the radius of convergence of the series.

5.19. Find the Taylor series expansion of

$$f(z) = \sin^{-1} z = \int_0^z \frac{1}{\sqrt{1 - \xi^2}} \, d\xi.$$

Determine the circle of convergence.

Hint: Consider

$$\frac{1}{\sqrt{1-z^2}} = \sum_{n=0}^{\infty} \frac{(2n)!}{2^{2n}(n!)^2} z^{2n}, \quad \text{valid for } |z| < 1.$$

5.20. Suppose the rational polynomial

$$\frac{a_0 + a_1 z + a_2 z^2}{b_0 + b_1 z + b_2 z^2 + b_3 z^3}$$

is expanded in a Taylor series in the form $\sum_{n=0}^{\infty} c_n z^n$; find c_0, c_1, c_2 and the recurrence relation between c_n, c_{n-1}, c_{n-2} and c_{n-3}, $n \geq 3$.

5.21. Suppose $f(z)$ admits a power series of the form

$$f(z) = \sum_{n=0}^{\infty} c_n z^n ;$$

prove that

(a) the coefficients of the odd powers of z vanish if $f(z)$ is even;
(b) the coefficients of the even powers of z vanish if $f(z)$ is odd.

5.22. The function $\frac{1}{1+x^2}$ is differentiable of all orders for real values of x, but its Taylor series expansion at $x = 0$

$$\frac{1}{1+x^2} = 1 - x^2 + x^4 - \cdots + (-1)^n x^{2n} + \cdots ,$$

converges only for $|x| < 1$. Explain why.

5.23. Supposing $f(z) = \sum_{n=0}^{\infty} c_n z^n$ is analytic inside $|z| < R$, show that

$$\frac{1}{2\pi} \int_0^{2\pi} |f(re^{i\theta})|^2 \, d\theta = \sum_{n=0}^{\infty} |c_n|^2 r^{2n}, \quad 0 < r < R.$$

5.24. Consider the following Taylor series

$$\sum_{n=1}^{\infty} \frac{z^{n+1}}{n(n+1)}.$$

Show that the circle of convergence is $|z| < 1$, and the series converges to the sum function

$$S(z) = (1-z)\text{Log}(1-z) + z,$$

where $\text{Log } 1 = 0$. At $z = 1$, show that the corresponding series

$$\sum_{n=1}^{\infty} \frac{1}{n(n+1)}$$

converges but $S(z)$ fails to be analytic. Quote another example of a Taylor series where the series is divergent at a point on the circle of convergence but the sum function is analytic at that point.

5.25. Suppose $f(z) = \sum_{n=0}^{\infty} c_n z^n$ is analytic inside the unit disk $|z| < 1$, together with Re $f(z) \geq 0$ and $f(0) = 1$; show that

(a) $|c_n| \leq 2, \; n \geq 0$; (b) $\dfrac{1 - |z|}{1 + |z|} \leq |f(z)| \leq \dfrac{1 + |z|}{1 - |z|}$.

5.26. Suppose the radii of convergence of the series

$$f(z) = \sum_{n=0}^{\infty} a_n z^n \quad \text{and} \quad g(z) = \sum_{n=0}^{\infty} b_n z^n$$

are greater than 1. Show that

$$\frac{1}{2\pi} \int_0^{2\pi} f(e^{i\theta}) g(e^{-i\theta}) \, d\theta = \sum_{n=0}^{\infty} a_n b_n.$$

5.27. A function $F(t, z)$ is called the *generating function* of the sequence of complex functions $\{f_n(z)\}$ if $F(t, z)$ admits a Taylor series expansion in powers of t of the form

$$F(t, z) = \sum_{n=0}^{\infty} f_n(z) t^n \qquad \text{for } |t| < R.$$

Some useful properties of the sequence of functions $\{f_n(z)\}$ can be derived via its generating function. Consider the generating function of the *Bernoulli polynomials* $B_n(z)$ defined by

$$t \, \frac{e^{tz} - 1}{e^t - 1} = \sum_{n=1}^{\infty} \frac{B_n(z)}{n!} \, t^n.$$

Prove the following properties of the Bernoulli polynomials:

(a) $B_n(z + 1) - B_n(z) = n z^{n-1}$;

(b) if m is a natural number, then

$$\frac{B_{n+1}(m)}{n + 1} = 1 + 2^n + 3^n + \cdots + (m - 1)^n;$$

(c) $B_n(z) = \displaystyle\sum_{k=0}^{n-1} \binom{n}{k} B_k z^{n-k}$, where B_k are the Bernoulli numbers defined in Example 5.3.4.

5.28. Find all the possible Laurent series expansions at the given point z_0 of each of the following functions:

(a) $f(z) = \dfrac{1}{(z-1)(z-2)}$, $z_0 = 0$;

(b) $f(z) = \dfrac{1}{z^2(z-i)}$, $z_0 = i$.

5.29. For each of the following functions, find the Laurent series at z_0 which is convergent within the annulus A as indicated:

(a) $\dfrac{1}{(z-i)(z-2)}$ where

    (i) $z_0 = 0$, $A = \{z: 1 < |z| < 2\}$;

    (ii) $z_0 = 0$, $A = \{z: |z| > 2\}$;

    (iii) $z_0 = i$, $A = \{z: 0 < |z - i| < \sqrt{5}\}$.

Then deduce the value of

$$\oint_C \frac{1}{(z-i)(z-2)}\, dz,$$

where C is a simple closed contour enclosing i but excluding 2.

(b) $\dfrac{1}{z(z^2 - 1)}$ where $z_0 = 1$ and $A = \{z : 1 < |z - 1| < 2\}$.

(c) $e^{z + \frac{1}{z}}$ where $z_0 = 0$ and $A = \mathbb{C}\backslash\{0\}$.

   Hint: Find the Cauchy product of e^z and $e^{\frac{1}{z}}$.

5.30. Find the Laurent series expansion of

$$f(z) = \sin\frac{z}{z-1},$$

valid for the annulus: $0 < |z - 1| < \infty$.

   Hint: Consider

$$\sin\frac{z}{z-1} = \sin\left(1 + \frac{1}{z-1}\right) = \sin 1 \cos\frac{1}{z-1} + \cos 1 \sin\frac{1}{z-1}.$$

5.31. Show that the coefficient c_n in the Laurent expansion at $z = 0$ of

$$f(z) = \cos\left(z + \frac{1}{z}\right)$$

is given by

$$c_n = \frac{1}{2\pi}\int_0^{2\pi} \cos(2\cos\theta)\cos n\theta\, d\theta, \qquad n = 0, \pm 1, \pm 2, \ldots.$$

Deduce the corresponding Laurent expansion at $z = 0$ of

$$f(z) = \cosh\left(z + \frac{1}{z}\right).$$

5.32. Show that the function represented by the Laurent series

$$\sum_{n=1}^{\infty} \frac{1}{n! z^n}$$

is analytic everywhere except at $z = 0$. Evaluate the contour integral of the above function around the unit circle $|z| = 1$ in the positive sense.

5.33. Consider a uniform horizontal potential flow past the perturbed circle $r = 1 - \epsilon \sin^2 \theta$. We would like to solve for the stream function $\psi(r, \theta)$ that is harmonic outside the perturbed circle and satisfies the boundary conditions

(a) $\dfrac{\partial \psi}{\partial y} \to 1$ and $\dfrac{\partial \psi}{\partial x} \to 0$ as $r \to \infty$,

(b) $\psi(1 - \epsilon \sin^2 \theta, \theta) = 0$.

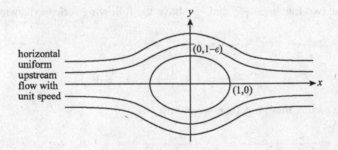

horizontal
uniform
upstream
flow with
unit speed

$(0, 1-\epsilon)$

$(1,0)$

The perturbation series expansion of $\psi(r, \theta)$ in powers of the parameter ϵ is given by

$$\psi(r, \theta) = \psi_0(r, \theta) + \epsilon \psi_1(r, \theta) + \epsilon^2 \psi_2(r, \theta) + \cdots.$$

Explain why the nth order solution $\psi_n(r, \theta)$ takes the form

$$\psi_n(r, \theta) = \sum_{j=0}^{n} A_{nj} \frac{\sin(2j + 1)\theta}{r^{2j+1}}, \quad n \geq 1.$$

Suppose we write

$$(1 - \epsilon \sin^2 \theta)^{-(2j+1)} = \sum_{k=0}^{\infty} C_{jk} \sin^{2k} \theta;$$

show that the surface boundary condition can be expressed in the form

$$\psi(1 - \epsilon \sin^2 \theta, \theta) = (1 - \epsilon \sin^2 \theta) \sin \theta$$
$$+ \sum_{j=0}^{\infty} \left[\sin(2j+1)\theta \left(\sum_{k=0}^{\infty} C_{jk} \sin^{2k} \theta \right) \left(\sum_{\ell=0}^{\infty} \epsilon^{j+\ell} A_{j+\ell, j} \right) \right].$$

Using the trigonometric relation

$$\sin(2j+1)\theta \sin^{2k} \theta = \sum_{m=0}^{j+k} B_m^{jk} \sin(2m+1)\theta,$$

show that the recurrence relations for the coefficients A_{nj} are given by

$$\sum_{j=0}^{n} A_{nj} C_{j_0} \sin(2j+1)\theta = - \sum_{j=0}^{n-1} \sum_{k=1}^{n-j} \sum_{m=0}^{j+k} C_{jk} A_{n-k, j} B_m^{jk} \sin(2m+1)\theta.$$

Use the above relations to show that

$$\psi_3(r, \theta) \doteq -\frac{1}{16} \frac{\sin \theta}{r} - \frac{9}{16} \frac{\sin 3\theta}{r^3} + \frac{21}{32} \frac{\sin 5\theta}{r^5} - \frac{7}{32} \frac{\sin 7\theta}{r^7}.$$

5.34. The two functions $\frac{z}{1-z}$ and $\frac{z}{z-1}$ have the following series expansions:

$$\frac{z}{1-z} = z + z^2 + z^3 + \cdots$$

$$\frac{z}{z-1} = 1 + \frac{1}{z} + \frac{1}{z^2} + \frac{1}{z^3} + \cdots.$$

By adding them together, we obtain

$$0 = \frac{z}{1-z} + \frac{z}{z-1} = \cdots + \frac{1}{z^3} + \frac{1}{z^2} + \frac{1}{z} + 1 + z + z^2 + z^3 + \cdots.$$

What is wrong with the above argument?

5.35. (a) Show that $\frac{1}{z(z+1)}$ is an analytic continuation of

$$f(z) = \sum_{k=0}^{\infty} (-1)^k \left(1 - \frac{1}{2^{k+1}} \right) (z-1)^k$$

$$= \frac{1}{2} - \frac{3}{4}(z-1) + \frac{7}{8}(z-1)^2 - \cdots + \cdots.$$

(b) Show that $\frac{1}{z^2}$ is an analytic continuation of

$$f(z) = \sum_{k=0}^{\infty} (k+1)(z+1)^k.$$

5.36. Show that the following two series are analytic continuations of each other:

$$f(z) = ? - \frac{1}{2} z^? + \frac{1}{3} z^3 - \cdots$$

$$g(z) = \ln 2 - \frac{1-z}{2} - \frac{1}{2} \frac{(1-z)^2}{2^2} - \frac{1}{3} \frac{(1-z)^3}{2^3} - \cdots.$$

5.37. For each of the following pairs of functions, show that they form analytic continuations:

(a) $f_1(z) = \displaystyle\sum_{n=0}^{\infty} z^n$, $|z| < 1$ and $f_2(z) = \frac{1}{1-z}$ for Re $z < \frac{1}{2}$;

(b) $f_1(z) = \text{Log } z$ and $f_2(z) = \ln|z| + i \arg z$ for $0 \le \arg z < 2\pi$.

5.38. Given that

$$\sin^2 x + \cos^2 x = 1$$

is valid for all real values x; using the corollary to Theorem 5.5.1, explain why

$$\sin^2 z + \cos^2 z = 1$$

is valid for any complex number z in the whole (unextended) complex plane.

5.39. Suppose

$$f(\bar{z}) = \overline{f(z)}$$

in a domain $\mathcal{D}$ which is symmetric about the real axis; show that the coefficients of the Laurent series expansion of $f(z)$ expanded about the origin are real.

5.40. A function $f(z)$ is analytic inside the annulus $r < |z| < \frac{1}{r}, r < 1$, and satisfies

$$f\left(\frac{1}{\bar{z}}\right) = \overline{f(z)}.$$

Let the Laurent series expansion of $f(z)$ at $z = 0$ be expressed in the form

$$f(z) = \sum_{k=-\infty}^{\infty} a_k z^k.$$

Show that $\overline{a_k} = a_{-k}$ and $f(z)$ is real on the unit circle $|z| = 1$.

6

Singularities and Calculus of Residues

This chapter begins with a discussion of the classification of isolated singularities of complex functions. The classification can be done effectively by examining the Laurent series expansion of a complex function in a deleted neighborhood around an isolated singularity. An isolated singularity can be either a pole, a removable singularity or an essential singularity. The various forms of behavior of a complex function near an isolated singularity are examined. Next, we introduce the definition of the residue of a complex function at an isolated singularity. We show how to apply residue calculus to the evaluation of different types of integral. The Fourier transform and Fourier integrals are considered, and the effective use of residue calculus for the analytic evaluation of these integrals is illustrated. The concept of the Cauchy principal value of an improper integral is introduced. We also consider the application of residue calculus to solving fluid flow problems.

6.1 Classification of singular points

By definition, a singularity or a singular point of a function $f(z)$ is a point at which $f(z)$ is not analytic. A point at which $f(z)$ is analytic is called a *regular point* of $f(z)$. A point z_0 is called an *isolated singularity* of $f(z)$ if there exists a neighborhood of z_0 inside which z_0 is the only singular point of $f(z)$. For example, $\pm i$ are isolated singularities of $f(z) = \frac{1}{z^2+1}$. More generally, the singularities of a rational function $\frac{P(z)}{Q(z)}$ are the zeros of $Q(z)$ and they are all isolated. The singularities of $\operatorname{cosec} z$ are all isolated and they are simply the zeros of $\sin z$, namely, $z = k\pi$, k is any integer.

Certainly we may have non-isolated singularities. For example, every point on the negative real axis (the branch cut of $\operatorname{Log} z$) is a non-isolated singularity of $\operatorname{Log} z$. The function $f(z) = \overline{z}$ is nowhere analytic, so every point in the complex plane is a non-isolated singularity. An interesting example is the function $f(z) = \operatorname{cosec} \frac{\pi}{z}$: all the points $z = \frac{1}{n}$, $n = \pm 1, \pm 2, \ldots$, are isolated singularities. Interestingly, the origin is a non-isolated singularity since every neighborhood of the origin contains other singularities.

In this section, we confine our discussion to isolated singularities only. Suppose z_0 is an isolated singularity; then there exists a positive number r such that $f(z)$ is analytic inside the deleted neighborhood $0 < |z - z_0| < r$. This forms an annular domain within which the Laurent series theorem is applicable. Suppose $f(z)$ has the Laurent series expansion at z_0 in the deleted neighborhood of the form

$$f(z) = \sum_{n=0}^{\infty} a_n (z - z_0)^n + \sum_{n=1}^{\infty} b_n (z - z_0)^{-n}, \quad 0 < |z - z_0| < r; \quad (6.1.1)$$

the isolated singularity z_0 is classified according to the principal part of the Laurent series.

Removable singularity

In this case, the principal part vanishes altogether and the Laurent series is essentially a Taylor series. The series represents an analytic function in the solid disk $|z - z_0| < r$. As there is no negative power term in the Laurent series, the limit $\lim_{z \to z_0} f(z)$ exists and is equal to a_0. The singularity z_0 is said to be removable since we can remove this singularity z_0 by defining $f(z_0) = a_0$.

For example, the function $\frac{\sin z}{z}$ is undefined at $z = 0$. The Laurent series of $\frac{\sin z}{z}$ in a deleted neighborhood of $z = 0$ is given by

$$f(z) = \frac{\sin z}{z} = 1 - \frac{z^2}{3!} + \frac{z^4}{5!} - \frac{z^6}{7!} + \cdots,$$

where the Laurent series has no negative power term. The singularity of $f(z) = \frac{\sin z}{z}$ at $z = 0$ can be removed by defining $f(0) = 1$.

Essential singularity

Here, the principal part has infinitely many nonzero terms. For example, $z = 0$ is an essential singularity of the function $z^2 e^{1/z}$. Inside the annular domain $0 < |z| < \infty$, the Laurent series of $z^2 e^{1/z}$ is found to be

$$f(z) = z^2 + z + \frac{1}{2!} + \frac{1}{3!} \frac{1}{z} + \frac{1}{4!} \frac{1}{z^2} + \cdots.$$

Pole of order k

In this case, the principal part has only a finite number of non-vanishing terms and the last non-vanishing coefficient is b_k; that is, the Laurent series in the deleted neighborhood of z_0 takes the form

$$f(z) = \sum_{n=0}^{\infty} a_n(z - z_0)^n + b_1(z - z_0)^{-1}$$
$$+ b_2(z - z_0)^{-2} + \cdots + b_k(z - z_0)^{-k}, \quad b_k \neq 0. \quad (6.1.2)$$

It is called a simple pole when $k = 1$. For example, $\frac{1}{z^2+1}$ has simple poles at $z = i$ and $z = -i$; $\frac{\sin z}{z^3}$ has a pole of order 2 at $z = 0$ since

$$\frac{\sin z}{z^3} = \frac{1}{z^2} - \frac{1}{3!} + \frac{z^2}{5!} - \cdots + \cdots, \quad 0 < |z| < \infty.$$

The above three cases are mutually exclusive, that is, an isolated singularity must be either a removable singularity, an essential singularity or a pole. A complex function is said to be *meromorphic* if it is analytic everywhere in the finite plane except at isolated poles. A meromorphic function has no essential singularities in the finite plane, though it may have an essential singularity at infinity. The number of poles in the finite complex plane can be infinite. An example of a meromorphic function is $\tan z$, where the isolated poles (infinitely many) are the zeros of $\cos z$.

It may be instructive to examine the behavior of $f(z)$ around an isolated singularity at $z = z_0$:

(i) When z_0 is a removable singularity, $\lim\limits_{z \to z_0} f(z)$ exists and $|f(z)|$ is bounded near z_0.

(ii) When z_0 is a pole of finite order, it is obvious that

$$\lim_{z \to z_0} f(z) = \infty.$$

(iii) When z_0 is an essential singularity, $f(z)$ has a complicated behavior around z_0. In fact, in every neighborhood of an essential singularity z_0, the function $f(z)$ comes arbitrarily close to any complex value. A more precise description of the limiting behavior of $f(z)$ at z_0 is given by the Weierstrass–Casorati theorem.

Theorem 6.1.1 (Weierstrass–Casorati theorem) *Let z_0 be an isolated essential singularity of a function $f(z)$ and let λ be any given complex value. For any positive number ϵ, however small, there exists a point z, distinct from z_0, in every neighborhood of z_0 such that*

$$|f(z) - \lambda| < \epsilon.$$

Proof Assuming the contrary, there exist $\epsilon > 0$ and $\delta > 0$ such that

$$|f(z) - \lambda| \geq \epsilon \quad \text{for} \quad 0 < |z - z_0| < \delta.$$

Define

$$g(z) = \frac{1}{f(z) - \lambda};$$

by the above hypothesis, g is analytic inside the deleted δ-neighborhood at z_0 and $|g(z)| \leq \frac{1}{\epsilon}$. We can then conclude that z_0 is a removable singularity of $g(z)$. Accordingly, we define

$$g(z_0) = \lim_{z \to z_0} g(z).$$

Now, $g(z)$ admits a Taylor series expansion inside the neighborhood $|z - z_0| < \delta$, where

$$g(z) = (z - z_0)^m \sum_{k=m}^{\infty} a_k (z - z_0)^k,$$

where m is a finite non-negative integer and $a_m \neq 0$. Consider the following two cases:

(i) When $m = 0$, $g(z_0) = a_0 \neq 0$ so

$$\frac{1}{g(z)} = f(z) - \lambda$$

is analytic at z_0, a contradiction to "z_0 being an isolated essential singularity of f".

(ii) When $m > 0$, $\frac{1}{g(z)}$ has a pole of finite order m. Again, a similar contradiction as in part (i) occurs.

Remark Indeed, there is a stronger result describing the behavior of f near an isolated essential singularity beyond that of the Weierstrass–Casorati theorem. The Picard theorem states that $f(z)$ assumes every finite value, with one possible exception, an infinite number of times in any neighborhood of an isolated essential singularity.

As an example, we examine the behavior of the function $f(z) = e^{1/z}$ around its isolated essential singularity at $z = 0$. We let z approach zero in different directions. First, if z approaches zero along the positive x-axis, we observe that

$$\lim_{z \to 0} e^{1/z} = \lim_{x \to 0^+} e^{1/x} = \infty, \quad z = x, \ x > 0.$$

However, when z approaches zero along the negative x-axis, we have

$$\lim_{z \to 0} e^{1/z} = \lim_{x \to 0^-} e^{1/x} = 0, \quad z = x, \; x < 0.$$

In the general case, if z approaches zero along the ray $\mathrm{Arg}\, z = \theta$, then

$$\lim_{z \to 0} e^{1/z} = \lim_{r \to 0} e^{\cos\theta/r} e^{-i\sin\theta/r}, \quad z = re^{i\theta}.$$

Given any value of r (equivalently, any neighborhood around $z = 0$) and any complex number λ, we can find θ such that $e^{\cos\theta/r} e^{-i\sin\theta/r}$ comes arbitrarily close to λ.

Singularity at complex infinity

The behavior of a complex function f at complex infinity ∞ in the extended complex plane $\mathbb{C}$ can be reviewed by examining the behavior of $f\left(\frac{1}{z}\right)$ in the neighborhood of the point 0. For example, the complex exponential function $f(z) = e^z$ is known to be entire. Its behavior at complex infinity can be explored by considering $f\left(\frac{1}{z}\right) = e^{1/z}$, which is known to have an isolated essential singularity at $z = 0$. Hence, the complex exponential function has an isolated essential singularity at complex infinity.

Simple method of finding the order of a pole

From eq. (6.1.2), we observe that if z_0 is a pole of order k, then

$$\lim_{z \to z_0} (z - z_0)^k f(z) = b_k, \quad b_k \neq 0.$$

In general, if we multiply $f(z)$ by $(z - z_0)^m$ and take the limit $z \to z_0$, we obtain

$$\lim_{z \to z_0} (z - z_0)^m f(z) = \begin{cases} b_k & m = k \\ 0 & m > k \\ \infty & m < k \end{cases}. \tag{6.1.3}$$

This formula provides the basis for a simple method to find the order k of a pole. We start by multiplying $f(z)$ by a factor $(z - z_0)^m$ for some integer m and check the limit of the product $(z - z_0)^m f(z)$ as $z \to z_0$. If m is too low, the limit does not exist; and if m is too high, the limit equals zero. Only when the correct value $m = k$ is hit does the limit exist and the resulting limit is a finite nonzero value.

Example 6.1.1 Suppose $f(z)$ has a pole of order k at z_0. Find a polynomial $p(z)$ such that $f(z) - \dfrac{p(z)}{(z-z_0)^k}$ is analytic (or at most has a removable singularity) at z_0.

Solution Let $\mathcal{D}$ be a deleted neighborhood of z_0 throughout which $f(z)$ is analytic. Since $f(z)$ has a pole of order k at z_0, the Laurent series expansion is of the form

$$f(z) = \sum_{n=-k}^{\infty} f_n(z - z_0)^n, \quad z \in \mathcal{D},$$

where f_n are the Laurent coefficients. We define the function

$$g(z) = f(z) - \frac{p(z)}{(z - z_0)^k}$$

for some polynomial $p(z)$ yet to be determined. Since $p(z)$ is analytic everywhere in the finite complex plane, $g(z)$ is then analytic in the same deleted neighborhood $\mathcal{D}$. Suppose $p(z)$ is a polynomial of degree $N \geq k - 1$. The Taylor series expansion of $p(z)$ at z_0 can be written as

$$p(z) = \sum_{n=0}^{N} \frac{p^{(n)}(z_0)}{n!} (z - z_0)^n,$$

where $p^{(n)}(z_0)$ is the nth-order derivative of $p(z)$ evaluated at z_0. Inside the deleted neighborhood $\mathcal{D}$, the Laurent series expansion of $g(z)$ at z_0 is given by

$$g(z) = \frac{f_{-k} - p(z_0)}{(z - z_0)^k} + \frac{f_{-k+1} - p'(z_0)}{(z - z_0)^{k-1}} + \cdots + \frac{f_{-1} - \frac{p^{(k-1)}(z_0)}{(k-1)!}}{z - z_0}$$
$$+ \sum_{n=0}^{N-k} \left[f_n - \frac{p^{(n+k)}(z_0)}{(n + k)!} \right] (z - z_0)^n + \sum_{n=N-k+1}^{\infty} f_n(z - z_0)^n.$$

Suppose $p(z)$ is chosen such that it satisfies the following condition

$$\frac{p^{(n)}(z_0)}{n!} = f_{-k+n}, \quad 0 \leq n \leq k - 1;$$

then the principal part of the Laurent series expansion of $g(z)$ in the deleted neighborhood $\mathcal{D}$ vanishes. In other words, any polynomial $p(z)$ of degree $N \geq k - 1$ together with the satisfaction of the above condition would have a corresponding $g(z)$ that is analytic (or at most has a removable singularity) at z_0.

Example 6.1.2   The Laurent series

$$\sum_{n=1}^{\infty} z^{-n} + \sum_{n=0}^{\infty} \frac{z^n}{2^{n+1}}$$

contains infinitely many negative power terms of z. Check whether the point $z = 0$ is an essential singular point of the function represented by the series.

Solution   The first sum converges to

$$\frac{\frac{1}{z}}{1 - \frac{1}{z}} = \frac{1}{z - 1} \quad \text{for } |z| > 1$$

and the second sum converges to

$$\frac{\frac{1}{2}}{1 - \frac{z}{2}} = \frac{1}{2 - z} \quad \text{for } |z| < 2.$$

Therefore, the Laurent series converges to the function

$$f(z) = \frac{1}{z - 1} + \frac{1}{2 - z} = -\frac{1}{z^2 - 3z + 2}$$

inside the annulus $\mathcal{A}: 1 < |z| < 2$. Obviously, $\mathcal{A}$ is not a deleted neighborhood of $z = 0$. It would be incorrect to claim that $z = 0$ is an essential singularity of $f(z)$.

Example 6.1.3   The function

$$f(z) = e^{1/(1-z)^2}$$

has an essential singularity at $z = 1$. Discuss the behavior of the function as $z \to 1$:

(i) along the circumference of the unit circle $|z| = 1$;
(ii) rectilinearly from the interior of the unit circle $|z| = 1$.

Solution

(i) On the circumference of the unit circle, we set $z = e^{i\theta}$, $-\pi < \theta \leq \pi$. Now,

$$\frac{1}{(1 - z)^2} = \frac{1}{\left[-2i \sin \frac{\theta}{2} \left(\cos \frac{\theta}{2} + i \sin \frac{\theta}{2} \right) \right]^2} = -\frac{1}{4} \frac{\cos \theta - i \sin \theta}{\sin^2 \frac{\theta}{2}},$$

so

$$|f(z)| = \left| \exp\left(-\frac{1}{4}\frac{\cos\theta}{\sin^2\frac{\theta}{2}}\right) \exp\left(\frac{i}{2}\cot\frac{\theta}{2}\right) \right| = \exp\left(-\frac{1}{4}\frac{\cos\theta}{\sin^2\frac{\theta}{2}}\right).$$

As $z \to 1$, we have $\theta \to 0$. We then observe that $|f(z)| \to 0$, so $f(z) \to 0$.

(ii) We represent the radial rays from the interior of the circle towards $z = 1$ by $z = 1 + re^{i\theta}$, $\frac{\pi}{2} < |\theta| \le \pi$. Along these rays,

$$f(z) = e^{(\cos 2\theta - i \sin 2\theta)/r^2},$$

so

$$|f(z)| = e^{\cos 2\theta/r^2}.$$

Next, we consider the following three separate cases. When $\frac{\pi}{2} < |\theta| < \frac{3}{4}\pi$, $|f(z)| \to 0$ as $r \to 0$, implying that $f(z) \to 0$. On the other hand, when $\frac{3}{4}\pi < |\theta| \le \pi$, $|f(z)|$ is unbounded as $r \to 0$, implying that the limit of $f(z)$ does not exist. In particular, when $\theta = \pm\frac{3}{4}\pi$, we have

$$f(z) = e^{\mp i/r^2}.$$

As $r \to 0$, the limit of $f(z)$ does not exist.

6.2 Residues and the Residue Theorem

Consider the contour integral

$$\oint_C f(z)\, dz;$$

if $f(z)$ has no singularity inside the closed contour C, then the value of the integral is zero by virtue of the Cauchy–Goursat integral theorem. How do we deal with the usual case where singularities of f are included inside the closed contour C? In this section, we illustrate how to apply the method of residues to the evaluation of the integral without resorting to direct integration.

Let z_0 be an isolated singularity of $f(z)$; then there exists a certain deleted neighborhood $N_\epsilon = \{z: 0 < |z - z_0| < \epsilon\}$ such that f is analytic everywhere inside N_ϵ. The residue of f at z_0 is defined by

$$\text{Res}(f, z_0) = \frac{1}{2\pi i} \oint_C f(z)\, dz, \tag{6.2.1}$$

where C is any simple closed contour around z_0 and inside N_ϵ.

A simple method of evaluating the residue is to examine the Laurent series of $f(z)$ at z_0 inside N_ϵ. Suppose $f(z)$ admits the Laurent series expansion inside N_ϵ, where

$$f(z) = \sum_{n=0}^{\infty} a_n (z - z_0)^n + \sum_{n=1}^{\infty} b_n (z - z_0)^{-n}; \qquad (6.2.2)$$

then by eq. (5.4.6),

$$b_1 = \frac{1}{2\pi i} \oint_C f(z)\, dz = \text{Res}(f, z_0). \qquad (6.2.3)$$

Sometimes the residue value can be found in a straightforward manner. Several examples are shown below.

(i) $\text{Res}\left(\frac{1}{(z-z_0)^k}, z_0 \right) = \begin{cases} 1 & \text{if } k = 1 \\ 0 & \text{if } k \neq 1 \end{cases}.$

(ii) $\text{Res}(e^{\frac{1}{z}}, 0) = 1$ as deduced from the following Laurent series expansion

$$e^{\frac{1}{z}} = 1 + \frac{1}{1!}\frac{1}{z} + \frac{1}{2!}\frac{1}{z^2} + \frac{1}{3!}\frac{1}{z^3} + \cdots, \quad |z| > 0.$$

(iii) $\text{Res}\left(\frac{1}{(z-1)(z-2)}, 1 \right) = \frac{1}{1-2} = -1$ by the Cauchy integral formula.

Theorem 6.2.1 *(Cauchy residue theorem) Let C be a simple closed contour inside which $f(z)$ is analytic everywhere except at the isolated singularities $z_1, z_2, \ldots, z_n$ (see Figure 6.1). The contour integral of $f(z)$ around C is given by*

$$\oint_C f(z)\, dz = 2\pi i\, [Res(f, z_1) + Res(f, z_2) + \cdots + Res(f, z_n)]. \quad (6.2.4)$$

The validity of the theorem can be shown easily using Corollary 3 of the Cauchy–Goursat integral theorem in Section 4.2.

In general the value of the residue of $f(z)$ at an isolated singularity z_0 is computed either by direct integration or by finding the coefficient b_1 in the appropriate Laurent series of f expanded inside the deleted neighborhood N_ϵ of z_0. When the isolated singularity z_0 is a pole, it is possible to derive some efficient formulas for computing the residue at the pole.

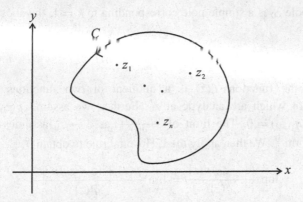

Figure 6.1. The closed contour C encircles isolated singularities of $f(z)$ at $z_1, z_2, \ldots, z_n$.

6.2.1 Computational formulas for evaluating residues

We now derive several efficient computational formulas for finding the residue of $f(z)$ at a pole z_0. First, it is necessary to know the exact order of the pole. This can be found by the method discussed earlier [see eq. (6.1.3)].

Suppose z_0 is a pole of order k of $f(z)$. In a deleted neighborhood of z_0, the Laurent series expansion takes the form

$$f(z) = \sum_{n=0}^{\infty} a_n(z - z_0)^n + b_1(z - z_0)^{-1} + \cdots + b_k(z - z_0)^{-k}, b_k \neq 0.$$

Consider the function

$$g(z) = (z - z_0)^k f(z)$$

whose principal part vanishes altogether. The power series expansion of $g(z)$ in the same deleted neighborhood of z_0 is given by

$$g(z) = b_k + b_{k-1}(z - z_0) + \cdots + b_1(z - z_0)^{k-1} + \sum_{n=0}^{\infty} a_n(z - z_0)^{n+k}.$$

$$(6.2.5)$$

Now $g(z)$ has a removable singularity at z_0, which can be removed by defining $g(z_0) = b_k$. The series in eq. (6.2.5) is essentially the Taylor series of $g(z)$ inside a neighborhood of z_0. The value of $\operatorname{Res}(f, z_0)$, which equals b_1, can be seen to be

$$b_1 = \frac{g^{(k-1)}(z_0)}{(k - 1)!} = \lim_{z \to z_0} \frac{d^{k-1}}{dx^{k-1}} \frac{[(z - z_0)^k f(z)]}{(k - 1)!}.$$

When the pole z_0 is a simple pole corresponding to $k = 1$, the above formula is reduced to

$$\text{Res}(f, z_0) = \lim_{z \to z_0} (z - z_0) f(z). \tag{6.2.6}$$

Suppose the function $f(z)$ is a quotient of two functions $p(z)$ and $q(z)$, both of which are analytic at z_0. Further, we assume $p(z_0) \neq 0$ but $q(z_0) = 0$, $q'(z_0) \neq 0$. The limit of $\frac{(z-z_0)p(z)}{q(z)}$ as $z \to z_0$ assumes the indeterminate form $\frac{0}{0}$. We then apply the L'Hospital rule to obtain

$$\lim_{z \to z_0} \frac{(z - z_0)p(z)}{q(z)} = \lim_{z \to z_0} \frac{p(z) + (z - z_0)p'(z)}{q'(z)}$$

$$= \frac{p(z_0)}{q'(z_0)} \neq 0.$$

Hence, z_0 is a simple pole of $f(z)$. To find $\text{Res}(f, z_0)$, we expand $p(z)$ and $q(z)$ in Taylor series at z_0. Using eq. (6.2.6) and observing that $q(z_0) = 0$, we obtain

$$\text{Res}(f, z_0) = \lim_{z \to z_0} (z - z_0) f(z)$$

$$= \lim_{z \to z_0} (z - z_0) \frac{p(z_0) + p'(z_0)(z - z_0) + \cdots}{q'(z_0)(z - z_0) + \frac{1}{2}q''(z_0)(z - z_0)^2 + \cdots}$$

$$= \frac{p(z_0)}{q'(z_0)}. \tag{6.2.7}$$

Example 6.2.1   Evaluate the integral

$$\oint_{|z|=1} \frac{z + 1}{z^2} \, dz$$

using

   (i) direct line integration,
   (ii) the calculus of residues,
   (iii) the primitive function $\log z - \dfrac{1}{z}$.

Solution

   (i) On the unit circle, $z = e^{i\theta}$ and $dz = ie^{i\theta} \, d\theta$. We then have

$$\oint_{|z|=1} \frac{z + 1}{z^2} \, dz = \int_0^{2\pi} (e^{-i\theta} + e^{-2i\theta}) ie^{i\theta} \, d\theta$$

$$= i \int_0^{2\pi} (1 + e^{-i\theta}) \, d\theta = 2\pi i.$$

(ii) The integrand $\frac{z+1}{z^2}$ has a double pole at $z = 0$. The Laurent series expansion in a deleted neighborhood of $z = 0$ is simply $\frac{1}{z} + \frac{1}{z^2}$, where the coefficient of $\frac{1}{z}$ is seen to be 1. We obtain

$$\text{Res}\left(\frac{z+1}{z^2}, 0\right) = 1,$$

and so

$$\oint_{|z|=1} \frac{z+1}{z^2}\, dz = 2\pi i\, \text{Res}\left(\frac{z+1}{z^2}, 0\right) = 2\pi i.$$

(iii) When a closed contour moves around the origin (which is the branch point of the function $\log z$) in the anticlockwise direction, the increase in the value of $\arg z$ equals 2π. Therefore,

$$\oint_{|z|=1} \frac{z+1}{z^2}\, dz = \text{change in value of } \ln|z| + i \arg z - \frac{1}{z} \text{ in}$$

$$\text{traversing one complete loop around the origin}$$

$$= 2\pi i.$$

Example 6.2.2 Evaluate the following integral

$$\oint_{|z|=2} \frac{\tanh z}{z}\, dz.$$

Solution The isolated singularities of the integrand function $\frac{\tanh z}{z}$ inside the closed contour $|z| = 2$ are $z = 0$, $\frac{i\pi}{2}$ and $-\frac{i\pi}{2}$. The singularity at $z = 0$ is removable since

$$\lim_{z \to 0} \frac{\tanh z}{z} = \lim_{z \to 0} \frac{\sinh z}{z} \lim_{z \to 0} \frac{1}{\cosh z} = 1.$$

By the Cauchy residue theorem, we then have

$$\oint_{|z|=2} \frac{\tanh z}{z}\, dz = 2\pi i \left[\text{Res}\left(\frac{\tanh z}{z}, \frac{i\pi}{2}\right) + \text{Res}\left(\frac{\tanh z}{z}, -\frac{i\pi}{2}\right) \right].$$

Since $z = \frac{i\pi}{2}$ is a simple zero of $\cosh z$, it is a simple pole of $\frac{\tanh z}{z}$; similarly, $z = -\frac{i\pi}{2}$ is a simple pole of $\frac{\tanh z}{z}$. We use two different methods to compute $\text{Res}\left(\frac{\tanh z}{z}, \frac{i\pi}{2}\right)$.

Method one

First, we observe that $\cosh z$ admits the following Taylor series at $z = \frac{i\pi}{2}$:

$$\cosh z = \left.\frac{\sinh z}{1!}\right|_{z=\frac{i\pi}{2}} \left(z - \frac{i\pi}{2}\right) + \left.\frac{\sinh z}{3!}\right|_{z=\frac{i\pi}{2}} \left(z - \frac{i\pi}{2}\right)^3 + \cdots$$

$$= i\left[\left(z - \frac{i\pi}{2}\right) + \frac{1}{3!}\left(z - \frac{i\pi}{2}\right)^3 + \cdots\right], \quad \left|z - \frac{i\pi}{2}\right| < \pi.$$

Since $z = \frac{i\pi}{2}$ is a simple pole of $\tanh z$, using eq. (6.2.6), we obtain

$$\text{Res}\left(\frac{\tanh z}{z}, \frac{i\pi}{2}\right) = \lim_{z \to \frac{i\pi}{2}} \left(z - \frac{i\pi}{2}\right) \frac{\tanh z}{z}$$

$$= \lim_{z \to \frac{i\pi}{2}} \frac{\left(z - \frac{i\pi}{2}\right)}{i\left[\left(z - \frac{i\pi}{2}\right) + \frac{1}{3!}\left(z - \frac{i\pi}{2}\right)^3 + \cdots\right]} \frac{\sinh z}{z}$$

$$= \frac{1}{\frac{i\pi}{2}} = -\frac{2}{\pi}i.$$

Method two

We write $\frac{\tanh z}{z}$ as $\frac{p(z)}{q(z)}$, where $p(z) = \frac{\sinh z}{z}$ and $q(z) = \cosh z$; and observe $p\left(\frac{i\pi}{2}\right) \neq 0, q\left(\frac{i\pi}{2}\right) = 0$ and $q'\left(\frac{i\pi}{2}\right) \neq 0$. Using eq. (6.2.7), we obtain

$$\text{Res}\left(\frac{\tanh z}{z}, \frac{i\pi}{2}\right) = \frac{p\left(\frac{i\pi}{2}\right)}{q'\left(\frac{i\pi}{2}\right)} = \left.\frac{1}{z}\right|_{z=\frac{i\pi}{2}} = -\frac{2}{\pi}i.$$

In a similar manner, we obtain

$$\text{Res}\left(\frac{\tanh z}{z}, -\frac{i\pi}{2}\right) = \left.\frac{1}{z}\right|_{z=-\frac{i\pi}{2}} = \frac{2}{\pi}i.$$

The sum of residues at the two isolated singularities is seen to be zero, so

$$\oint_{|z|=2} \frac{\tanh z}{z} \, dz = 0.$$

Remark Since $\frac{\tanh z}{z}$ is an even function, part (b) of Problem 6.15 reveals that

$$\text{Res}\left(\frac{\tanh z}{z}, \frac{i\pi}{2}\right) = -\text{Res}\left(\frac{\tanh z}{z}, -\frac{i\pi}{2}\right).$$

This is consistent with the result obtained in the above calculations.

Example 6.2.3 Suppose $f(z)$ and $g(z)$ are analytic in some domain containing $z = a$ and $z = a$ is a double zero of $g(z) = 0$ and $f(a) \neq 0$. Show that the residue of $\frac{f(z)}{g(z)}$ at $z = a$ is given by

$$\text{Res}\left(\frac{f(z)}{g(z)}, a\right) = \frac{6f'(a)g''(a) - 2f(a)g'''(a)}{3[g''(a)]^2}.$$

Solution Since $z = a$ is a double zero of $g(z) = 0$, we then have $g(a) = g'(a) = 0$. The Taylor series of $f(z)$ and $g(z)$ expanded in some neighborhood centered at $z = a$ are, respectively,

$$f(z) = f(a) + f'(a)(z - a) + \frac{f''(a)}{2!}(z - a)^2 + \cdots,$$

$$g(z) = \frac{g''(a)}{2!}(z - a)^2 + \frac{g'''(a)}{3!}(z - a)^3 + \frac{g''''(a)}{4!}(z - a)^4 + \cdots.$$

Their quotient can be expressed in the form

$$\frac{f(z)}{g(z)} = \frac{f(a) + f'(a)(z - a) + \frac{f''(a)}{2!}(z - a)^2 + \cdots}{\frac{g''(a)}{2}(z - a)^2\left[1 + \frac{g'''(a)}{3g''(a)}(z - a) + \frac{g''''(a)}{12g''(a)}(z - a)^2 + \cdots\right]}$$

$$= \frac{f(a) + \left[f'(a) - \frac{f(a)g'''(a)}{3g''(a)}\right](z - a) + \cdots}{\frac{g''(a)}{2}(z - a)^2}.$$

From the above expansion, we deduce that the coefficient of $\frac{1}{z-a}$ in the Laurent series expansion of $\frac{f(z)}{g(z)}$ at $z = a$ is given by

$$\frac{f'(a) - \frac{f(a)g'''(a)}{3g''(a)}}{\frac{g''(a)}{2}};$$

therefore,

$$\text{Res}\left(\frac{f(z)}{g(z)}, a\right) = \frac{6f'(a)g''(a) - 2f(a)g'''(a)}{3\,[g''(a)]^2}.$$

Example 6.2.4 Suppose $p_n(z)$ is a polynomial of degree n and α is a point inside the unit circle centered at the origin; show that

$$\oint_{|z|=1} \frac{p_n(z)}{z^{n+1}(z - \alpha)}\, dz = 0.$$

Solution Inside the unit circle $|z| = 1$, the integrand has two poles, namely, $z = 0$ as a pole of order $n + 1$ and $z = \alpha$ as a simple pole. By the Cauchy residue theorem, we have

$$\oint_{|z|=1} \frac{p_n(z)}{z^{n+1}(z-\alpha)} dz$$
$$= 2\pi i \left[\text{Res}\left(\frac{p_n(z)}{z^{n+1}(z-\alpha)}, 0 \right) + \text{Res}\left(\frac{p_n(z)}{z^{n+1}(z-\alpha)}, \alpha \right) \right].$$

To evaluate the residue at $z = 0$, it is necessary to expand the integrand in a Laurent series inside a deleted neighborhood around $z = 0$. Suppose we write

$$p_n(z) = a_0 + a_1 z + \cdots + a_n z^n.$$

The Laurent series expansion of the integrand takes the form

$$\frac{p_n(z)}{z^{n+1}(z-\alpha)}$$
$$= \left(\frac{a_0 + a_1 z + \cdots + a_n z^n}{z^{n+1}} \right) \left(-\frac{1}{\alpha} \right) \left(1 + \frac{z}{\alpha} + \cdots + \frac{z^n}{\alpha^n} + \cdots \right),$$

which is valid inside the annular domain $0 < |z| < |\alpha|$. The residue at $z = 0$ is given by

$$\text{Res}\left(\frac{p_n(z)}{z^{n+1}(z-\alpha)}, 0 \right) = \text{coefficient of } \frac{1}{z} \text{ in the above Laurent series}$$
$$= \left(-\frac{1}{\alpha} \right) \left(\frac{a_0}{\alpha^n} + \frac{a_1}{\alpha^{n-1}} + \cdots + a_n \right)$$
$$= -\frac{1}{\alpha^{n+1}} (a_0 + a_1 \alpha + \cdots + a_n \alpha^n)$$
$$= -\frac{1}{\alpha^{n+1}} p_n(\alpha).$$

Since $z = \alpha$ is a simple pole, the residue at α is found to be [see eq. (6.2.6)]

$$\text{Res}\left(\frac{p_n(z)}{z^{n+1}(z-\alpha)}, \alpha \right) = \lim_{z \to \alpha} \frac{p_n(z)}{z^{n+1}} = \frac{p_n(\alpha)}{\alpha^{n+1}}.$$

Combining the above results, we obtain

$$\oint_{|z|=1} \frac{p_n(z)}{z^{n+1}(z-\alpha)} dz = 2\pi i \left[-\frac{p_n(\alpha)}{\alpha^{n+1}} + \frac{p_n(\alpha)}{\alpha^{n+1}} \right] = 0.$$

Example 6.2.5　Suppose α is a simple pole of a meromorphic function $f(z)$. Show that the line integral

$$\int_\Gamma f(z) \, dz$$

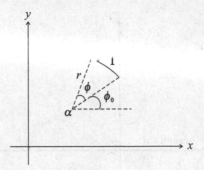

Figure 6.2. The circular arc Γ subtends an angle ϕ around the center α with an infinitesimal radius r.

equals $i\phi \operatorname{Res}(f(z), \alpha)$, where Γ is a circular arc subtended at an angle ϕ around the point α with an infinitesimal radius.

Solution Since $f(z)$ has a simple pole at $z = \alpha$; in a deleted neighborhood of α, $f(z)$ admits a Laurent series expansion of the form

$$f(z) = \frac{a_{-1}}{z - \alpha} + \sum_{n=0}^{\infty} a_n (z - \alpha)^n.$$

On the circular arc around α, $z = \alpha + re^{i\theta}$, $\phi_0 \leq \theta \leq \phi_0 + \phi$, where r is the radius of the arc (see Figure 6.2). The contour integral becomes

$$\int_{\Gamma} f(z)\, dz = \int_{\phi_0}^{\phi_0 + \phi} f(\alpha + re^{i\theta}) i r e^{i\theta}\, d\theta$$

$$= \int_{\phi_0}^{\phi_0 + \phi} \left(\frac{a_{-1}}{re^{i\theta}} + \sum_{n=0}^{\infty} a_n r^n e^{in\theta} \right) i r e^{i\theta}\, d\theta.$$

On taking the limit $r \to 0$, we obtain

$$\lim_{r \to 0} \int_{\Gamma} f(z)\, dz = \int_{\phi_0}^{\phi_0 + \phi} a_{-1} i\, d\theta = i\phi a_{-1} = i\phi \operatorname{Res}(f(z), \alpha).$$

Example 6.2.6 Find the residue of $f(z) = \frac{z^\alpha}{(z-i)^2}$, $0 < \alpha < 1$, at each of its isolated singularities, where the principal branch of the power function z^α is chosen.

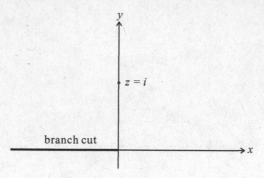

Figure 6.3. The function $f(z) = \frac{z^\alpha}{(z-i)^2}$ has only one isolated singularity at $z = i$. The branch cut of $f(z)$ is the whole negative real axis, including the origin. All the points along the branch cut of z^α (including the end point $z = 0$) are non-isolated singularities.

Solution   The only isolated singularity of

$$f(z) = \frac{z^\alpha}{(z-i)^2}, \quad 0 < \alpha < 1,$$

is at $z = i$; all the points along the branch cut of z^α (the whole negative real axis, including $z = 0$) are non-isolated singularities (see Figure 6.3). The principal branch of z^α is taken to be $e^{\alpha \mathrm{Log}\, z}$, where $\mathrm{Log}\, z$ is the principal branch of the logarithm function. Obviously, $z = i$ is a pole of order 2 since

$$\lim_{z \to i}(z - i)^2 \frac{z^\alpha}{(z-i)^2} = \lim_{z \to i} e^{\alpha \mathrm{Log}\, z} = e^{i\alpha\pi/2},$$

which is a finite nonzero value. The residue of f at $z = i$ is then given by

$$\mathrm{Res}\left(\frac{z^\alpha}{(z-i)^2}, i\right)$$

$$= \lim_{z \to i} \frac{d}{dz} z^\alpha = \lim_{z \to i} \frac{\alpha}{z} e^{\alpha \mathrm{Log}\, z} = \alpha e^{i(\alpha-1)\pi/2}, \quad 0 < \alpha < 1.$$

Example 6.2.7   Let $f(z)$ be analytic in a domain containing the whole real axis, and let $z_n = \left(n + \frac{1}{2}\right)\pi$, n is any integer. Show that

$$\mathrm{Res}\left(\frac{f(z)}{\cos^2 z}, z_n\right) = f'(z_n), \quad n \text{ is any integer.}$$

Solution   First, we observe that $z_n = \left(n + \frac{1}{2}\right)\pi$ is a double pole of $\frac{1}{\cos^2 z}$, for any integer n. In some deleted ϵ-neighborhood around the point

$z_n = \left(n + \frac{1}{2}\right)\pi$, $\frac{1}{\cos^2 z}$ admits the following Laurent series expansion:

$$\frac{1}{\cos^2 z} - \frac{c_{-2}}{(z - z_n)^2} + \frac{c_{-1}}{z - z_n} + \sum_{k=0}^{\infty} c_k (z - z_n)^k \qquad 0 < |z - z_n| < \epsilon.$$

On the other hand, since $f(z)$ is analytic in a domain containing the whole real axis, $f(z)$ admits the following Taylor series expansion inside the ϵ-neighborhood centered at $z = z_n$:

$$f(z) = \sum_{m=0}^{\infty} a_m (z - z_n)^m, \quad |z - z_n| < \epsilon.$$

Now, we consider the Laurent series expansion of

$$\frac{f(z)}{\cos^2 z} = \left[\sum_{m=0}^{\infty} a_m (z - z_n)^m\right]\left[\frac{c_{-2}}{(z - z_n)^2} + \frac{c_{-1}}{z - z_n} + \sum_{k=0}^{\infty} c_k (z - z_n)^k\right].$$

Since $\mathrm{Res}\left(\frac{f(z)}{\cos^2 z}, z_n\right)$ is equal to the coefficient of $\frac{1}{z - z_n}$ in the above Laurent series expansion, we obtain

$$\mathrm{Res}\left(\frac{f(z)}{\cos^2 z}, z_n\right) = a_1 c_{-2} + a_0 c_{-1},$$

where $a_0 = f(z_n)$ and $a_1 = f'(z_n)$. The Taylor series of $\cos^2 z$ at $z = z_n$ contains only even power terms, so the Laurent series of $\frac{1}{\cos^2 z}$ at $z = z_n$ has only even power terms. This gives $c_{-1} = 0$. Lastly, we compute c_{-2} by evaluating the following limit using the L'Hospital rule, where

$$c_{-2} = \lim_{z \to z_n} \frac{(z - z_n)^2}{\cos^2 z} = \lim_{z \to z_n} \frac{2(z - z_n)}{-2\sin z \cos z} = \lim_{z \to z_n} \frac{2}{-2\cos 2z} = 1.$$

Therefore, we have

$$\mathrm{Res}\left(\frac{f(z)}{\cos^2 z}, z_n\right) = a_1 = f'(z_n).$$

Example 6.2.8 Suppose $f(z)$ is analytic everywhere in the finite complex plane except for a finite number of isolated singularities interior to a positively oriented simple closed contour C. Show that

$$\oint_C f(z)\,dz = 2\pi i\, \mathrm{Res}\left(\frac{1}{z^2} f\left(\frac{1}{z}\right), 0\right).$$

Use the result to evaluate

$$\oint_{|z|=3} \frac{6z + 4}{z^3 + 3z^2 + 2z}\,dz.$$

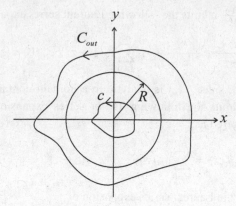

Figure 6.4. The function $f(z)$ is analytic everywhere outside the simple closed contour C. The contour C lies completely inside the circle $|z| = R$. The exterior simple closed contour C_{out} encloses the circle $|z| = R$.

Solution   Given the simple closed contour C that encircles all isolated singularities of $f(z)$, we choose a circle $|z| = R$ such that the contour C lies completely inside the circle. An exterior simple closed contour C_{out} is chosen such that it encloses the circle $|z| = R$ (see Figure 6.4).

Let us consider the Laurent series expansion of $f(z)$ in the annulus $R < |z| < \infty$, where

$$f(z) = \sum_{n=-\infty}^{\infty} c_n z_n, \quad R < |z| < \infty.$$

From the Laurent series theorem, the Laurent coefficients are given by

$$c_n = \frac{1}{2\pi i} \oint_{C_{out}} \frac{f(z)}{z^{n+1}} \, dz, \quad n = 0, \pm 1, \pm 2, \dots.$$

In particular, we are interested in computing

$$c_{-1} = \frac{1}{2\pi i} \oint_{C_{out}} f(z) \, dz.$$

The above annulus $R < |z| < \infty$ is not a deleted neighborhood about $z = 0$, so c_{-1} is not equal to $\mathrm{Res}\,(f, 0)$. Suppose we replace z by $\frac{1}{z}$ in the above Laurent series expansion; we obtain

$$\frac{1}{z^2} f\left(\frac{1}{z}\right) = \sum_{n=-\infty}^{\infty} \frac{c_n}{z^{n+2}} = \sum_{n=-\infty}^{\infty} \frac{c_{n-2}}{z^n},$$

which is valid in the annulus $0 < |z| < \frac{1}{R}$. This new annulus is now a deleted neighborhood about $z = 0$. We then deduce that

$$\text{Res}\left(\frac{1}{z^2} f\left(\frac{1}{z}\right), 0\right) = c_{-1} = \frac{1}{2\pi i} \oint_{C_{out}} f(z)\,dz.$$

Since there is no singularity lying between C_{out} and C, we can deform the exterior contour C_{out} to C, so we obtain

$$\oint_C f(z)\,dz = 2\pi i \,\text{Res}\cdot\left(\frac{1}{z^2} f\left(\frac{1}{z}\right), 0\right).$$

The integrand function

$$f(z) = \frac{6z + 4}{z^3 + 3z^2 + z}$$

has isolated singularities at $z = 0, z = -1$ and $z = -2$ inside the contour $|z| = 3$. By the Cauchy residue theorem, we have

$$\oint_{|z|=3} f(z)\,dz = 2\pi i[\text{Res}\,(f, 0) + \text{Res}\,(f, -1) + \text{Res}\,(f, -2)].$$

Instead of evaluating the above three residue values, a more efficient evaluation method can be derived by computing $\text{Res}\left(\frac{1}{z^2} f\left(\frac{1}{z}\right), 0\right)$. Now, we consider

$$\frac{1}{z^2} f\left(\frac{1}{z}\right) = \frac{1}{z^2} \frac{\frac{6}{z} + 4}{\frac{1}{z^3} + \frac{3}{z^2} + \frac{1}{z}} = \frac{4z + 6}{2z^2 + 3z + 1},$$

which is seen to be analytic at $z = 0$, so

$$\oint_{|z|=3} f(z)\,dz = 2\pi i \,\text{Res}\left(\frac{1}{z^2} f\left(\frac{1}{z}\right), 0\right) = 0.$$

Remark   By convention, we define the residue at complex infinity by

$$\text{Res}\,(f, \infty) = -\frac{1}{2\pi i} \oint_C f(z)\,dz = -\text{Res}\left(\frac{1}{z^2} f\left(\frac{1}{z}\right), 0\right),$$

where the contour C is chosen such that all singularities of $f(z)$ in the finite plane are included inside C. The negative sign is deliberately chosen in the above definition so that

$$\sum_{all} \text{Res}(f, z_n) + \text{Res}\,(f, \infty) = 0,$$

where the summation is taken over all singularities z_n inside C. That is, if $f(z)$ is analytic in the extended complex plane except for a finite number of singularities, then the sum of all residues of $f(z)$, including the residue at complex infinity, is zero.

6.3 Evaluation of real integrals by residue calculus

A wide variety of real definite integrals can be evaluated effectively by the calculus of residues. The techniques of evaluation can be split into different categories as discussed below.

6.3.1 Integrals of trigonometric functions over $[0, 2\pi]$

We consider a real integral involving trigonometric functions of the form

$$\int_0^{2\pi} R(\cos\theta, \sin\theta)\, d\theta,$$

where $R(x, y)$ is a rational function defined inside the unit circle $|z| = 1$, $z = x + iy$. The real integral can be converted into a contour integral around the unit circle by the following substitutions:

$$z = e^{i\theta},\ dz = ie^{i\theta}d\theta = iz\, d\theta,$$

$$\cos\theta = \frac{e^{i\theta} + e^{-i\theta}}{2} = \frac{1}{2}\left(z + \frac{1}{z}\right),$$

$$\sin\theta = \frac{e^{i\theta} - e^{-i\theta}}{2i} = \frac{1}{2i}\left(z - \frac{1}{z}\right).$$

The real integral can then be transformed:

$$\int_0^{2\pi} R(\cos\theta, \sin\theta)\, d\theta$$

$$= \oint_{|z|=1} \frac{1}{iz} R\left(\frac{z + \frac{1}{z}}{2}, \frac{z - \frac{1}{z}}{2i}\right) dz$$

$$= 2\pi i\left[\text{sum of residues of } \frac{1}{iz} R\left(\frac{z + \frac{1}{z}}{2}, \frac{z - \frac{1}{z}}{2i}\right) \text{inside} |z| = 1\right]. \quad (6.3.1)$$

Example 6.3.1   Evaluate the real definite integral

$$I = \int_0^{\pi} \frac{1}{a - b\cos\theta}\, d\theta,\ a > b > 0.$$

Solution Since the integrand is symmetric about $\theta = \pi$, it is desirable to extend the integration interval to $[0, 2\pi]$ so that

$$I = \frac{1}{2} \int_0^{2\pi} \frac{1}{a - b\cos\theta} \, d\theta = \int_0^{2\pi} \frac{e^{i\theta}}{2ae^{i\theta} - b(e^{2i\theta} + 1)} \, d\theta.$$

We let $z = e^{i\theta}$ and as θ increases from 0 to 2π, z moves around the unit circle $|z| = 1$. By eq. (6.3.1), the real integral can be transformed into the following contour integral

$$I = -\frac{1}{i} \oint_{|z|=1} \frac{1}{bz^2 - 2az + b} \, dz,$$

where the path of integration is the unit circle $|z| = 1$. The integrand has two simple poles, which are given by the zeros of the denominator. The product of the poles is seen to be 1. Let α denote the pole that is inside the unit circle; then the other pole will be $\frac{1}{\alpha}$. The two poles are found to be

$$\alpha = \frac{a - \sqrt{a^2 - b^2}}{b} \quad \text{and} \quad \frac{1}{\alpha} = \frac{a + \sqrt{a^2 - b^2}}{b}.$$

Since $a > b > 0$, the two poles are distinct. We then have

$$I = -\frac{1}{ib} \oint_{|z|=1} \frac{1}{(z - \alpha)\left(z - \frac{1}{\alpha}\right)} \, dz$$

$$= -\frac{2\pi i}{ib} \operatorname{Res}\left(\frac{1}{(z - \alpha)\left(z - \frac{1}{\alpha}\right)}, \alpha\right)$$

$$= -\frac{2\pi i}{ib\left(\alpha - \frac{1}{\alpha}\right)} = \frac{\pi}{\sqrt{a^2 - b^2}}.$$

6.3.2 Integrals of rational functions

We consider a real integral involving a rational function of the form

$$\int_{-\infty}^{\infty} f(x) \, dx,$$

where

(i) $f(z)$ is a rational function with no singularity on the real axis,
(ii) $\lim_{z \to \infty} zf(z) = 0$.

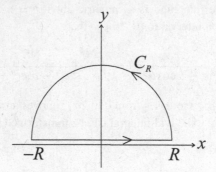

Figure 6.5. The contour of integration is the union of the upper semi-circle C_R and the line segment from $-R$ to R along the real axis.

It can be shown that

$$\int_{-\infty}^{\infty} f(x)\,dx$$

$$= 2\pi i \ [\text{sum of residues at the poles of } f \text{ in the upper half-plane}]. \quad (6.3.2)$$

To verify the claim, we integrate $f(z)$ around a closed contour C that consists of the upper semi-circle C_R and the line segment from $-R$ to R along the real axis (see Figure 6.5). By the Cauchy residue theorem, we have

$$\oint_C f(z)\,dz = \int_{-R}^{R} f(x)\,dx + \int_{C_R} f(z)\,dz$$

$$= 2\pi i \ [\text{sum of residues at the poles of } f \text{ inside } C].$$

As $R \to \infty$, all the poles of f in the upper half-plane will be enclosed inside C. To establish eq. (6.3.2), it suffices to show that as $R \to \infty$,

$$\lim_{R\to\infty} \int_{C_R} f(z)\,dz = 0. \qquad (6.3.3)$$

The modulus of the above integral is estimated by the modulus inequality (4.1.6) as follows:

$$\left| \int_{C_R} f(z)\,dz \right| \leq \int_0^\pi |f(Re^{i\theta})|\, R\, d\theta$$

$$\leq \max_{0\leq\theta\leq\pi} |f(Re^{i\theta})|\, R \int_0^\pi d\theta$$

$$= \max_{z\in C_R} |zf(z)|\pi,$$

which goes to zero as $R \to \infty$, since $\lim_{z\to\infty} zf(z) = 0$. Hence, the result in eq. (6.3.3) is established.

Example 6.3.2 Evaluate the real integral

$$\int_{-\infty}^{\infty} \frac{x^4}{1 + x^6} \, dx$$

by the residue method.

Solution The complex function

$$f(z) = \frac{z^4}{1 + z^6}$$

has simple poles at i, $\frac{\sqrt{3}+i}{2}$ and $\frac{-\sqrt{3}+i}{2}$ in the upper half-plane, and it has no singularity on the real axis. Furthermore, the integrand observes the property $\lim_{z \to \infty} zf(z) = 0$. By eq. (6.3.2), we have

$$\int_{-\infty}^{\infty} f(x) \, dx = 2\pi i \left[\text{Res}(f, i) + \text{Res}\left(f, \frac{\sqrt{3}+i}{2} \right) + \text{Res}\left(f, \frac{-\sqrt{3}+i}{2} \right) \right].$$

By eq. (6.2.7), the residue values at the simple poles are found to be

$$\text{Res}(f, i) = \frac{z^4}{\frac{d}{dz}(1 + z^6)} \bigg|_{z=i} = \frac{1}{6z} \bigg|_{z=i} = -\frac{i}{6},$$

$$\text{Res}\left(f, \frac{\sqrt{3}+i}{2} \right) = \frac{1}{6z} \bigg|_{z=\frac{\sqrt{3}+i}{2}} = \frac{\sqrt{3}-i}{12},$$

and

$$\text{Res}\left(f, \frac{-\sqrt{3}+i}{2} \right) = \frac{1}{6z} \bigg|_{z=\frac{-\sqrt{3}+i}{2}} = -\frac{\sqrt{3}+i}{12}.$$

Combining the results, we obtain

$$\int_{-\infty}^{\infty} \frac{x^4}{1 + x^6} \, dx = 2\pi i \left(-\frac{i}{6} + \frac{\sqrt{3}-i}{12} - \frac{\sqrt{3}+i}{12} \right) = \frac{2\pi}{3}.$$

6.3.3 Integrals involving multi-valued functions

We would like to evaluate a real integral involving a fractional power function of the form

$$\int_0^{\infty} \frac{f(x)}{x^\alpha} \, dx, \qquad \text{where } 0 < \alpha < 1.$$

The required properties of $f(z)$ are:

1. $f(z)$ is a rational function with no singularity on the positive real axis, including the origin;
2. $\lim\limits_{z \to \infty} f(z) = 0$.

Remark   We may relax the first requirement to allow the function $f(z)$ to have a finite number of simple poles along the positive real axis. In this case, the Cauchy principal value of the integral should be considered (see Section 6.5).

We integrate $\phi(z) = \frac{f(z)}{z^\alpha}$ along the closed contour C as shown in Figure 6.6. The contour C consists of four parts:

(i) the line segment from ϵ to R along the upper side of the positive real axis, that is, $z = x, \epsilon \le x \le R$;

(ii) the large circle C_R traversing in the anticlockwise direction, that is, $z = Re^{i\theta}, 0 < \theta < 2\pi$;

(iii) the line segment from R to ϵ along the lower side of the positive real axis, that is, $z = xe^{2\pi i}, \epsilon \le x \le R$;

(iv) the infinitesimal circle C_ϵ traversing in the clockwise direction, that is, $z = \epsilon e^{i\theta}, 0 < \theta < 2\pi$.

Subsequently, we take the limits $R \to \infty$ and $\epsilon \to 0$. We first establish the following two results:

$$\lim_{R \to \infty} \int_{C_R} \phi(z) = 0 \qquad \text{and} \qquad \lim_{\epsilon \to 0} \int_{C_\epsilon} \phi(z) = 0. \qquad (6.3.4)$$

The bounds on the moduli of the above two integrals are found to be

$$\left| \int_{C_R} \phi(z)\, dz \right| \le \int_0^{2\pi} |\phi(Re^{i\theta})|\, R\, d\theta \le 2\pi \max_{z \in C_R} |z\phi(z)|,$$

and

$$\left| \int_{C_\epsilon} \phi(z)\, dz \right| \le \int_0^{2\pi} |\phi(\epsilon e^{i\theta})|\, \epsilon\, d\theta \le 2\pi \max_{z \in C_\epsilon} |z\phi(z)|.$$

It suffices to show that

$$z\phi(z) \to 0 \quad \text{as either} \quad z \to \infty \quad \text{or} \quad z \to 0.$$

As $\lim\limits_{z \to \infty} f(z) = 0$ and $f(z)$ is a rational function, the degree of the denominator of $f(z)$ is at least one higher than that of its numerator. Since $1 - \alpha < 1$, we have $z\phi(z) = z^{1-\alpha} f(z) \to 0$ as $z \to \infty$. Also, we deduce that $f(z)$ is

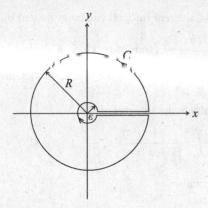

Figure 6.6. The closed contour C consists of an infinitely large circle and an infinitesimal circle joined by line segments along the positive x-axis.

continuous at $z = 0$ since $f(z)$ has no singularity at the origin. Therefore, $z\phi(z) = z^{1-\alpha} f(z) \sim 0 \cdot f(0) = 0$ as $z \to 0$.

Lastly, we observe that the term z^α in the integrand is multi-valued since

$$z^\alpha = e^{\alpha \log z} = e^{\alpha(\ln |z| + i\theta)} = |z|^\alpha e^{i\alpha\theta}, \text{ where } \theta = \arg z.$$

The choice of the contour of integration as shown in Figure 6.6 dictates that the argument of the principal branch of z^α is chosen to be $0 \le \theta < 2\pi$. Note that the origin is a branch point and the positive real axis is chosen to be the branch cut.

Now the contour integral around C can be expressed as the sum of four contour integrals:

$$\oint_C \phi(z) \, dz = \int_{C_R} \phi(z) \, dz + \int_{C_\epsilon} \phi(z) \, dz$$
$$+ \int_\epsilon^R \frac{f(x)}{x^\alpha} \, dx + \int_R^\epsilon \frac{f(xe^{2\pi i})}{x^\alpha e^{2\alpha\pi i}} \, dx$$
$$= 2\pi i \, [\text{sum of residues of all the isolated singularities}$$
$$\text{of } f \text{ enclosed inside the closed contour } C]. \quad (6.3.5)$$

By taking the limits $\epsilon \to 0$ and $R \to \infty$, the first two integrals vanish. The last integral can be expressed as

$$-\int_0^\infty \frac{f(x)}{x^\alpha e^{2\alpha\pi i}} \, dx = -e^{-2\alpha\pi i} \int_0^\infty \frac{f(x)}{x^\alpha} \, dx.$$

Combining the results, the real integral can be evaluated by the formula:

$$\int_0^\infty \frac{f(x)}{x^\alpha}\,dx = \frac{2\pi i}{1 - e^{-2\alpha\pi i}} \text{ [sum of residues at all the isolated singularities}$$

$$\text{of } f \text{ in the finite complex plane].} \qquad (6.3.6)$$

Example 6.3.3 Evaluate the integral

$$\int_0^\infty \frac{1}{(1+x)x^\alpha}\,dx, \qquad 0 < \alpha < 1.$$

Solution The integrand

$$f(z) = \frac{1}{(1+z)z^\alpha}$$

is multi-valued and has an isolated singularity (a simple pole) at $z = -1$. We choose the branch cut to be along the positive real axis with the origin as a branch point. We now consider the contour integral

$$\oint_C \frac{1}{(1+z)z^\alpha}\,dz$$

around the closed contour C as shown in Figure 6.6.

The contour integrals along the upper line segment and the lower line segment along the positive real axis can be expressed as

$$\int_\epsilon^R \frac{1}{(1+x)x^\alpha}\,dx \quad \text{and} \quad -e^{-2\alpha\pi i}\int_\epsilon^R \frac{1}{(1+x)x^\alpha}\,dx,$$

respectively. By the Cauchy residue theorem, the contour integral is found to be

$$\oint_C \frac{1}{(1+z)z^\alpha}\,dz$$

$$= (1 - e^{-2\alpha\pi i})\int_\epsilon^R \frac{dx}{(1+x)x^\alpha} + \int_{C_R} \frac{dz}{(1+z)z^\alpha} + \int_{C_\epsilon} \frac{dz}{(1+z)z^\alpha}$$

$$= 2\pi i \operatorname{Res}\left(\frac{1}{(1+z)z^\alpha}, -1\right) = \frac{2\pi i}{e^{\alpha\pi i}},$$

where C_R and C_ϵ denote the outer large circle and the inner small circle, respectively. Given $0 < \alpha < 1$, the moduli of the second and third integrals can be estimated as follows:

$$\left|\int_{C_R} \frac{1}{(1+z)z^\alpha}\,dz\right| \leq \frac{2\pi R}{(R-1)R^\alpha} \sim R^{-\alpha} \to 0 \qquad \text{as} \quad R \to \infty;$$

$$\left|\int_{C_\epsilon} \frac{1}{(1+z)z^\alpha}\,dz\right| \leq \frac{2\pi\epsilon}{(1-\epsilon)\epsilon^\alpha} \sim \epsilon^{1-\alpha} \to 0 \qquad \text{as} \quad \epsilon \to 0.$$

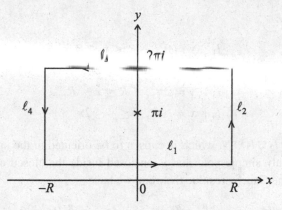

Figure 6.7. The closed rectangular contour encloses only one simple pole at $z = \pi i$.

On taking the limits $R \to \infty$ and $\epsilon \to 0$, we obtain

$$(1 - e^{-2\alpha\pi i}) \int_0^\infty \frac{1}{(1+x)x^\alpha}\, dx = \frac{2\pi i}{e^{\alpha\pi i}};$$

so

$$\int_0^\infty \frac{1}{(1+x)x^\alpha}\, dx = \frac{2\pi i}{e^{\alpha\pi i}\,(1 - e^{-2\alpha\pi i})} = \frac{\pi}{\sin\alpha\pi}.$$

6.3.4 Miscellaneous types of integral

The success of applying the residue calculus techniques to evaluation of different types of real integral relies on the ingenious choice of the contour, possibly together with an appropriate modification of the integrand. Additional examples are presented below to illustrate the high level of ingenuity exhibited in the residue calculus approach.

Example 6.3.4 Evaluate the real integral

$$\int_{-\infty}^\infty \frac{e^{\alpha x}}{1 + e^x}\, dx, \quad 0 < \alpha < 1.$$

Solution The integrand function in its complex extension has infinitely many poles in the complex plane, namely, at $z = (2k + 1)\pi i$, where k is any integer. Therefore, it is not advisable to choose a contour that includes an infinitely large outer circle. Instead, we choose the rectangular contour as depicted in Figure 6.7. The contour consists of four line segments:

$$l_1 : y = 0, \ -R \le x \le R,$$
$$l_2 : x = R, \ 0 \le y \le 2\pi,$$
$$l_3 : y = 2\pi, \ -R \le x \le R,$$
$$l_4 : x = -R, \ 0 \le y \le 2\pi.$$

Let $C = l_1 \cup l_2 \cup l_3 \cup l_4$, which is chosen to be oriented in the anticlockwise sense. The only simple pole that is enclosed inside the closed contour C is $z = \pi i$. By the Cauchy residue theorem, we have

$$\oint_C \frac{e^{\alpha z}}{1 + e^z} \, dz = \int_{-R}^{R} \frac{e^{\alpha x}}{1 + e^x} \, dx + \int_0^{2\pi} \frac{e^{\alpha(R+iy)}}{1 + e^{R+iy}} \, i\, dy$$
$$+ \int_R^{-R} \frac{e^{\alpha(x+2\pi i)}}{1 + e^{x+2\pi i}} \, dx + \int_{2\pi}^0 \frac{e^{\alpha(-R+iy)}}{1 + e^{-R+iy}} \, i\, dy$$
$$= 2\pi i \operatorname{Res} \left(\frac{e^{\alpha z}}{1 + e^z}, \pi i \right)$$
$$= 2\pi i \left. \frac{e^{\alpha z}}{e^z} \right|_{z=\pi i} = -2\pi i e^{\alpha \pi i}.$$

The above residue is evaluated using the formula in eq. (6.2.7). We estimate the bounds on the moduli of the second and the fourth integrals as follows:

$$\left| \int_0^{2\pi} \frac{e^{\alpha(R+iy)}}{1 + e^{R+iy}} \, i\, dy \right| \le \int_0^{2\pi} \frac{e^{\alpha R}}{e^R - 1} \, dy \sim O(e^{-(1-\alpha)R}),$$

$$\left| \int_{2\pi}^0 \frac{e^{\alpha(-R+iy)}}{1 + e^{-R+iy}} \, i\, dy \right| \le \int_0^{2\pi} \frac{e^{-\alpha R}}{1 - e^{-R}} \, dy \sim O(e^{-\alpha R}).$$

As $0 < \alpha < 1$, both $e^{-(1-\alpha)R}$ and $e^{-\alpha R}$ tend to zero as $R \to \infty$. Therefore, the second and the fourth integrals tend to zero as $R \to \infty$. Taking the limit $R \to \infty$, the sum of the first and the third integrals becomes

$$(1 - e^{2\alpha \pi i}) \int_{-\infty}^{\infty} \frac{e^{\alpha x}}{1 + e^x} \, dx = -2\pi i e^{\alpha \pi i};$$

so

$$\int_{-\infty}^{\infty} \frac{e^{\alpha x}}{1 + e^x} \, dx = \frac{2\pi i}{e^{\alpha \pi i} - e^{-\alpha \pi i}} = \frac{\pi}{\sin \alpha \pi}.$$

Example 6.3.5   Evaluate the real integral

$$\int_0^\infty \frac{1}{1+x^3}\,dx.$$

Solution   If the interval of integration is $(-\infty, \infty)$ instead of $(0, \infty)$, then the corresponding integral can be evaluated using the technique discussed in Subsection 6.3.2. Here, the integrand is not an even function, so it serves no purpose to extend the interval of integration to $(-\infty, \infty)$. Alternatively, we consider the branch cut integral

$$\oint_C \frac{\log z}{1+z^3}\,dz,$$

where C is the closed contour shown in Figure 6.6. The inner small circle C_ϵ has radius ϵ, $\epsilon \to 0$, and the outer large circle C_R has radius R, $R \to \infty$. The positive real axis is taken to be the branch cut of $\log z$ so that $0 \le \arg z < 2\pi$. From the relation $\log z = \ln|z| + i \arg z$, we have

$$\log z = \ln|z| \qquad \text{along the upper side of the positive } x\text{-axis,}$$
$$\log z = \ln|z| + 2\pi i \quad \text{along the lower side of the positive } x\text{-axis.}$$

The asymptotic order of the modulus of the contour integral around the inner small circle C_ϵ is estimated to be

$$\left| \oint_{C_\epsilon} \frac{\log z}{1+z^3}\,dz \right| = O(2\pi\epsilon \cdot \ln \epsilon) \to 0 \text{ as } \epsilon \to 0.$$

Therefore, the contribution from the contour integral around C_ϵ is zero. Similarly, the contribution from the contour integral around the outer large circle C_R is also zero since

$$\left| \oint_{C_R} \frac{\log z}{1+z^3}\,dz \right| = O\left(2\pi R \cdot \frac{\ln R}{R^3}\right) \to 0 \text{ as } R \to \infty.$$

Taking the limits $\epsilon \to 0$ and $R \to \infty$, the contour integral around C becomes

$$\lim_{\substack{R \to \infty \\ \epsilon \to 0}} \oint_C \frac{\log z}{1+z^3}\,dz = \int_0^\infty \frac{\ln x}{1+x^3}\,dx + \int_\infty^0 \frac{\ln x + 2\pi i}{1+x^3}\,dx$$

$$= -2\pi i \int_0^\infty \frac{1}{1+x^3}\,dx$$

$$= 2\pi i \left[\text{sum of residues of } \frac{\log z}{1+z^3} \text{ at the} \right.$$

$$\left. \text{poles in the finite complex plane} \right].$$

Interestingly, the two terms with $\ln x$ cancel. The integrand $\frac{\log z}{1+z^3}$ has simple poles at $\alpha = e^{i\pi/3}$, $\beta = e^{i\pi}$ and $\gamma = e^{5\pi i/3}$. The sum of the residues at the three poles is found to be

$$\text{Res}\left(\frac{\log z}{1+z^3}, \alpha\right) + \text{Res}\left(\frac{\log z}{1+z^3}, \beta\right) + \text{Res}\left(\frac{\log z}{1+z^3}, \gamma\right)$$

$$= \frac{\log \alpha}{(\alpha - \beta)(\alpha - \gamma)} + \frac{\log \beta}{(\beta - \alpha)(\beta - \gamma)} + \frac{\log \gamma}{(\gamma - \alpha)(\gamma - \beta)}$$

$$= -i \frac{\left[\frac{\pi}{3}(\beta - \gamma) + \pi(\gamma - \alpha) + \frac{5\pi}{3}(\alpha - \beta)\right]}{(\alpha - \beta)(\beta - \gamma)(\gamma - \alpha)} = -\frac{2\pi}{3\sqrt{3}}.$$

Combining the results, we obtain

$$\int_0^\infty \frac{1}{1+x^3}\, dx = \frac{2\pi}{3\sqrt{3}}.$$

6.4 Fourier transforms

A real function is said to be *piecewise continuous* in an interval if it is continuous everywhere in the interval except at a finite number of points where the function is allowed to have jump discontinuities. Let $u(x)$ be a piecewise continuous real function over $(-\infty, \infty)$ satisfying the condition

$$\int_{-\infty}^\infty |u(x)|\, dx < \infty.$$

The Fourier transform of $u(x)$ is defined by

$$\mathcal{F}\{u(x)\} = U(\omega) = \int_{-\infty}^\infty e^{i\omega x} u(x)\, dx, \quad -\infty < \omega < \infty. \quad (6.4.1)$$

The following properties of Fourier transforms can be deduced immediately from the basic rules in calculus:

(i) $\mathcal{F}\{\alpha_1 u_1(x) + \alpha_2 u_2(x)\} = \alpha_1 \mathcal{F}\{u_1(x)\} + \alpha_2 \mathcal{F}\{u_2(x)\}$,
　　　where α_1 and α_2 are complex constants;
(ii) $\mathcal{F}\{u(x - a)\} = e^{ia\omega} \mathcal{F}\{u(x)\}$, a is real;
(iii) $\mathcal{F}\{u(bx)\} = \frac{1}{b} U\left(\frac{\omega}{b}\right)$;
(iv) $\mathcal{F}\{e^{-i\omega_0 x} u(x)\} = U(\omega - \omega_0)$;
(v) $\frac{d}{d\omega} \mathcal{F}\{u(x)\} = i\omega \mathcal{F}\{u(x)\}$;
(vi) $\mathcal{F}\{u'(x)\} = -i\omega \mathcal{F}\{u(x)\}$.

Convolution product

The convolution product of two functions $f(x)$ and $g(x)$ is defined by

$$(f * g)(x) = \int_{-\infty}^{\infty} f(\xi) g(x - \xi) \, d\xi \tag{6.4.2}$$

The Fourier transform of $f * g$ is given by

$$\mathcal{F}\{f * g\} = \int_{-\infty}^{\infty} e^{i\omega x} (f * g)(x) \, dx$$

$$= \int_{-\infty}^{\infty} e^{i\omega x} \int_{-\infty}^{\infty} f(\xi) g(x - \xi) \, d\xi dx. \tag{6.4.3}$$

Suppose f and g are absolutely integrable; we apply the interchange of the order of integration to give

$$\mathcal{F}\{f * g\} = \int_{-\infty}^{\infty} e^{i\omega(x-\xi)} g(x - \xi) \, dx \int_{-\infty}^{\infty} e^{i\omega\xi} f(\xi) \, d\xi.$$

The first and second integrals are recognized as $\mathcal{F}\{g(x)\}$ and $\mathcal{F}\{f(x)\}$, respectively, so we obtain

$$\mathcal{F}\{f * g\} = \mathcal{F}\{f\}\mathcal{F}\{g\}. \tag{6.4.4}$$

6.4.1 Fourier inversion formula

We would like to establish the relation

$$u(x) = \frac{1}{2\pi} \int_{-\infty}^{\infty} e^{-i\omega x} \int_{-\infty}^{\infty} e^{i\omega t} u(t) \, dt d\omega, \tag{6.4.5}$$

from which we can deduce the Fourier inversion formula

$$u(x) = \frac{1}{2\pi} \int_{-\infty}^{\infty} e^{-i\omega x} U(\omega) \, d\omega. \tag{6.4.6}$$

This is an elegant result since the Fourier transform formula and its Fourier inversion resemble each other closely, except with the flip of the sign in the exponent of the integration kernel and the appearance of the factor $\frac{1}{2\pi}$.

To establish eq. (6.4.5), we start from the Fourier series representation of a periodic function $u(x)$ defined over the finite interval $[-L, L]$. The Fourier series expansion of $u(x)$ takes the form

$$u(x) = \frac{a_0}{2} + \sum_{n=1}^{\infty} a_n \cos \frac{n\pi x}{L} + \sum_{n=1}^{\infty} b_n \sin \frac{n\pi x}{L}. \tag{6.4.7}$$

The validity of the above expansion requires $f(x)$ to have only a finite number of finite discontinuities and only a finite number of maxima and minima. These conditions are sufficient but not necessary, and they are commonly called the *Dirichlet conditions*. By applying the orthogonality properties of the component functions over $[-L, L]$, the Fourier coefficients a_n and b_n are given by

$$a_n = \frac{1}{L} \int_{-L}^{L} u(t) \cos \frac{n\pi t}{L} \, dt, \quad n = 0, 1, \ldots$$

$$b_n = \frac{1}{L} \int_{-L}^{L} u(t) \sin \frac{n\pi t}{L} \, dt, \quad n = 1, 2, \ldots.$$

Writing out all the terms in full, eq. (6.4.7) can be written as

$$u(x) = \frac{1}{2L} \int_{-L}^{L} u(t) \, dt + \frac{1}{L} \sum_{n=1}^{\infty} \cos \frac{n\pi x}{L} \int_{-L}^{L} u(t) \cos \frac{n\pi t}{L} \, dt$$

$$+ \frac{1}{L} \sum_{n=1}^{\infty} \sin \frac{n\pi x}{L} \int_{-L}^{L} u(t) \sin \frac{n\pi t}{L} \, dt$$

$$= \frac{1}{2L} \int_{-L}^{L} u(t) \, dt + \frac{1}{L} \sum_{n=1}^{\infty} \int_{-L}^{L} u(t) \cos \frac{n\pi}{L} (t - x) \, dt. \quad (6.4.8)$$

Suppose we let L approach infinity so that $[-L, L]$ becomes $(-\infty, \infty)$. Further, we set

$$\frac{n\pi}{L} = \omega \quad \text{and} \quad \frac{\pi}{L} = \Delta\omega.$$

As $L \to \infty$, the first term in eq. (6.4.8) vanishes since the value of the integral is finite given that $u(t)$ is absolutely integrable. The second term becomes the series

$$\frac{1}{\pi} \sum_{n=1}^{\infty} \Delta\omega \int_{-\infty}^{\infty} u(t) \cos \omega(t - x) \, dt.$$

In the limit $\Delta\omega \to 0$, the above infinite sum is replaced by an integral with integration variable ω. We thus formally obtain

$$u(x) = \frac{1}{\pi} \int_{0}^{\infty} \int_{-\infty}^{\infty} u(t) \cos \omega(t - x) \, dt \, d\omega.$$

To arrive at the form shown in eq. (6.4.5), we observe that $\cos \omega(t - x)$ is an even function of ω and $\sin \omega(t - x)$ is an odd function of ω. We then have

$$u(x) = \frac{1}{2\pi} \int_{-\infty}^{\infty} \int_{-\infty}^{\infty} u(t) \cos \omega(t - x) \, dt \, d\omega \quad (6.4.9a)$$

and

$$0 = \frac{1}{2\pi} \int_{-\infty}^{\infty} \int_{-\infty}^{\infty} u(t) \sin \omega(t - x) \, dt \, d\omega. \quad (6.4.9b)$$

Adding eq. (6.4.9a) and i times eq. (6.4.9b) together, the result in eq. (6.4.5) is finally established.

Remark Suppose $u_e(x)$ is an even function, where $u_e(-x) = u_e(x)$, $x > 0$. The Fourier cosine transform of $u_e(x)$ is defined by

$$U_c(\omega) = \mathcal{F}_c\{u_e(x)\} = \int_0^\infty \cos \omega x \, u_e(x) \, dx, \quad x > 0. \qquad (6.4.10)$$

The corresponding inversion formula can be deduced to be

$$u_e(x) = \frac{2}{\pi} \int_0^\infty \cos \omega x \, U_c(\omega) \, d\omega, \quad \omega > 0. \qquad (6.4.11)$$

Hints for the proof of eq. (6.4.11) are given in Problem 6.24.

Similarly, suppose $u_o(x)$ is an odd function, where $u_o(-x) = -u_o(x)$, $x > 0$. The Fourier sine transform of $u_o(x)$ and its inversion formula are given by

$$U_s(\omega) = \mathcal{F}_s\{u_o(x)\} = \int_0^\infty \sin \omega x \, u_o(x) \, dx, \quad x > 0 \qquad (6.4.12)$$

and

$$u_o(x) = \frac{2}{\pi} \int_0^\infty \sin \omega x \, U_s(\omega) \, d\omega, \quad \omega > 0, \qquad (6.4.13)$$

respectively.

Parseval identity

The Parseval identity states that

$$\int_{-\infty}^\infty |u|^2(x) \, dx = \frac{1}{2\pi} \int_{-\infty}^\infty |U(\omega)|^2 \, d\omega. \qquad (6.4.14)$$

To prove the identity, we apply the Fourier inversion formula to $\mathcal{F}\{f * g\}$ so that

$$(f * g)(x) = \int_{-\infty}^\infty f(\xi) g(x - \xi) \, d\xi = \frac{1}{2\pi} \int_{-\infty}^\infty e^{-i\omega x} \mathcal{F}\{f * g\} \, d\omega.$$

Suppose we choose $f(\xi) = u(\xi)$ and $g(\xi) = \bar{u}(-\xi)$, and set $x = 0$. We then obtain

$$\int_{-\infty}^\infty u(\xi) \bar{u}(\xi) \, d\xi = \frac{1}{2\pi} \int_{-\infty}^\infty \mathcal{F}\{u(x)\} \mathcal{F}\{\bar{u}(-x)\} \, d\omega$$

$$= \frac{1}{2\pi} \int_{-\infty}^\infty U(\omega) \overline{U(\omega)} \, d\omega;$$

hence the result.

Dirac function

The construction of the Dirac function may be motivated by attempts to find a function $\psi(x)$ such that the convolution of ψ with any function f always gives f. That is,

$$\int_{-\infty}^{\infty} f(\xi)\psi(x - \xi)\,d\xi = f * \psi = \psi * f$$

$$= \int_{-\infty}^{\infty} \psi(\xi)f(x - \xi)\,d\xi = f(x). \quad (6.4.15)$$

If we set $x = 0$, then $\psi(x)$ has to observe

$$\int_{-\infty}^{\infty} \psi(\xi)f(-\xi)\,d\xi = f(0). \qquad (6.4.16)$$

The above result indicates that the values of $f(x)$ for $x \neq 0$ do not have any impact on the value of the integral of the product $\psi(\xi)f(-\xi)$. This occurs only when $\psi(x) = 0$ for all nonzero values of x. On the other hand, by taking $f(x) = 1$, we obtain

$$\int_{-\infty}^{\infty} \psi(\xi)\,d\xi = 1. \qquad (6.4.17)$$

Apparently, there is no function defined in the usual pointwise sense that possesses the above two properties. Actually, $\psi(x)$ belongs to the class of *generalized functions*.

Since no such function exists in the usual pointwise sense, can we approximate the generalized function $\psi(x)$ in some form as the limit of a sequence of functions? Consider the sequence of functions $\delta_n(x), n = 1, 2, \ldots,$ defined by

$$\delta_n(x) = \begin{cases} n & -\frac{1}{2n} < x < \frac{1}{2n} \\ 0 & \text{otherwise} \end{cases}. \qquad (6.4.18)$$

As n increases, $\delta_n(x)$ resembles a high and narrow spike, like an impulse with infinite magnitude. Also, $\{\delta_n(x)\}$ is a sequence of functions that satisfy the normalized condition:

$$\int_{-\infty}^{\infty} \delta_n(x)\,dx = 1.$$

The limit

$$\lim_{n \to \infty} \delta_n(x)$$

does not exist; however, the sequence of integrals

$$\int_{-\infty}^{\infty} \delta_n(x)f(x)\,dx$$

for any continuous function $f(x)$ exists and the corresponding limit is equal to

$$\lim_{n \to \infty} \int_{m}^{\infty} \delta_n(x) f(x) \, dx = f(0).$$

The Dirac function $\delta(x)$ is formally defined by

$$\int_{-\infty}^{\infty} \delta(x) f(x) \, dx = \lim_{n \to \infty} \int_{-\infty}^{\infty} \delta_n(x) f(x) \, dx = f(0), \qquad (6.4.19a)$$

where the left-hand integral is the limit of a sequence of integrals. Here, $\delta(x)$ is a distribution, which means that it has effect on continuous functions but is not itself a function. With $f(x) = 1$, we obtain

$$\int_{-\infty}^{\infty} \delta(x) \, dx = 1. \qquad (6.4.19b)$$

Though $\delta(x)$ cannot be defined in the usual pointwise sense, sometimes it may be convenient to think of $\delta(x)$ as a function of x with the property

$$\delta(x) = \begin{cases} \infty & x = 0 \\ 0 & x \neq 0 \end{cases}. \qquad (6.4.19c)$$

The choice of the sequence of limiting functions $\{\delta_n(x)\}$ is not unique. Some other possible choices are

$$\delta_n^{(1)}(x) = \frac{n}{\sqrt{\pi}} e^{-n^2 x^2}, \qquad\qquad n = 1, 2, \ldots; \qquad (6.4.20a)$$

$$\delta_n^{(2)}(x) = \frac{n}{\pi} \frac{1}{1 + n^2 x^2}, \qquad\qquad n = 1, 2, \ldots; \qquad (6.4.20b)$$

$$\delta_n^{(3)}(x) = \frac{\sin nx}{\pi x} = \frac{1}{2\pi} \int_{-n}^{n} e^{-ixt} \, dt, \quad n = 1, 2, \ldots. \qquad (6.4.20c)$$

These sequences of functions all satisfy the properties:

$$\lim_{n \to \infty} \int_{-\infty}^{\infty} \delta_n(x) f(x) \, dx = f(0) \quad \text{and} \quad \int_{-\infty}^{\infty} \delta_n(x) \, dx = 1. \quad (6.4.21)$$

They invariably become infinitely high and narrow spikes as n increases to infinity. Readers should be wary that we may have a sequence of functions that satisfy the normalized condition and exhibit the property

$$\lim_{n \to \infty} \delta_n(x) = \begin{cases} \infty & x \neq 0 \\ 0 & x = 0 \end{cases};$$

however the limit of which is not the distribution of $\delta(x)$.

Suppose we translate $\delta(x)$ by x_0 to give another generalized function defined as $\delta(x - x_0)$, and this new generalized function observes the property:

$$\int_{-\infty}^{\infty} \delta(x - x_0) f(x) \, dx = f(x_0). \tag{6.4.22}$$

In a loose sense, we may think of $\delta(x - x_0)$ as a generalized function which assumes zero value when $x \neq x_0$ and infinite value when $x = x_0$.

The Dirac function is closely related to the Heaviside function $H(x)$ defined in eq. (3.3.13). By observing

$$\int_{-\infty}^{x} \delta(x) \, dx = 0, \quad x < 0,$$

$$\int_{x}^{\infty} \delta(x) \, dx = 0, \quad x > 0,$$

we deduce that

$$\int_{-\infty}^{x} \delta(\xi) \, d\xi = \begin{cases} 1 & \text{if } x > 0 \\ 0 & \text{if } x < 0 \end{cases}.$$

By formally differentiating with respect to x on both sides of the above equation, the relation between $H(x)$ and $\delta(x)$ is found to be

$$\frac{d}{dx} H(x) = \delta(x). \tag{6.4.23}$$

Suppose we take $f(x) = e^{i\omega x}$ in eq. (6.4.22); we obtain

$$\int_{-\infty}^{\infty} e^{i\omega x} \delta(x - x_0) \, dx = e^{i\omega x_0}.$$

Setting $x_0 = 0$ in the above equation, we obtain the following formula for the Fourier transform of $\delta(x)$:

$$\mathcal{F}\{\delta(x)\} = 1. \tag{6.4.24}$$

Taking the Fourier inversion of $U(\omega) = 1$, we have

$$\delta(x) = \frac{1}{2\pi} \int_{-\infty}^{\infty} e^{-i\omega x} \, d\omega. \tag{6.4.25}$$

Suppose we rewrite the above integral formally as

$$\frac{1}{2\pi} \int_{-\infty}^{\infty} e^{-i\omega x} \, d\omega = \lim_{n \to \infty} \int_{-n}^{n} \frac{1}{2\pi} e^{-i\omega x} \, d\omega = \lim_{n \to \infty} \frac{\sin nx}{\pi x}. \tag{6.4.26}$$

Therefore, the sequence $\{\delta_n^{(3)}(x)\}$ defined in eq. (6.4.20c) is seen to be a sequence of functions that approximate the distribution of $\delta(x)$.

Interestingly, eq. (6.4.5) can be established readily using eqs. (6.4.22) and (6.4.25), assuming that interchanging order of integration is valid. The proof is shown as follows:

$$u(x) = \int_{-\infty}^{\infty} u(t)\delta(x-t) \, dt$$

$$= \int_{-\infty}^{\infty} u(t) \left(\frac{1}{2\pi} \int_{-\infty}^{\infty} e^{-i\omega(x-t)} d\omega \right) dt$$

$$= \frac{1}{2\pi} \int_{-\infty}^{\infty} e^{-i\omega x} \int_{-\infty}^{\infty} e^{i\omega t} u(t) \, dt d\omega.$$

6.4.2 Evaluation of Fourier integrals

We would like to evaluate a Fourier integral of the form

$$\int_{-\infty}^{\infty} e^{imx} f(x) \, dx, \qquad m > 0,$$

where

(i) $f(z)$ has a finite number of isolated singularities,
(ii) $\lim_{z \to \infty} f(z) = 0$,
(iii) $f(z)$ has no singularity along the real axis.

Remark

(i) The assumption $m > 0$ is not strictly essential. With some minor modifications, the method proposed below also works even when m is negative or imaginary.
(ii) When $f(z)$ has singularities on the real axis, the Cauchy principal value of the integral is considered (see Section 6.5 and Example 6.5.3).

Jordan lemma

Suppose $f(z) \to 0$ as $z \to \infty$; the Jordan lemma states that

$$\lim_{R \to \infty} \int_{C_R} e^{i\lambda z} f(z) \, dz = 0, \tag{6.4.27}$$

where λ is a positive real number and the contour C_R is the upper semi-circle with radius R, $R \to \infty$. The proof of the lemma requires the following inequality:

$$\sin \theta \geq \frac{2\theta}{\pi} \qquad \text{for} \qquad 0 \leq \theta \leq \frac{\pi}{2}. \tag{6.4.28}$$

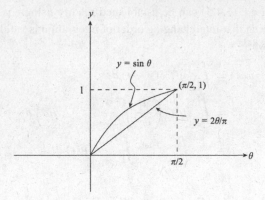

Figure 6.8. Geometric visualization of inequality (6.4.28).

The inequality can be visualized by observing that the curve for $y = \sin\theta$ is concave over the interval $[0, \frac{\pi}{2}]$, so it always lies on or above the line segment $y = \frac{2\theta}{\pi}$ (see Figure 6.8).

We consider the modulus of the integral

$$
\left| \int_{C_R} f(z) e^{i\lambda z}\, dz \right| \leq \int_0^\pi |f(Re^{i\theta})|\, |e^{i\lambda Re^{i\theta}}|\, R\, d\theta
$$

$$
\leq \max_{z \in C_R} |f(z)|\, R \int_0^\pi e^{-\lambda R \sin\theta}\, d\theta
$$

$$
= 2R \max_{z \in C_R} |f(z)| \int_0^{\frac{\pi}{2}} e^{-\lambda R \sin\theta}\, d\theta
$$

$$
\leq 2R \max_{z \in C_R} |f(z)| \int_0^{\frac{\pi}{2}} e^{-\lambda R \frac{2\theta}{\pi}}\, d\theta
$$

$$
= 2R \max_{z \in C_R} |f(z)|\, \frac{\pi}{2R\lambda}\, (1 - e^{-\lambda R}),
$$

which tends to 0 as $R \to \infty$, given that $f(z) \to 0$ uniformly as $R \to \infty$ for $z \in C_R$.

Remark We may relax the restriction that $\lambda > 0$. For $\lambda < 0$, the same result can be obtained if C_R is chosen to be the lower semi-circle. When $\lambda = im$, $m > 0$, we choose C_R to be the right semi-circle; and when $\lambda = -im$, $m > 0$, C_R is chosen to be the left semi-circle.

To evaluate the Fourier integral, we integrate $e^{imz} f(z)$ along the closed contour C that consists of the upper semi-circle C_R and the line segment from

$-R$ to R along the real axis (see Figure 6.8). We then have

$$\oint_L e^{imz} f(z) \, dz = \int_R^R e^{imx} f(x) \, dx + \int_{C_R} e^{imz} f(z) \, dz. \quad (6.4.29)$$

Taking the limit $R \to \infty$, the integral along C_R vanishes by virtue of the Jordan lemma. Lastly, we apply the Cauchy residue theorem to obtain

$$\int_{-\infty}^{\infty} e^{imx} f(x) \, dx = 2\pi i \text{ [sum of residues at all the isolated}$$

$$\text{singularities of } f \text{ in the upper half-plane} \quad (6.4.30)$$

since C encloses all the singularities of f in the upper half-plane as $R \to \infty$.

Example 6.4.1 Evaluate the Fourier integral

$$\int_{-\infty}^{\infty} \frac{\sin 2x}{x^2 + x + 1} \, dx.$$

Solution It is easy to check that

$$f(z) = \frac{1}{z^2 + z + 1}$$

has no singularity along the real axis and

$$\lim_{z \to \infty} \frac{1}{z^2 + z + 1} = 0.$$

The integrand has two simple poles, namely, $z = e^{2\pi i/3}$ in the upper half-plane and $e^{-2\pi i/3}$ in the lower half-plane. By virtue of the Jordan lemma, we have

$$\int_{-\infty}^{\infty} \frac{\sin 2x}{x^2 + x + 1} \, dx = \text{Im} \int_{-\infty}^{\infty} \frac{e^{2ix}}{x^2 + x + 1} \, dx = \text{Im} \oint_C \frac{e^{2iz}}{z^2 + z + 1} \, dz,$$

where C is the union of the infinitely large upper semi-circle and its diameter along the real axis. By eq. (6.4.30), we have

$$\oint_C \frac{e^{2iz}}{z^2 + z + 1} \, dz = 2\pi i \, \text{Res}\left(\frac{e^{2iz}}{z^2 + z + 1}, e^{\frac{2\pi i}{3}}\right)$$

$$= 2\pi i \left.\frac{e^{2iz}}{2z + 1}\right|_{z = e^{\frac{2\pi i}{3}}} = 2\pi i \frac{e^{2ie^{\frac{2\pi i}{3}}}}{2e^{\frac{2\pi i}{3}} + 1}.$$

Hence, the Fourier integral is found to be

$$\int_{-\infty}^{\infty} \frac{\sin 2x}{x^2 + x + 1} \, dx = \text{Im}\left(2\pi i \frac{e^{2ie^{\frac{2\pi i}{3}}}}{2e^{\frac{2\pi i}{3}} + 1}\right) = -\frac{2}{\sqrt{3}} \pi e^{-\sqrt{3}} \sin 1.$$

6.5 Cauchy principal value of an improper integral

Suppose a real function $f(x)$ is continuous everywhere in the interval $[a, b]$ except at a point x_0 inside the interval. The integral of $f(x)$ over the interval $[a, b]$ is an improper integral, which may be defined as

$$\int_a^b f(x)\, dx = \lim_{\epsilon_1, \epsilon_2 \to 0} \left[\int_a^{x_0 - \epsilon_1} f(x)\, dx + \int_{x_0 + \epsilon_2}^b f(x)\, dx \right], \quad \epsilon_1,\ \epsilon_2 > 0. \tag{6.5.1}$$

In many cases, the above limit exists only when $\epsilon_1 = \epsilon_2$, and does not exist otherwise. For example, the function $y = \frac{1}{x}$ is not continuous at $x = 0$, and in a strict sense the integral

$$\int_{-1}^2 \frac{1}{x}\, dx$$

does not exist. If we define the improper integral according to eq. (6.5.1), we obtain

$$\int_{-1}^2 \frac{1}{x}\, dx$$

$$= \lim_{\epsilon_1 \to 0} \int_{-1}^{-\epsilon_1} \frac{1}{x}\, dx + \lim_{\epsilon_2 \to 0} \int_{\epsilon_2}^2 \frac{1}{x}\, dx, \quad \epsilon_1 > 0 \text{ and } \epsilon_2 > 0$$

$$= \lim_{\epsilon_1, \epsilon_2 \to 0} (\ln \epsilon_1 + \ln 2 - \ln \epsilon_2) = \ln 2 + \lim_{\epsilon_1, \epsilon_2 \to 0} \left(\ln \frac{\epsilon_1}{\epsilon_2} \right). \tag{6.5.2}$$

The limit does not exist if $\epsilon_1 \to 0$ and $\epsilon_2 \to 0$ in an arbitrary manner.

The *Cauchy principal value* of the improper integral in eq. (6.5.1) (symbolized by the letter P) is defined by taking $\epsilon_1 = \epsilon_2 = \epsilon$ so that

$$P \int_a^b f(x)\, dx = \lim_{\epsilon \to 0} \left[\int_a^{x_0 - \epsilon} f(x)\, dx + \int_{x_0 + \epsilon}^b f(x)\, dx \right], \quad \epsilon > 0, \tag{6.5.3}$$

provided that the limit exists. For the improper integral in eq. (6.5.2), the corresponding Cauchy principal value is found to be

$$P \int_{-1}^2 \frac{1}{x}\, dx = \ln 2 + \lim_{\epsilon \to 0} \left(\ln \frac{\epsilon}{\epsilon} \right) = \ln 2. \tag{6.5.4}$$

In contour integration, consideration of the principal value of an improper integral is necessary when the integration contour passes through an isolated singularity of the integrand. We illustrate the evaluation of the Cauchy principal value of an improper integral through the following examples.

Example 6.5.1 Find the principal value of the following definite integral:

$$I_n = \int_0^\pi \frac{\cos n\phi}{\cos \phi - \cos \theta} \, d\phi, \quad 0 < \theta < \pi \text{ and } n \text{ is any non-negative integer}$$

Solution By definition, the principal value of I_n is given by

$$I_n = \lim_{\epsilon \to 0} \left(\int_0^{\theta - \epsilon} \frac{\cos n\phi}{\cos \phi - \cos \theta} \, d\phi + \int_{\theta + \epsilon}^{\pi} \frac{\cos n\phi}{\cos \phi - \cos \theta} \, d\phi \right).$$

When $n = 0$, we have

$$\int_0^{\theta - \epsilon} \frac{1}{\cos \phi - \cos \theta} \, d\phi = \frac{1}{\sin \theta} \ln \frac{\sin \frac{1}{2}(\theta + \phi)}{\sin \frac{1}{2}(\theta - \phi)} \Big|_0^{\theta - \epsilon}$$

and

$$\int_{\theta + \epsilon}^{\pi} \frac{1}{\cos \phi - \cos \theta} \, d\phi = \frac{1}{\sin \theta} \ln \frac{\sin \frac{1}{2}(\phi + \theta)}{\sin \frac{1}{2}(\phi - \theta)} \Big|_{\theta + \epsilon}^{\pi} \, ;$$

so

$$I_0 = \lim_{\epsilon \to 0} \frac{1}{\sin \theta} \ln \frac{\sin \left(\theta - \frac{\epsilon}{2} \right)}{\sin \left(\theta + \frac{\epsilon}{2} \right)} = 0.$$

When $n = 1$, by observing $I_0 = 0$, we obtain

$$I_1 = I_1 - I_0 \cos \theta = \int_0^{\pi} \frac{\cos \phi - \cos \theta}{\cos \phi - \cos \theta} \, d\phi = \pi.$$

To evaluate I_n, $n > 1$, we write

$$I_n = I_n - I_0 \cos n\theta = \int_0^{\pi} \frac{\cos n\phi - \cos n\theta}{\cos \phi - \cos \theta} \, d\phi.$$

Using the trigonometric relation

$$\cos (n + 1)\phi - \cos (n + 1)\theta + \cos (n - 1)\phi - \cos (n - 1)\theta$$
$$= 2 \cos n\phi (\cos \phi - \cos \theta) + 2 \cos \theta (\cos n\phi - \cos n\theta),$$

and observing

$$\int_0^{\pi} \cos n\phi \, d\phi = 0,$$

we divide each term by $\cos \phi - \cos \theta$ and integrate with resepct to ϕ from 0 to π. This gives the reduction formula

$$I_{n+1} + I_{n-1} = 2 \cos \theta \, I_n.$$

The general solution to the above reduction formula is given by

$$I_n = A \sin n\theta + B \cos n\theta,$$

where A and B are arbitrary constants. The constants A and B can be determined using the initial values $I_0 = 0$ and $I_1 = \pi$. We obtain $A = \frac{\pi}{\sin\theta}$ and $B = 0$. The final result is

$$P\left(\int_0^\pi \frac{\cos n\phi}{\cos\phi - \cos\theta} \, d\phi \right) = \pi \frac{\sin n\theta}{\sin\theta}, \qquad n \geq 0.$$

Example 6.5.2 Show that

$$\int_{-\infty}^\infty \frac{e^{px} - e^{qx}}{1 - e^x} \, dx = \pi(\cot p\pi - \cot q\pi),$$

provided that $0 < p < 1$ and $0 < q < 1$. Hence, deduce that

$$\int_{-\infty}^\infty \frac{\sinh ax}{\sinh bx} \, dx = \frac{\pi}{b} \tan \frac{a\pi}{2b}, \qquad \text{if } b > |a|.$$

Solution   The point $z = 0$ is a simple pole of the two functions

$$\frac{e^{pz}}{1 - e^z} \quad \text{and} \quad \frac{e^{qz}}{1 - e^z};$$

however it is not so for the integrand

$$\frac{e^{pz} - e^{qz}}{1 - e^z}.$$

Therefore,

$$\int_{-\infty}^\infty \frac{e^{px}}{1 - e^x} \, dx \quad \text{and} \quad \int_{-\infty}^\infty \frac{e^{qx}}{1 - e^x} \, dx$$

are improper integrals, and it can be shown readily that their principal values exist. To simplify the valuation procedure, we choose to split the integral as follows:

$$\int_{-\infty}^\infty \frac{e^{px} - e^{qx}}{1 - e^x} \, dx = P\int_{-\infty}^\infty \frac{e^{px}}{1 - e^x} \, dx - P\int_{-\infty}^\infty \frac{e^{qx}}{1 - e^x} \, dx.$$

The integrand functions

$$\frac{e^{pz}}{1 - e^z} \quad \text{and} \quad \frac{e^{qz}}{1 - e^z}$$

have infinitely many simple poles at $z = 2n\pi i$, where n is any integer. The two simple poles that are of interest in the present problem are $z = 0$ and $z = 2\pi i$. The following closed rectangular contour C is chosen: (i) it has two

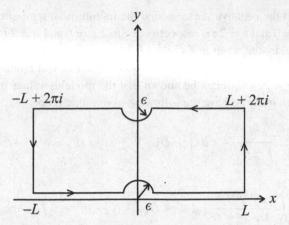

Figure 6.9. The contour is a closed rectangle with infinitesimal indentations at $z = 0$ and $z = 2\pi i$.

infinitesimal indentations at $z = 0$ and $z = 2\pi i$; (ii) the left and right vertical lines are, respectively, $z = -L + iy$ and $z = L + iy$, where $0 \le y \le 2\pi$; (iii) the upper and lower horizontal lines are, respectively, $z = x + 2\pi i$ and $z = x$, where $x \in [-L, L] \backslash (-\epsilon, \epsilon)$ (see Figure 6.9). The indentations are infinitesimal semi-circles with common radius ϵ. They are adopted here so as to avoid the passage of the integration path through the poles at $z = 0$ and $z = 2\pi i$. As a result, there is no inclusion of the poles inside the contour. The contour C is seen to consist of eight parts, so the contour integral around C can be split into eight line integrals. Consider the integration of $\frac{e^{pz}}{1-e^z}$ around the closed contour C; since there is no pole enclosed inside the contour C, we have

$$
\begin{aligned}
0 &= \oint_C \frac{e^{pz}}{1 - e^z}\, dz \\
&= \int_{-L}^{-\epsilon} \frac{e^{px}}{1 - e^x}\, dx - \frac{2\pi i}{2} \mathrm{Res}\left(\frac{e^{pz}}{1 - e^z}, 0 \right) \\
&\quad + \int_{\epsilon}^{L} \frac{e^{px}}{1 - e^x}\, dx + \int_0^{2\pi} \frac{e^{p(L+iy)}}{1 - e^{L+iy}}\, i\, dy \\
&\quad + \int_L^{\epsilon} \frac{e^{p(x+2\pi i)}}{1 - e^{x+2\pi i}}\, dx - \frac{2\pi i}{2} \mathrm{Res}\left(\frac{e^{pz}}{1 - e^z}, 2\pi i \right) \\
&\quad + \int_{-\epsilon}^{-L} \frac{e^{p(x+2\pi i)}}{1 - e^{x+2\pi i}}\, dx + \int_{2\pi}^0 \frac{e^{p(-L+iy)}}{1 - e^{-L+iy}}\, i\, dy.
\end{aligned}
$$

Subsequently, we take the limits $\epsilon \to 0$ and $L \to \infty$ in the above integrals. The second and sixth terms in the right-hand expression arise from the contour

integration in the negative sense around the infinitesimal semi-circular indentations at $z = 0$ and $z = 2\pi i$, respectively. Since $z = 0$ and $z = 2\pi i$ are simple
poles of $\frac{e^{pz}}{1-e^z}$, by the result in Example 6.2.5, the value of the contour integral
around the infinitesimal semi-circle is equal to half of that around the whole
circle. For $0 < p < 1$, it can be shown that the modulus values of the fourth
and the eighth integrals have the following asymptotic properties:

$$\left| \int_0^{2\pi} \frac{e^{p(L+iy)}}{1 - e^{L+iy}} \, i \, dy \right| \sim O(e^{-(1-p)L}) \to 0 \quad \text{as} \quad L \to \infty,$$

$$\left| \int_{2\pi}^0 \frac{e^{p(-L+iy)}}{1 - e^{-L+iy}} \, i \, dy \right| \sim O(e^{-pL}) \to 0 \quad \text{as} \quad L \to \infty.$$

Moreover, the respective residue values at the simple poles $z = 0$ and $z = 2\pi i$
are found to be

$$\text{Res}\left(\frac{e^{pz}}{1 - e^z}, 0 \right) = -1 \quad \text{and} \quad \text{Res}\left(\frac{e^{pz}}{1 - e^z}, 2\pi i \right) = -e^{2p\pi i}.$$

Combining the results, we have

$$(1 - e^{2p\pi i}) \lim_{\substack{\epsilon \to 0 \\ L \to \infty}} \left[\int_{-L}^{-\epsilon} \frac{e^{px}}{1 - e^x} \, dx + \int_{\epsilon}^L \frac{e^{px}}{1 - e^x} \, dx \right]$$

$$= (1 - e^{2p\pi i}) P \int_{-\infty}^{\infty} \frac{e^{px}}{1 - e^x} \, dx$$

$$= \pi i \left[\text{Res}\left(\frac{e^{pz}}{1 - e^z}, 0 \right) + \text{Res}\left(\frac{e^{pz}}{1 - e^z}, 2\pi i \right) \right] = -\pi i (1 + e^{2p\pi i}),$$

and so

$$P \int_{-\infty}^{\infty} \frac{e^{px}}{1 - e^x} \, dx = \pi \frac{(e^{p\pi i} + e^{-p\pi i})/2}{(e^{p\pi i} - e^{-p\pi i})/2i} = \pi \frac{\cos p\pi}{\sin p\pi} = \pi \cot p\pi.$$

For $0 < q < 1$, we deduce a similar result

$$P \int_{-\infty}^{\infty} \frac{e^{qx}}{1 - e^x} \, dx = \pi \cot q\pi.$$

Finally, we obtain

$$\int_{-\infty}^{\infty} \frac{e^{px} - e^{qx}}{1 - e^x} \, dx = \pi (\cot p\pi - \cot q\pi).$$

To evaluate the second integral in the problem, we first rewrite the integrand
into the form

$$\frac{\sinh ax}{\sinh bx} = \frac{e^{ax} - e^{-ax}}{e^{bx} - e^{-bx}} = \frac{-e^{(a+b)x}}{1 - e^{2bx}} + \frac{e^{(b-a)x}}{1 - e^{2bx}}$$

When $b > |a|$, we have

$$0 < \frac{b-a}{2b} < 1 \quad \text{and} \quad 0 < \frac{a+b}{2b} < 1.$$

By taking $p = \frac{b-a}{2b}$ and $q = \frac{a+b}{2b}$, we have

$$\int_{-\infty}^{\infty} \frac{\sinh ax}{\sinh bx}\, dx = P\int_{-\infty}^{\infty} \frac{e^{(b-a)x}}{1 - e^{2bx}}\, dx - P\int_{-\infty}^{\infty} \frac{e^{(a+b)x}}{1 - e^{2bx}}\, dx$$

$$= \frac{1}{2b}P\int_{-\infty}^{\infty} \frac{e^{\frac{b-a}{2b}x}}{1 - e^{x}}\, dx - \frac{1}{2b}P\int_{-\infty}^{\infty} \frac{e^{\frac{b+a}{2b}x}}{1 - e^{x}}\, dx$$

$$= \frac{\pi}{2b}\left[\cot\left(\frac{b-a}{2b}\right)\pi - \cot\left(\frac{a+b}{2b}\right)\pi\right]$$

$$= \frac{\pi}{2b}\left[\cot\left(\frac{\pi}{2} - \frac{a\pi}{2b}\right) - \cot\left(\frac{\pi}{2} + \frac{a\pi}{2b}\right)\right]$$

$$= \frac{\pi}{b}\tan\frac{a\pi}{2b}.$$

Example 6.5.3   Find the Cauchy principal value of

$$\int_{-\infty}^{\infty} \frac{e^{ix}}{x}\, dx,$$

and use the result to deduce that

$$\int_{0}^{\infty} \frac{\sin x}{x}\, dx = \frac{\pi}{2}.$$

Solution   The integrand $\frac{e^{ix}}{x}$ has an isolated singularity at $x = 0$. The Cauchy
principal value of

$$\int_{-\infty}^{\infty} \frac{e^{ix}}{x}\, dx$$

is defined as

$$P\int_{-\infty}^{\infty} \frac{e^{ix}}{x}\, dx = \lim_{\substack{R\to\infty \\ \epsilon\to 0}} \left[\int_{-R}^{-\epsilon} \frac{e^{ix}}{x}\, dx + \int_{\epsilon}^{R} \frac{e^{ix}}{x}\, dx\right].$$

We consider the contour integral

$$\oint_{C} \frac{e^{iz}}{z}\, dz,$$

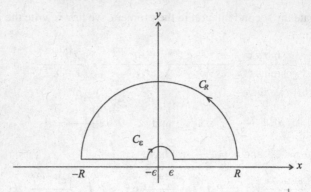

Figure 6.10. The closed contour C consists of the infinitely large upper semi-circle C_R of radius R, the line segments $(-R, -\epsilon)$ and (ϵ, R) along the x-axis, and the infinitesimal upper semi-circle C_ϵ of radius ϵ around $z = 0$.

where C is the closed contour shown in Figure 6.10. An infinitesimal upper semi-circle C_ϵ around $z = 0$ is appended so that the integration path does not cross the pole of the integrand. With this construction of the closed contour C, the pole $z = 0$ is not included inside the contour.

Since $\frac{e^{iz}}{z}$ has no singularity inside C, we have

$$\oint_C \frac{e^{iz}}{z}\, dz = 0.$$

On the other hand, the above contour integral can be split into three parts, namely,

$$\oint_C \frac{e^{iz}}{z}\, dz = \int_{C_R} \frac{e^{iz}}{z}\, dz + P \int_{-\infty}^{\infty} \frac{e^{ix}}{x}\, dx + \int_{C_\epsilon} \frac{e^{iz}}{z}\, dz.$$

Since $z = 0$ is a simple pole of $\frac{e^{iz}}{z}$ and C_ϵ is looped in the negative sense, using the result in Example 6.2.5, we obtain

$$\int_{C_\epsilon} \frac{e^{iz}}{z}\, dz = -\pi i \operatorname{Res}\left(\frac{e^{iz}}{z}, 0\right) = -\pi i.$$

By virtue of the Jordan lemma, the first integral

$$\int_{C_R} \frac{e^{iz}}{z}\, dz$$

is seen to be zero as $R \to \infty$. Combining the results, we obtain

$$P \int_{-\infty}^{\infty} \frac{e^{ix}}{x}\, dx = -\int_{C_\epsilon} \frac{e^{iz}}{z}\, dz = \pi i.$$

For the integral

$$\int_0^\infty \frac{\sin x}{x}\, dx,$$

the point $x = 0$ is a removable singularity of the integrand since $\frac{\sin x}{x} \to 1$ as $x \to 0$. We observe that

$$\mathrm{Im}\left(P \int_{-\infty}^\infty \frac{e^{ix}}{x}\, dx\right) = \lim_{\substack{R \to \infty \\ \epsilon \to 0}} \left(\int_{-R}^{-\epsilon} \frac{\sin x}{x}\, dx + \int_\epsilon^R \frac{\sin x}{x}\, dx\right)$$

$$= 2 \lim_{\substack{R \to \infty \\ \epsilon \to 0}} \int_\epsilon^R \frac{\sin x}{x}\, dx = 2 \int_0^\infty \frac{\sin x}{x}\, dx,$$

so

$$\int_0^\infty \frac{\sin x}{x}\, dx = \frac{1}{2}\, \mathrm{Im}\left(P \int_{-\infty}^\infty \frac{e^{ix}}{x}\, dx\right) = \mathrm{Im}\left(\frac{i\pi}{2}\right) = \frac{\pi}{2}.$$

6.6 Hydrodynamics in potential fluid flows

In this section, we would like to illustrate the application of contour integration and residue calculus to compute the hydrodynamic lifting force and the moment on a body immersed in a potential flow field.

Consider a two-dimensional body immersed in a potential flow field whose complex potential is given by $w = f(z)$, $z = x + iy$. Let the boundary of the body be represented by the closed curve γ_b, where the boundary itself is a streamline. As the velocity values V along the boundary of the body are not uniform, this causes the pressure P to vary along the boundary. This physical property follows from the Bernoulli law of hydrodynamics, which states that $P + \frac{\rho}{2}|V|^2$ is constant along any streamline. Here, ρ is the density of the fluid, which is assumed to be constant in the present discussion. Bernoulli's law is a rephrasing of the principle of conservation of energy. The non-uniformity of pressure values along the boundary of the body leads to hydrodynamic force and moment exerted on the body by the potential flow motion.

6.6.1 Blasius laws of hydrodynamic force and moment

The Blasius laws of hydrodynamic force F and the moment M exerted on a body immersed in the potential flow field as characterized by the complex

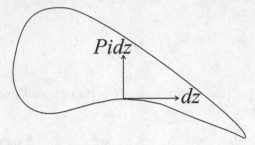

Figure 6.11. The differential pressure force $Pi\,dz$ acts normal to the differential segment dz on the boundary of an immersed body.

potential $w = f(z)$ are given by

$$\overline{F} = \frac{\rho i}{2} \oint_{\gamma} [f'(z)]^2 dz, \tag{6.6.1}$$

$$M = \mathrm{Re}\left(-\frac{\rho}{2} \oint_{\gamma} z[f'(z)]^2 dz\right), \tag{6.6.2}$$

respectively. Here, γ is any curve surrounding the body, provided that there is no singularity included between γ and the body's boundary γ_b.

To prove the above formulas, we consider the differential pressure force exerted on the body normal to the differential length dz on the body's boundary γ_b. Since dz is along the tangent to the body, $i\,dz$ lies along the interior normal to γ_b (see Figure 6.11); so the differential pressure force dF is given by $Pi\,dz$.

Let B denote the constant in the Bernoulli law. The differential pressure force dF acting along the boundary of the body γ_b is then given by

$$dF = Pi\,dz = \left[B - \frac{\rho}{2}|V(z)|^2\right] i\,dz,$$

where $V(z) = \overline{f'(z)}$ is the velocity of the flow at the point $z = x + iy$. The total force $F = F_x + iF_y$ exerted on the immersed body is given by the contour integral around γ_b, where

$$F = F_x + iF_y = \oint_{\gamma_b} Pi\,dz$$

$$= \oint_{\gamma_b} Bi\,dz - \oint_{\gamma_b} \frac{\rho i}{2}|V(z)|^2\,dz. \tag{6.6.3}$$

The first integral is zero since the integrand Bi is a constant. It is more preferable to express the second integral in terms of the complex potential $f(z)$. Since the body is a streamline, the velocity along the body is tangential to the body. When $V(z)$ and dz are treated as vector quantities, they should have the same angle of inclination. We then obtain

$$\text{Arg } V(z) = \text{Arg } \overline{f'(z)} = \text{Arg } dz.$$

Let θ denote the common value of the above two argument quantities. The integrand in the second integral can be expressed as

$$-\frac{\rho i}{2}[\overline{f'(z)}]^2 e^{-2i\theta} \, dz = -\frac{\rho i}{2}\overline{[f'(z)]^2 dz} = \overline{\frac{\rho i}{2}[f'(z)]^2 dz},$$

so

$$\overline{F} = F_x - iF_y = \frac{\rho i}{2} \oint_{\gamma_b} [f'(z)]^2 \, dz. \tag{6.6.4}$$

By the Cauchy–Goursat integral theorem, we can replace γ_b by any simple closed contour γ surrounding the body, provided that there is no singularity enclosed between γ and γ_b.

Next, we would like to compute the hydrodynamic moment exerted on the body immersed in the potential flow field. Let us consider the differential moment dM due to the differential pressure force $dF = dF_x + i\, dF_y$ at $z = x + iy$. Since the pressure force acts along the interior normal to γ_b, we have

$$\text{Arg } dF = \frac{\pi}{2} + \text{Arg } dz = \frac{\pi}{2} + \theta.$$

The differential moment dM can then be expressed as

$$\begin{aligned}
dM &= x\, dF_y - y\, dF_x \\
&= P|dz|\left[x \sin\left(\frac{\pi}{2} + \theta\right) - y\cos\left(\frac{\pi}{2} + \theta\right)\right] \\
&= P|dz|(x\cos\theta + y\sin\theta) = P\,\text{Re}(ze^{-2i\theta}\, dz).
\end{aligned}$$

The total moment exerted on the body is given by

$$\begin{aligned}
M &= \oint_{\gamma_b} \left[B - \frac{\rho}{2}|V(z)|^2\right]\text{Re}(ze^{-2i\theta}\, dz) \\
&= \text{Re} \oint_{\gamma_b} -\frac{\rho}{2}[f'(z)]^2 e^{2i\theta} z e^{-2i\theta}\, dz \\
&= \text{Re}\left(-\frac{\rho}{2}\oint_{\gamma_b} z[f'(z)]^2\, dz\right). \tag{6.6.5}
\end{aligned}$$

Again, provided that there is no singularity included between γ and γ_b, we may replace γ_b by γ. Thus we obtain the desired result in (6.6.2).

Example 6.6.1 Consider the potential flow field in the right half-plane, $\text{Re } z > 0$, due to a field source of strength m located at $z = a$, $a > 0$. An

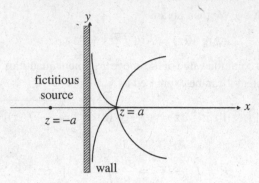

Figure 6.12. The combination of the real source at $z = a$ and the fictitious source at $z = -a$ produces a flow field with the y-axis as a streamline.

infinite wall is placed along the y-axis. Find the total pressure force exerted on the wall due to the fluid source.

Solution   Consider the effect of adding a fictitious image source of source strength m at $z = -a$. The complex potential of the flow field in the whole complex plane due to the two sources is given by

$$f(z) = \frac{m}{2\pi}[\text{Log}(z - a) + \text{Log}(z + a)].$$

When z assumes either real values or imaginary values, Im $f(z)$ becomes constant, indicating that both the real and imaginary axes are streamlines of the flow field. Effectively, we may replace the streamlines by non-penetrable walls without distorting the flow field since the fluid particles flow tangentially along the streamlines, as they would flow along the walls. Since we are seeking the solution in the right half-plane, it is appropriate to place the wall along the y-axis. By limiting the domain of definition to Re $z > 0$, the above $f(z)$ is already the solution for the complex potential of the given flow field. This method is called the *method of images* since the fictitious source can be considered as the mirror image of the physical source with respect to the wall (considered as the mirror). Several streamlines of the flow field are sketched in Figure 6.12.

The square of the speed of the fluid flow is given by

$$\frac{df}{dz}\frac{\overline{df}}{dz} = \frac{m^2}{4\pi^2}\left(\frac{1}{z+a} + \frac{1}{z-a}\right)\left(\frac{1}{\bar{z}+a} + \frac{1}{\bar{z}-a}\right)$$

$$= \frac{m^2}{\pi^2}\frac{x^2 + y^2}{(x^2 + y^2)^2 - 2a^2(x^2 - y^2) + a^4}.$$

According to the Blasius law of hydrodynamic force, the pressure force exerted on the wall is obtained by integrating along the y-axis from $y = \infty$ to $y = -\infty$ (following the anticlockwise sense of the integration contour) $\frac{\rho i}{2} \frac{df}{dz} \frac{\overline{df}}{dz}$ evaluated at $x = 0$. This gives

$$\overline{F} = \frac{\rho i}{2} \int_{\infty}^{-\infty} \left(\frac{df}{dz} \frac{\overline{df}}{dz} \right) \bigg|_{x=0} i \, dy, \quad dz = i \, dy$$

$$= \frac{\rho}{2} \int_{-\infty}^{\infty} \frac{m^2}{\pi^2} \frac{y^2}{(y^2 + a^2)^2} \, dy = \frac{\rho m^2}{4\pi a}.$$

As F is a real quantity, the direction of the net force is horizontal. The positive sign indicates that the net force is to the right, that is, the wall is sucked towards the source.

6.6.2 *Kutta–Joukowski's lifting force theorem*

Consider an airfoil immersed in an otherwise uniform potential flow with the free stream velocity U_∞ (see Figure 5.3), and a circulation of magnitude Γ (can be either positive or negative) developed around the airfoil. Using the Blasius law of hydrodynamic force, it can be shown that the airfoil experiences a lift of magnitude $\rho|\Gamma U_\infty|$ exerted by the potential flow field with fluid density ρ. The direction of the lifting force vector is obtained by rotating the free stream velocity vector through a right angle in the sense opposite to that of the circulation.

First, we observe that the general representation of the complex potential $f(z)$ of the flow field due to the presence of the airfoil together with circulation Γ around the airfoil is given by

$$f(z) = \overline{U_\infty} z + \frac{\Gamma}{2\pi i} \text{Log } z + \frac{b_1}{z} + \frac{b_2}{z^2} + \cdots, \tag{6.6.6}$$

given that the origin is placed inside the airfoil. To justify the above claim, the first term is seen to correspond to the free stream U_∞ at $|z| \to \infty$ since

$$\lim_{|z| \to \infty} \overline{f'(z)} = U_\infty. \tag{6.6.7}$$

The second term is seen to account for the circulation around the airfoil. This is because the circulation around the airfoil is given by

$$\text{Re} \oint_{\gamma_b} f'(z) \, dz = \text{Re}\big(2\pi i \, \text{Res}(f'(z), 0)\big)$$

$$= \text{Re}\left(2\pi i \, \frac{\Gamma}{2\pi i} \right) = \Gamma. \tag{6.6.8}$$

Applying the Blasius law of hydrodynamic lifting force, we use eq. (6.6.1) to compute the lifting force exerted on the airfoil by the potential fluid flow. To evaluate the contour integral, it suffices to consider the Laurent series expansion of the integrand function $[f'(z)]^2$ up to the first negative power term. The Laurent series expansion of $[f'(z)]^2$ is seen to be

$$[f'(z)]^2 = \overline{U_\infty}^2 + \frac{\Gamma \overline{U_\infty}}{\pi i} \frac{1}{z} + \cdots .$$

Accordingly, the lifting force exerted on the airfoil is given by

$$F = \frac{\rho i}{2} \oint_{\gamma_b} \overline{\left[\overline{U_\infty}^2 + \frac{\Gamma \overline{U_\infty}}{\pi i} \frac{1}{z} + \cdots \right] dz}$$

$$= 2\pi i \left(\frac{\rho}{2\pi} \Gamma \overline{U_\infty} \right) = -i\rho \Gamma U_\infty. \tag{6.6.9}$$

The magnitude of the lifting force equals $\rho |\Gamma U_\infty|$. Due to the factor $-i$, the direction of the lifting force is at right-angles to U_∞, rotating in the sense opposite to that of the circulation.

6.7 Problems

6.1. Suppose the radius of convergence of the Taylor series

$$\sum_{n=0}^{\infty} c_n z^n, \qquad c_n \ge 0,$$

equals 1; explain why $z = 1$ is a singular point.

6.2. Discuss the nature of the isolated singularity $z = 0$ of the following functions:

(a) $f_1(z) = \begin{cases} \sin z & z \ne 0 \\ 1 & z = 0 \end{cases}$;

(b) $f_2(z) = \begin{cases} \sin z & z \ne 0 \\ \infty & z = 0 \end{cases}$;

(c) $f_3(z) = \begin{cases} \sin \frac{1}{z} & z \ne 0 \\ 1 & z = 0 \end{cases}$.

6.3. Suppose $z = 0$ is a pole of orders n and m, respectively, of the functions $f_1(z)$ and $f_2(z)$. For the following functions, find the possible order of the pole $z = 0$:

(a) $f_1(z) + f_2(z)$; (b) $f_1(z) f_2(z)$; (c) $f_1(z) / f_2(z)$.

6.4. Suppose $z = z_0$ is an essential singularity of $f_1(z)$ and $f_2(z)$. Can the point $z = z_0$ become a removable singularity or a pole of the function $f_1(z) + f_2(z)$? If yes, construct such examples.

6.5. Find all the isolated singularities of each of the following functions in the unextended complex plane and classify the nature of these singularities:

(a) $\dfrac{z^2 + 1}{e^z}$; (b) $\dfrac{1}{e^z - 1} - \dfrac{1}{z}$; (c) $e^{-z} \cos \dfrac{1}{z}$; (d) $\sin \dfrac{1}{\sin \frac{1}{z}}$.

6.6. The function

$$f(z) = \frac{1}{z(z-1)^2}$$

has a pole of order 2 at $z = 1$, and it admits the following Laurent series expansion:

$$\frac{1}{z(z-1)^2} = \frac{1}{(z-1)^3} - \frac{1}{(z-1)^4} + \frac{1}{(z-1)^5} - \cdots + \cdots, \ |z-1| > 1.$$

We then conclude that $z = 1$ is an essential singularity since the Laurent series expansion has infinitely many negative power terms. Also, since the coefficient of $\frac{1}{z-1}$ is zero in the above expansion, we claim that

$$\text{Res}(f(z), 1) = 0.$$

Explain why the above claims on the nature of the singularity and residue value are incorrect.

6.7. For each of the following functions, find all the isolated singularities in the unextended complex plane and evaluate the residue value at each of these singularities.

(a) $\dfrac{z^2 - 1}{z^3(z^2 + 1)}$; (b) $\dfrac{\tan z}{1 - e^z}$; (c) $\dfrac{e^{i\alpha x}}{z^4 + \beta^4}$, α and β are real;

(d) $\dfrac{1}{z \sin z}$; (e) $\dfrac{e^{1/z}}{z}$.

6.8. Evaluate $\text{Res}\left(\dfrac{f'(z)}{f(z)}, \alpha\right)$ if

(a) α is a zero of order n of $f(z)$,
(b) α is a pole of order n of $f(z)$.

Let $g(z)$ be analytic at α; evaluate $\text{Res}(g(z) \dfrac{f'(z)}{f(z)}, \alpha)$ subject to the same conditions as in (a) and (b).

6.9. Classify the nature of the isolated singularity of each of the following functions:

(a) $\sin \dfrac{1}{z-1}$ at $z = 1$; (b) $\dfrac{1}{z^3(e^{z^3}-1)}$ at $z = 0$.

6.10. Suppose $f(z)$ and $g(z)$ are analytic at $z = z_0$, $f(z_0) \neq 0$, and $z = z_0$ is a zero of order 2 of $g(z)$. Express $\mathrm{Res}\left(\dfrac{f(z)}{g(z)}, z_0\right)$ in terms of the Taylor coefficients of $f(z)$ and $g(z)$ at $z = z_0$.

6.11. Locate the isolated singularities of the function $f(z) = \pi \cot \pi z$, and determine whether each is a removable singularity, a pole or an essential singularity. If the singularity is removable, give the limit of the function at the point. If the singularity is a pole, give the order of the pole, and compute the residue at the singularity.

6.12. Compute the residue at each of the isolated singularities of

$$f(z) = \frac{\cos z}{z^2(z-\pi)^3}.$$

6.13. What is the order of the pole of

$$f(z) = \frac{1}{(2\cos z - 2 + z^2)^2}$$

at $z = 0$? Compute $\mathrm{Res}\,(f, 0)$.

6.14. Compute

(a) $\mathrm{Res}\left(\tan z, \left(k + \tfrac{1}{2}\right)\pi\right)$, k is any integer;

(b) $\mathrm{Res}\left(\dfrac{z^{2n}}{(z-1)^n}, 1\right)$.

6.15. Suppose α is an isolated singularity of $f(z)$; show that

(a) $\mathrm{Res}(f, \alpha) = \mathrm{Res}(f, -\alpha)$ if $f(z)$ is odd;

(b) $\mathrm{Res}(f, \alpha) = -\mathrm{Res}(f, -\alpha)$ if $f(z)$ is even.

6.16. Let $a_1, a_2, \ldots, a_n$ be distinct points in the complex plane and suppose all of them lie within the circle $|z| = R$. Suppose $f(z)$ is analytic on and inside the circle; evaluate the integral

$$\oint_{|z|=R} \frac{f(z)}{(z-a_1)(z-a_2)\cdots(z-a_n)}\, dz.$$

6.17. A proper rational function $f(z)$ with only poles of first order can be represented by

$$f(z) = \frac{b_0 z^n + \cdots + b_{n-1} z + b_n}{(z-z_1)(z-z_2)\cdots(z-z_k)},$$

where $n < k$, and $z_1, z_2, \ldots, z_k$ are distinct. Show that the corresponding partial fraction decomposition of $f(z)$ takes the form

$$f(z) \qquad \frac{c_1}{} \qquad \frac{c_2}{} \qquad \qquad \frac{c_k}{}$$

where $c_j = \mathrm{Res}(f, z_j)$, $j = 1, 2, \ldots, k$. Apply the result to find the partial fraction decomposition of each of the following functions:

(a) $\dfrac{z+1}{(z-1)(z-2)(z-3)}$; (b) $\dfrac{z^2}{z^5+1}$; (c) $\dfrac{1}{z^n-1}$.

6.18. Show that the value of each of the following integrals is zero.

(a) $\displaystyle\oint_{|z|=1} ze^{2z}\,dz$; (b) $\displaystyle\oint_{|z|=1} \tanh z\,dz$;

(c) $\displaystyle\oint_{|z|=1} \frac{1}{\cos z}\,dz$; (d) $\displaystyle\oint_{|z|=3} \frac{1}{z^2+1}\,dz$;

(e) $\displaystyle\oint_{|z|=2} \frac{\tan z}{z}\,dz$.

6.19. Use the Cauchy residue theorem to evaluate the following integrals:

(a) $\displaystyle\oint_{|z|=2} \frac{z^4+z}{(z-1)^2}\,dz$; (b) $\displaystyle\oint_{|z|=2} \frac{z^3+3z+1}{z^4-5z^2}\,dz$;

(c) $\displaystyle\oint_{|z|=2} \frac{\sinh^2 z}{z^4}\,dz$; (d) $\displaystyle\oint_{|z-i|=2} \frac{e^z+z}{(z-1)^4}\,dz$;

(e) $\displaystyle\oint_{|z-i|=2} \frac{e^{-z}\sin z}{z^2}\,dz$;

(f) $\displaystyle\oint_{|z-i|=2} \frac{\sin z}{(z-i)^n}\,dz$, n is any positive integer.

6.20. Evaluate the following definite integrals. When the integral is improper, find its principal value if it exists.

(a) $\displaystyle\int_0^{2\pi} \frac{d\theta}{(a+b\cos\theta)^2}$, $\quad a > b > 0$;

(b) $\displaystyle\int_0^{2\pi} \frac{d\theta}{(a+b\cos^2\theta)^2}$, $\quad a > b > 0$;

(c) $\displaystyle\int_0^{2\pi} \frac{d\theta}{1-2a\cos\theta+a^2}$, $\quad a$ is a complex number and $a \neq \pm 1$.

6.21. Evaluate the following integrals:

(a) $\int_{-\infty}^{\infty} \dfrac{x}{(x^2 + 4x + 13)^2}\, dx$;   (b) $\int_{0}^{\infty} \dfrac{x^2}{(x^2 + a^2)^2}\, dx$, $a > 0$;

(c) $\int_{0}^{\infty} \dfrac{1}{(x^2 + 1)^n}\, dx$, n is a positive integer;

(d) $\int_{-\infty}^{\infty} \dfrac{1}{(x^2 + a^2)(x^2 + b^2)}\, dx$, $a > 0, b > 0$;   (e) $\int_{0}^{\infty} \dfrac{x^2 + 1}{x^4 + 1}\, dx$;

(f) $\int_{0}^{\infty} \dfrac{1}{1 + x^n}\, dx$, $n \geq 2$ and n is a positive integer;

(g) $\int_{-\infty}^{\infty} \dfrac{x^4}{1 + x^8}\, dx$.

6.22. Show that

$$\int_{0}^{2\pi} \ln |e^{i\theta} - 1|\, d\theta = 0,$$

and use the result to deduce

$$\int_{0}^{\pi} \ln \sin \theta\, d\theta = -\pi \ln 2.$$

6.23. Find the Fourier transform of each of the following functions:

(a) $u(t) = \dfrac{1}{1 + t^2}$;   (b) $u(t) = e^{-t^2}$;   (c) $u(t) = \begin{cases} 1 & |t| < \beta \\ 0 & |t| > \beta \end{cases}$.

6.24. Suppose $u_e(x)$ is an even function, where $u_e(-x) = u_e(x)$, $x > 0$. Show that

$$\mathcal{F}\{u_e(x)\} = 2U_c(w) = 2\int_{0}^{\infty} \cos wx\, u_e(x)\, dx,$$

and hence deduce the Fourier cosine transform inversion formula given in eq. (6.4.11).

6.25. Use the Parseval identity to show that

$$\frac{1}{2\pi} \int_{-\infty}^{\infty} \frac{|e^{-ia\omega} - e^{-ib\omega}|^2}{\omega^2}\, d\omega = b - a,$$

and deduce that

$$\frac{1}{\pi} \int_{-\infty}^{\infty} \frac{\sin a\omega \sin b\omega}{\omega^2}\, d\omega = \min(a, b), \quad a > 0 \text{ and } b > 0.$$

6.26. Show that each of the following functions satisfies the defining properties of the Dirac function $\delta(x)$ when the limit $\epsilon \to 0$ is taken:

(a) $\dfrac{1}{\sqrt{\pi\epsilon}}e^{-x^2/\epsilon}$;   (b) $\dfrac{\epsilon}{\pi(x^2+\epsilon^2)}$.

6.27. Evaluate the following Fourier integrals:

(a) $\displaystyle\int_{-\infty}^{\infty}\dfrac{x\cos x}{x^2-2x+10}\,dx$;

(b) $\displaystyle\int_{-\infty}^{\infty}\dfrac{x\sin x}{x^2+4x+20}\,dx$;

(c) $\displaystyle\int_{-\infty}^{\infty}\dfrac{\cos ax}{x^2+b^2}\,dx$, a and b are positive;

(d) $\displaystyle\int_{-\infty}^{\infty}\dfrac{x\sin ax}{x^2+b^2}\,dx$, a and b are positive;

(e) $\displaystyle\int_{0}^{\infty}\dfrac{\sin ax}{x(x^2+1)}\,dx$.

6.28. Let $f(z) = e^{imz}F(z)$, $m > 0$, and the function $F(z)$ have the following properties:

(a) it has a finite number of isolated singularities $z_1,\ z_2,\ldots,z_n$ in the upper half-plane;
(b) it is analytic at all points on the real axis, except at the points a_1, $a_2,\ldots,a_m$, which are *simple* poles;
(c) $F(z) \to 0$ if $z \to \infty$ and $\mathrm{Im}\,z \geq 0$.

Show that

$$P\int_{-\infty}^{\infty} f(x)\,dx = 2\pi i\left[\sum_{k=1}^{n}\mathrm{Res}(f(z),z_k)+\frac{1}{2}\sum_{k=1}^{m}\mathrm{Res}(f(z),a_k)\right].$$

6.29. Evaluate

$$\int_{-\infty}^{\infty}\frac{xe^x}{e^{4x}+1}\,dx.$$

Hint:　Choose the closed rectangular contour whose four sides are

$$\ell_1 : \{(x, y) : -R \le x \le R, y = 0\}$$
$$\ell_2 : \left\{(x, y) : x = R, 0 \le y \le \frac{\pi}{2}\right\}$$
$$\ell_3 : \left\{(x, y) : -R \le x \le R, y = \frac{\pi}{2}\right\}$$
$$\ell_4 : \left\{(x, y) : x = -R, 0 \le y \le \frac{\pi}{2}\right\}.$$

6.30. Evaluate

$$\int_0^\infty \frac{\ln x}{x^2 + 4}\, dx.$$

Note that $\ln x$ has a singularity at $x = 0$. The improper integral is thus defined as

$$\lim_{\substack{\epsilon \to 0 \\ R \to \infty}} \int_\epsilon^R \frac{\ln x}{x^2 + 4}\, dx.$$

6.31. Show that

$$\int_0^\infty \frac{\ln(x^2 + 1)}{x^2 + 1}\, dx = \pi \ln 2.$$

Hint:　Use the relation

$$\text{Log}(i - x) + \text{Log}(i + x) = \text{Log}(i^2 - x^2) = \ln(x^2 + 1) + \pi i.$$

6.32. Suppose $a > 0$; show that

$$\frac{1}{2\pi i} \int_{a-i\infty}^{a+i\infty} \frac{e^{zt}}{z}\, dz = \begin{cases} 1 & t > 0 \\ 0 & t < 0 \end{cases}.$$

The path of contour integration is along the infinite vertical line $\text{Re}\, z = a$.

6.33. The following integral occurs in the quantum theory of atomic collisions:

$$I = \int_{-\infty}^\infty \frac{\sin t}{t}\, e^{ipt}\, dt,$$

where p is real. Show that

$$I = \begin{cases} 0 & |p| > 1 \\ \pi & |p| < 1 \end{cases}.$$

Discuss the cases when $p = \pm 1$.

6.34. Suppose the analytic functions $P(z)$ and $Q(z)$ both have a zero at z_0; prove that the L'Hospital rule is given by

$$\lim_{z \to z_0} \frac{P(z)}{Q(z)} = \lim_{z \to z_0} \frac{P'(z)}{Q'(z)}$$

6.35. Show that

$$\int_{-\infty}^{\infty} \frac{x^{2p} - x^{2q}}{1 - x^{2r}} \, dx = \frac{\pi}{r} \left(\cos \frac{2p+1}{r} \pi - \cos \frac{2q+1}{r} \pi \right),$$

where p, q, r are non-negative integers, $p < r$ and $q < r$.

Hint: Choose the closed contour of integration that includes the infinitely large semi-circle in the upper half-plane.

6.36. Suppose $f(z)$ has poles $\alpha_1, \alpha_2, \ldots, \alpha_n$, none of which lies on the positive real axis or equals zero, and there exists a real number p such that

$$\lim_{z \to 0} \left[z^{p+1} f(z) \right] = \lim_{z \to \infty} \left[z^{p+1} f(z) \right] = 0.$$

Prove that

$$\int_0^{\infty} x^p f(x) \, dx$$

$$= \begin{cases} -\dfrac{\pi}{\sin p\pi} e^{-p\pi i} \displaystyle\sum_{k=1}^n \mathrm{Res}(z^p f(z), \alpha_k) & p \text{ is not an integer} \\[2ex] -\displaystyle\sum_{k=1}^n \mathrm{Res}(z^p f(z) \log z, \alpha_k) & p \text{ is an integer.} \end{cases}$$

6.37. Suppose we integrate e^{iz^2} along the closed wedge-shaped contour shown in the figure below, and take the limit $R \to \infty$ subsequently; show that

$$\int_0^{\infty} \cos x^2 \, dx = \int_0^{\infty} \sin x^2 \, dx = \frac{\sqrt{2\pi}}{4}.$$

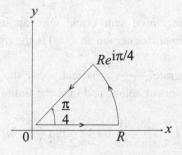

6.38. Show that

$$\int_0^\infty e^{-ax^2} \cos bx^2 \, dx = \frac{\sqrt{2\pi}}{4\sqrt{a^2+b^2}} \sqrt{a + \sqrt{a^2+b^2}}, \quad a > 0.$$

Hint: Use a wedge-shaped closed contour similar to that in Problem 6.37, where the angle of the wedge should be chosen appropriately according to the parameter values a and b.

6.39. Evaluate the Cauchy principal value of each of the following integrals:

(a) $\displaystyle\int_{-\infty}^\infty \frac{xe^{ix}}{x^2 - \pi^2} \, dx;$ (b) $\displaystyle\int_{-\infty}^\infty \frac{e^{imx}}{(x-1)(x-2)} \, dx, \quad m > 0.$

6.40. Show that

$$\int_0^\infty \frac{\cos(\ln x)}{1+x^2} \, dx = \frac{\pi}{2\cosh\frac{\pi}{2}}.$$

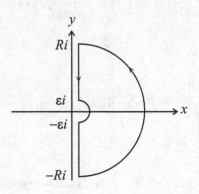

Hint: Integrate the function

$$f(z) = \frac{z^i}{z^2 - 1}$$

along the closed semi-circle with an infinite radius and an infinitesimal indentation at $z = 0$ in the right-hand plane.

6.41. Suppose we integrate $\frac{e^{az}}{\sinh \pi z}$, a is real, along the indented closed rectangular contour shown below, and take the limits $L \to \infty$ and $\epsilon \to 0$; show that

$$\int_0^\infty \frac{\sinh ax}{\sinh \pi x} \, dx = \frac{1}{2} \tan \frac{a}{2}, \quad |a| < \pi.$$

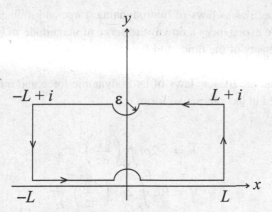

6.42. By considering the contour integral

$$\oint_C \frac{1 - e^{2iz}}{z^2} \, dz,$$

where C is the closed contour depicted in Figure 6.10, show that

$$\int_0^\infty \frac{\sin^2 x}{x^2} \, dx = \frac{\pi}{2}.$$

Use a similar technique to evaluate

$$\int_0^\infty \frac{\sin^3 x}{x^3} \, dx.$$

6.43. Consider the potential flow over the circle $|z| = a$, where the free stream velocity U_∞ is aligned with the positive x-axis and a circulation $\Gamma > 0$ is developed around the circle. Show that the complex potential of the flow field is given by

$$f(z) = U_\infty \left(z + \frac{a^2}{z} \right) + \frac{\Gamma}{2\pi i} \, \text{Log} \, \frac{z}{a}.$$

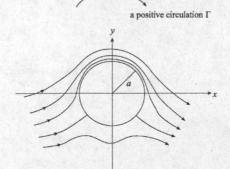

a positive circulation Γ

Using the Blasius laws of hydrodynamic force and moment, show that the circle experiences a downward force of magnitude $\rho U_\infty \Gamma$, where ρ is the density of the fluid, and the hydrodynamic moment acting on the circle is zero.

6.44. Show that the Blasius laws of hydrodynamic force and moment can be expressed in the alternative forms

$$\overline{F} = -2\rho i \oint_\gamma \left(\frac{\partial \psi}{\partial z}\right)^2 dz,$$

$$M = \text{Re} \oint_\gamma 2\rho z \left(\frac{\partial \psi}{\partial z}\right)^2 dz,$$

where $\psi(z)$ is the stream function of the fluid flow.

Hint: Use the relation

$$u - iv = -2i \frac{\partial \psi}{\partial z}$$

and

$$u + iv = 2i \frac{\partial \psi}{\partial \overline{z}},$$

where u and v are the x- and y-components of velocity, respectively. Further, on the boundary of the body, we have

$$0 = d\psi = \frac{\partial \psi}{\partial z} dz + \frac{\partial \psi}{\partial \overline{z}} d\overline{z}.$$

By combining these relations, show that

$$d\overline{F} = -2\rho i \left(\frac{\partial \psi}{\partial z}\right)^2 dz.$$

7

Boundary Value Problems and
Initial-Boundary Value Problems

In the earlier chapters, we have analyzed several prototype potential field problems, including potential fluid flows, steady state temperature distribution, electrostatics problems and gravitational potential problems. All of these potential field problems are governed by the Laplace equation. There is no time variable in these problems, and the characterization of individual physical problems is exhibited by the corresponding prescribed boundary conditions. The mathematical problem of finding the solution of a partial differential equation that satisfies the prescribed boundary conditions is called a *boundary value problem*, of which there are two main types: *Dirichlet problems* where the boundary values of the solution function are prescribed, and *Neumann problems* where the values of the normal derivative of the solution function along the boundary are prescribed. In other physical problems, like the heat conduction and wave propagation models, the time variable is also involved in the model. To describe fully the partial differential equations modeling these problems, one needs to prescribe both the associated boundary conditions and the initial conditions. The latter class is called an *initial-boundary value problem*. This chapter discusses some of the solution methodologies for solving boundary value problems and initial-boundary value problems using complex variables methods.

The link between analytic functions and harmonic functions is exhibited by the fact that both the real and imaginary parts of a complex function that is analytic inside a domain satisfy the Laplace equation in the same domain. The Gauss mean value theorem (see Subsection 4.3.3) states that the value of a harmonic function at the center of any circle inside the domain of harmonicity equals the average of the values of the function along the boundary of the circle. In Section 7.1, we generalize the above result by establishing two forms of integral representation of harmonic functions: the Poisson integral formula and the Schwarz integral formula. We also discuss the properties of solutions to the

311

Dirichlet and Neumann problems. In particular, the compatibility conditions required for the existence of solutions to Neumann problems are examined. In Section 7.2, we discuss the Laplace transform and its inversion method. Though the Laplace transform is applied to functions that are real, its inversion formula involves the evaluation of the Bromwich contour integral in the complex plane. The last section is devoted to the discussion of application of the Laplace transform method to solutions of two prototype initial-boundary value problems, namely, heat conduction in a thin rod and longitudinal oscillations of an elastic thin rod.

7.1 Integral formulas of harmonic functions

Recall that the Cauchy integral formula gives the value of an analytic function f inside a contour C in terms of its values on the contour:

$$ f(z) = \frac{1}{2\pi i} \oint_C \frac{f(\zeta)}{\zeta - z} \, d\zeta, \quad z \text{ lies inside } C. \tag{7.1.1} $$

This harmonic-analytic dualism leads us to expect that the solution to the Laplace equation inside a domain $\mathcal{D}$ can be expressed in terms of the values of the function along the boundary of $\mathcal{D}$ in the form of an integral. In this section, we derive the integral representation formulas of harmonic functions with the Dirichlet-type and Neumann-type boundary conditions. The types of domain are limited to circles and the upper half-plane.

7.1.1 Poisson integral formula

We would like to find the integral representation of a function $u(x, y)$ that is harmonic on and inside the circle $|z| = R$ together with the pre-scribed boundary condition $u(R, \theta)$ along the boundary of the circle. Let $f(z) = u(x, y) + i v(x, y)$, $z = x + iy$, be analytic on and inside the circle $|z| = R$. By the Cauchy integral formula, we have

$$ f(z) = \frac{1}{2\pi i} \oint_{|\zeta|=R} \frac{f(\zeta)}{\zeta - z} \, d\zeta. \tag{7.1.2} $$

The symmetry point of z with respect to the circle $|z| = R$ is given by (see Subsection 1.3.2)

$$ z_s = \frac{R^2}{\bar{z}} = \frac{\zeta \bar{\zeta}}{\bar{z}}, $$

where ζ is a point on the circle (see Figure 7.1). Since the symmetry point z_s is outside the circle, the function $\frac{f(\zeta)}{\zeta - z_s}$ is analytic on and inside $|\zeta| = R$; we then

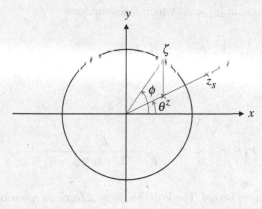

Figure 7.1. The pair of symmetry points z and z_s with respect to the circle $|\zeta| = R$ are on the same ray through the center. The polar forms of z and ζ are $re^{i\theta}$ and $Re^{i\phi}$, respectively, and $|\zeta - z|^2 = R^2 - 2Rr\cos(\phi - \theta) + r^2$.

have

$$f(z) = \frac{1}{2\pi i} \oint_{|\zeta|=R} f(\zeta) \left(\frac{1}{\zeta - z} - \frac{1}{\zeta - z_s} \right) d\zeta, \qquad (7.1.3)$$

by virtue of the Cauchy–Goursat integral theorem. Consider the quantity

$$\frac{\zeta}{\zeta - z} - \frac{\zeta}{\zeta - z_s} = \frac{\zeta}{\zeta - z} - \frac{\zeta}{\zeta - \frac{\zeta\bar{\zeta}}{\bar{z}}} = \frac{\zeta}{\zeta - z} + \frac{\bar{z}}{\bar{\zeta} - \bar{z}} = \frac{|\zeta|^2 - |z|^2}{|\zeta - z|^2},$$

which is manifestly real. We write $\zeta = Re^{i\phi}$ and $z = re^{i\theta}$; then the above quantity can be expressed as

$$\frac{|\zeta|^2 - |z|^2}{|\zeta - z|^2} = \frac{R^2 - r^2}{\zeta\bar{\zeta} - (\zeta\bar{z} + \bar{\zeta}z) + z\bar{z}} = \frac{R^2 - r^2}{R^2 - 2Rr\cos(\phi - \theta) + r^2}.$$

By observing that

$$d\zeta = iRe^{i\phi}\,d\phi = i\zeta\,d\phi,$$

the contour integral in eq. (7.1.3) can be expressed as

$$f(z) = \frac{1}{2\pi} \int_0^{2\pi} \frac{R^2 - r^2}{R^2 - 2Rr\cos(\phi - \theta) + r^2} f(Re^{i\phi})\,d\phi. \qquad (7.1.4)$$

Finally, we take the real parts of both sides of the above equation. As a result, we obtain the *Poisson integral formula* for the solution of the Dirichlet problem in the circular domain $|z| < R$, which takes the form

$$u(r, \theta) = \frac{1}{2\pi} \int_0^{2\pi} P(R, r, \phi - \theta) u(R, \phi) \, d\phi, \quad r < R. \tag{7.1.5}$$

The kernel

$$P(R, r, \phi - \theta) = \frac{R^2 - r^2}{R^2 - 2Rr \cos(\phi - \theta) + r^2} \tag{7.1.6}$$

is called the *Poisson kernel*. The Poisson integral formula resembles an integral transform of the boundary value $u(R, \phi)$, with $\frac{1}{2\pi} P(R, r, \phi - \theta)$ as the transform kernel. This integral formula implies that $u(r, \theta)$, $r < R$, is determined completely by its boundary value $u(R, \phi)$ on the circle.

In the above derivation of the Poisson integral formula, it has been assumed that f is analytic on and inside the circle $|z| = R$. This implies implicitly that the real part $u(R, \theta)$ would be a continuous function of θ on the circle. However, the continuity of the boundary value function $u(R, \theta)$ can be relaxed. It can be proved rigorously that the Poisson integral formula holds even when $u(R, \theta)$ is piecewise continuous.

Fourier series expansion

The Poisson kernel can be expressed as

$$P(R, r, \phi - \theta) = \frac{\zeta}{\zeta - z} + \frac{\bar{z}}{\bar{\zeta} - \bar{z}} = \frac{z}{\zeta - z} + \frac{\bar{\zeta}}{\bar{\zeta} - \bar{z}} = \frac{1}{2} \left(\frac{\zeta + z}{\zeta - z} + \frac{\bar{\zeta} + \bar{z}}{\bar{\zeta} - \bar{z}} \right)$$

$$= \operatorname{Re} \left(\frac{\zeta + z}{\zeta - z} \right) = \operatorname{Re} \left(-1 + \frac{2}{1 - \frac{z}{\zeta}} \right)$$

$$= \operatorname{Re} \left(-1 + \frac{2}{1 - \frac{r}{R} e^{-i(\phi - \theta)}} \right).$$

Since $r < R$, the Poisson kernel can be decomposed into the following Fourier cosine series:

$$P(R, r, \phi - \theta) = 1 + 2 \sum_{n=1}^{\infty} \left(\frac{r}{R} \right)^n \cos n(\phi - \theta).$$

The harmonic function $u(r, \theta)$ can then be expressed as

$$u(r, \theta) = \frac{1}{2\pi} \int_0^{2\pi} u(R, \phi) \, d\phi$$

$$+ \frac{1}{\pi} \sum_{n=1}^{\infty} \int_0^{2\pi} u(R, \phi) \left(\frac{r}{R}\right)^n \cos n(\phi - \theta) \, d\phi$$

$$= \frac{1}{2\pi} \int_0^{2\pi} u(R, \phi) \, d\phi$$

$$+ \sum_{n=1}^{\infty} \left(\frac{1}{\pi} \int_0^{2\pi} u(R, \phi) \cos n\phi \, d\phi\right) \left(\frac{r}{R}\right)^n \cos n\theta$$

$$+ \sum_{n=1}^{\infty} \left(\frac{1}{\pi} \int_0^{2\pi} u(R, \phi) \sin n\phi \, d\phi\right) \left(\frac{r}{R}\right)^n \sin n\theta.$$

We write

$$a_n = \frac{1}{\pi} \int_0^{2\pi} u(R, \phi) \cos n\phi \, d\phi, \quad n = 0, 1, 2, \ldots,$$

$$b_n = \frac{1}{\pi} \int_0^{2\pi} u(R, \phi) \sin n\phi \, d\phi, \quad n = 1, 2, \ldots.$$

Here, a_n and b_n are recognized as the Fourier coefficients in the Fourier expansion of the boundary value function $u(R, \phi)$. The Fourier series expansion of $u(r, \theta)$ can then be expressed as

$$u(r, \theta) = \frac{a_0}{2} + \sum_{n=1}^{\infty} a_n \left(\frac{r}{R}\right)^n \cos n\theta + \sum_{n=1}^{\infty} b_n \left(\frac{r}{R}\right)^n \sin n\theta. \quad (7.1.7)$$

Properties of the Poisson kernel

(i) Since $\zeta = Re^{i\phi}$, the function

$$\psi(z) = \frac{\zeta + z}{\zeta - z}$$

is an analytic function of z inside the circle $|z| = R$. Therefore, its real part $P(R, r, \phi - \theta)$ is harmonic inside the same circle.

(ii) Note that the function $u(r, \theta) = 1$ is harmonic inside the domain $|z| < R$ and satisfies the boundary condition $u(R, \theta) = 1$. By setting $u(r, \theta) = 1$ and $u(R, \theta) = 1$ in the Poisson integral formula, we then obtain

$$\frac{1}{2\pi} \int_0^{2\pi} P(R, r, \phi - \theta) \, d\phi = 1. \quad (7.1.8)$$

(iii) For fixed values of R, r and θ, the maximum value of $P(R, r, \phi - \theta)$ over $\phi \in [0, 2\pi]$ occurs when its denominator attains its minimum value. This occurs at $\phi = \theta$, for any value of r. The corresponding maximum value is found to be

$$\max_{0 \leq \phi \leq 2\pi} P(R, r, \phi - \theta) = \frac{R + r}{R - r},$$

which tends to infinity as $r \to R$.

Similarly, the minimum value of $P(R, r, \phi - \theta)$ occurs at either $\phi = \theta + \pi$ or $\phi = \theta - \pi$ (only one of these two values falls inside $[0, 2\pi]$), corresponding to the points where $R^2 - 2Rr \cos(\phi - \theta) + r^2$ attains its maximum value. We then have

$$\min_{0 \leq \phi \leq 2\pi} P(R, r, \phi - \theta) = \frac{R - r}{R + r} \to 0 \quad \text{as } r \to R.$$

(iv) Let $f(\theta)$ be piecewise continuous over $[0, 2\pi]$. For each $\theta \in [0, 2\pi]$, we state without proof the following result:

$$\lim_{r \to R} \frac{1}{2\pi} \int_0^{2\pi} P(R, r, \phi - \theta) f(\phi) \, d\phi$$

$$= \begin{cases} f(\theta) & \text{if } f \text{ is continuous at } \theta \\ \alpha f(\theta^-) + (1 - \alpha) f(\theta^+), \ 0 < \alpha < 1 & \text{if } f \text{ has a finite jump at } \theta. \end{cases}$$

$$(7.1.9)$$

The value α depends on the angle of approach to the boundary point $\zeta = Re^{i\phi}$ from the interior of the circle (see Problem 7.5 and Example 7.1.1). In particular, α equals $\frac{1}{2}$ when the boundary point $\zeta = Re^{i\phi}$ is approached radially.

Uniqueness of solution for Dirichlet problems

The integral representation of the solution to the Dirichlet problem for a circular domain was established in eq. (7.1.5). Is that solution unique? The uniqueness of the solution for Dirichlet problems with arbitrary domain can be established quite readily using the maximum principle for harmonic functions.

Let $\phi(z)$ be the boundary value prescribed along the boundary of the domain for a Dirichlet problem. Suppose the Dirichlet problem admits two solutions $u_1(z)$ and $u_2(z)$ inside the domain $\mathcal{D}$, and both solutions satisfy the prescribed boundary condition along the boundary $\partial \mathcal{D}$. Define $u = u_1 - u_2$; then by linearity of the Laplace equation, u is also harmonic in $\mathcal{D}$ since u_1 and u_2 are both harmonic in $\mathcal{D}$. The boundary condition for u becomes

$$u(z) = \phi(z) - \phi(z) = 0, \quad \text{for } z \in \partial \mathcal{D}.$$

Using the maximum principle for harmonic functions (see Subsection 4.3.3), the maximum value of $u(z)$ for any point z inside $\mathcal{D}$ must be zero since its boundary value is zero. We then have $u(z) = 0$, or equivalently $u_1 = u_2$, throughout $\mathcal{D}$. The uniqueness of the solution is then established. An alternative proof using Green's theorem is outlined in Problem 7.10.

Example 7.1.1 An infinitely long metal cylinder of radius R is cut into two halves. The upper half is grounded and the lower half is maintained at a fixed potential K. Find the electrostatic potential at points inside the cylinder.

Solution The infinite length of the metal cylinder fits well the two-dimensional consideration of the problem. The posed problem is equivalent to the Dirichlet problem of finding the harmonic function in the domain $|z| < R$ with boundary condition

$$f(\phi) = \begin{cases} 0 & 0 < \phi < \pi \\ K & \pi < \phi < 2\pi \end{cases}.$$

There is a discontinuity at both $\phi = 0$ and $\phi = \pi$. Using the Poisson integral formula, the solution to the electrostatic potential $u(r, \theta)$ is given by

$$u(r, \theta) = \frac{1}{2\pi} \int_{\pi}^{2\pi} K \frac{R^2 - r^2}{R^2 - 2Rr \cos(\phi - \theta) + r^2} \, d\phi.$$

To evaluate the integral, we set $t = \tan\frac{\phi - \theta}{2}$ so that $d\phi = \frac{2}{1+t^2} dt$. The integral is then reduced to

$$
\begin{aligned}
u(r, \theta) &= \frac{K}{2\pi} \int_{\tan(\frac{\pi}{2} - \frac{\theta}{2})}^{\tan(\pi - \frac{\theta}{2})} \frac{R^2 - r^2}{R^2 - 2Rr \frac{1-t^2}{1+t^2} + r^2} \frac{2}{1+t^2} \, dt \\
&= \frac{K}{\pi} \int_{\tan(\frac{\pi}{2} - \frac{\theta}{2})}^{\tan(\pi - \frac{\theta}{2})} \frac{(R+r)/(R-r)}{1 + (R+r)^2 t^2 / (R-r)^2} \, dt \\
&= \frac{K}{\pi} \tan^{-1} \left[\frac{R+r}{R-r} \tan\left(\frac{\phi - \theta}{2}\right) \right]_{\pi}^{2\pi} \\
&= \frac{K}{\pi} \tan^{-1} \left(\frac{R^2 - r^2}{2Rr \sin\theta} \right).
\end{aligned}
$$

The equation of the equipotential line $u(r, \theta) = u_0$, $0 < u_0 < K$, is given by

$$u_0 = \frac{K}{\pi} \tan^{-1} \left(\frac{R^2 - r^2}{2Rr \sin\theta} \right), \quad r < R.$$

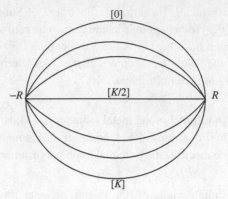

Figure 7.2. The equipotential lines of the charged cylinder are circular arcs passing through the two points $z = R$ and $z = -R$.

By substituting $r^2 = x^2 + y^2$, $r \sin \theta = y$, and rearranging the terms, we obtain

$$x^2 + y^2 + \left(2R \tan \frac{\pi u_0}{K}\right) y = R^2, \quad x^2 + y^2 < R^2.$$

The equipotential lines are circular arcs passing through the two points $z = R$ and $z = -R$. These two points exhibit a jump of discontinuity in the boundary value. The equipotential lines of the charged cylinder are shown in Figure 7.2. The pattern of the equipotential lines reveals that the electric potential values at points close to $z = R$ and $z = -R$ depend on the direction of approach to that boundary point [see also eq. (7.1.9)].

Example 7.1.2   Find the function $u(r, \theta)$ that is harmonic inside the domain $\{z : |z| < R \text{ and } \operatorname{Im} z > 0\}$ and satisfies the boundary condition

$$u(R, \theta) = \begin{cases} f(\theta) & 0 < \theta < \pi, \quad r = R \\ 0 & \theta = 0 \text{ or } \theta = \pi, \quad r < R \end{cases}.$$

Solution   The boundary value function $f(\theta)$ has a discontinuity at $\theta = 0$ and $\theta = \pi$. Since $u(R, 0) = u(R, \pi) = 0$, we may try to consider the odd extension of $f(\theta)$ over $(\pi, 2\pi)$ by defining

$$f(\theta) = -f(2\pi - \theta) \quad \text{for} \quad \pi < \theta < 2\pi.$$

Now, with the boundary value of the extended $f(\theta)$ prescribed along the whole circumference of the circle, the solution to the Dirichlet problem in the upper

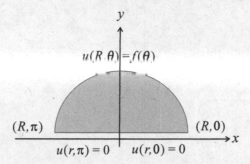

Figure 7.3. The Dirichlet problem in a semi-circular domain is formulated by prescribing the boundary condition along the upper semi-circle and the diameter.

half-disk is given by

$$u(r, \theta) = \frac{1}{2\pi} \int_0^\pi P(R, r, \phi - \theta) f(\phi) \, d\phi$$

$$- \frac{1}{2\pi} \int_\pi^{2\pi} P(R, r, \phi - \theta) f(2\pi - \phi) \, d\phi$$

$$= \frac{1}{2\pi} \int_0^\pi [P(R, r, \phi - \theta) - P(R, r, \phi + \theta)] f(\phi) \, d\phi,$$

$$r < R, \ 0 < \theta < \pi,$$

where $P(R, r, \phi - \theta)$ is the Poisson kernel.

Since $P(R, r, \phi - \theta)$ and $P(R, r, \phi + \theta)$ are harmonic inside $|z| < R$, $u(r, \theta)$ should also be harmonic in the same domain. One can check easily that

$$u(r, 0) = u(r, \pi) = 0, \quad r < R,$$

hence the solution satisfies the boundary condition along the real axis. Also, the boundary condition on the upper half-disc is satisfied, by virtue of eq. (7.1.9). Hence, $u(r, \theta)$ as defined above solves the given Dirichlet problem.

7.1.2 Schwarz integral formula

Next we consider the problem of finding the real function that is harmonic in the upper half-plane $\text{Im } z > 0$ and satisfies the prescribed Dirichlet boundary condition along the real axis. Again, we derive the integral representation of the solution in terms of the boundary values.

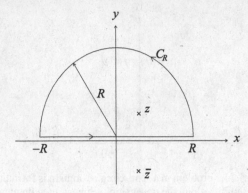

Figure 7.4. The point $\bar{z}$ lies outside the closed contour C since it is the mirror image of z with respect to the real axis.

Let $f(z)$ be analytic and satisfy the order property $|z^k f(z)| < M$, for some positive constants k and M, in the closed upper half-plane $\operatorname{Im} z \geq 0$. Let C denote the positively oriented closed contour which consists of the upper half semi-circle C_R with radius R and the line segment from $-R$ to R along the real axis (see Figure 7.4). Let z be any point inside C. By the Cauchy integral formula, we have

$$
\begin{aligned}
f(z) &= \frac{1}{2\pi i} \oint_C \left(\frac{1}{\zeta - z} - \frac{1}{\zeta - \bar{z}} \right) f(\zeta)\, d\zeta \\
&= \frac{1}{2\pi i} \left[\int_{C_R} \left(\frac{1}{\zeta - z} - \frac{1}{\zeta - \bar{z}} \right) f(\zeta)\, d\zeta \right. \\
&\quad \left. + \int_{-R}^{R} \left(\frac{1}{\zeta - z} - \frac{1}{\zeta - \bar{z}} \right) f(\zeta)\, d\zeta \right].
\end{aligned}
\tag{7.1.10}
$$

Note that the contribution from the second term in each of the above integrals is zero since $\bar{z}$ lies outside the contour C.

Since $|f(\zeta)| < \frac{M}{R^k}$ for points ζ lying on C_R, the asymptotic order of the modulus of the integral satisfies

$$
\left| \frac{1}{2\pi i} \int_{C_R} \left(\frac{1}{\zeta - z} - \frac{1}{\zeta - \bar{z}} \right) f(\zeta)\, d\zeta \right| \sim O\left(\frac{1}{R^k} \right), \quad k > 0,
\tag{7.1.11}
$$

which tends to zero as R goes to infinity. Combining eqs. (7.1.10) and (7.1.11), and taking the limit $R \to \infty$, we then obtain

$$
\begin{aligned}
f(z) &= \frac{1}{2\pi i} \int_{-\infty}^{\infty} \left(\frac{1}{t - z} - \frac{1}{t - \bar{z}} \right) f(t)\, dt \\
&= \frac{1}{2\pi i} \int_{-\infty}^{\infty} \frac{z - \bar{z}}{|t - z|^2} f(t)\, dt, \quad \operatorname{Im} z > 0.
\end{aligned}
$$

Suppose we write $f(z) = u(x, y) + iv(x, y)$, $z = x + iy$. By taking the real parts on both sides of the above equation, we obtain

$$u(x, y) = \frac{1}{\pi} \int_{-\infty}^{\infty} \frac{y}{(t - x)^2 + y^2} u(t, 0) \, dt$$

$$= \frac{1}{\pi} \int_{-\infty}^{\infty} S(t - x, y) \, u(t, 0) \, dt, \quad y > 0, \qquad (7.1.12)$$

where the kernel function

$$S(t - x, y) = \frac{y}{(t - x)^2 + y^2}$$

is called the *Schwarz kernel*. The above integral representation formula is called the *Schwarz integral formula*.

In the above proof, we require $f(z)$ to be analytic and satisfy the order property $|z^k f(z)| < M$ in the closed region Im $z \geq 0$. However, it can be proved rigorously that the Schwarz integral formula remains valid with less stringent requirements on the boundary value function $u(x, 0)$. The precise requirements are that $u(x, 0)$ is bounded for all x and continuous except for at most a finite number of finite jumps.

The Schwarz kernel possesses properties similar to those of the Poisson kernel. Some of the obvious ones are:

(i) $S(t - x, y)$ is harmonic inside the domain $y > 0$.
(ii) Since the Schwarz integral formula is satisfied by setting $u(x, y) = 1$ and $u(t, 0) = 1$, we have

$$\frac{1}{\pi} \int_{-\infty}^{\infty} S(t - x, y) \, dt = 1, \quad y > 0. \qquad (7.1.13)$$

(iii) The maximum value of $S(t - x, \ y)$ as a function of t over $(-\infty, \infty)$ occurs at $t = x$, and

$$\max_{-\infty < t < \infty} S(t - x, y) = S(0, y) = \frac{1}{y} \to \infty \text{ as } y \to 0. \qquad (7.1.14)$$

Example 7.1.3 Find the function $u(x, y)$ that is harmonic inside the first quadrant $x > 0$ and $y > 0$, and satisfies the boundary conditions

$$\begin{cases} u(x, 0) = f(x), & 0 < x < \infty \\ u(0, y) = g(y), & 0 < y < \infty \end{cases},$$

where $f(x)$ and $g(y)$ are bounded and continuous except for at most a finite number of finite jumps (see Figure 7.5).

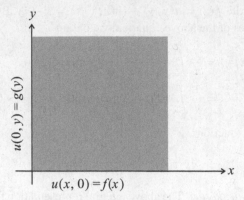

Figure 7.5. Dirichlet problem in the first quadrant, $x > 0$ and $y > 0$, with prescribed boundary conditions: $u(x, 0) = f(x)$ and $u(0, y) = g(y)$.

Solution   First, we find the harmonic functions $u_1(x, y)$ and $u_2(x, y)$ such that:

(i) $u_1(x, y)$ is harmonic in the first quadrant $x > 0$ and $y > 0$, and satisfies the boundary conditions

$$\begin{cases} u_1(x, 0) = f(x), & 0 < x < \infty \\ u_1(0, y) = 0, & 0 < y < \infty \end{cases};$$

(ii) $u_2(x, y)$ is harmonic in the first quadrant $x > 0$ and $y > 0$, and satisfies the boundary conditions

$$\begin{cases} u_2(x, 0) = 0, & 0 < x < \infty \\ u_2(0, y) = g(y), & 0 < y < \infty \end{cases}.$$

Since the Laplace equation is linear, by virtue of the principle of superposition of solutions, $u_1 + u_2$ is also harmonic. As the boundary conditions for $u_1(x, y) + u_2(x, y)$ are the same as those for $u(x, y)$, by virtue of uniqueness of solution, we then have

$$u(x, y) = u_1(x, y) + u_2(x, y).$$

To solve for $u_1(x, y)$, given the zero boundary condition $u_1(0, y) = 0$ along the positive y-axis, we perform the odd extension of the boundary value function $f(x)$ over the interval $(-\infty, 0)$ by defining

$$f(x) = -f(-x), \quad -\infty < x < 0.$$

By the Schwarz integral formula (7.1.12), the solution for $u_1(x, y)$ is given by

$$u_1(x, y) = \frac{1}{\pi} \int_0^\infty \frac{y}{(t-x)^2 + y^2} f(t)\, dt - \frac{1}{\pi} \int_{-\infty}^0 \frac{y}{(t-x)^2 + y^2} f(-t)\, dt \quad \text{(i)}$$

$$= \frac{y}{\pi} \int_0^\infty \left[\frac{1}{(t-x)^2 + y^2} - \frac{1}{(t+x)^2 + y^2} \right] f(t)\, dt, \quad x > 0 \text{ and } y > 0.$$

Obviously, the boundary condition $u_1(0, y) = 0$ is satisfied by the above solution; thus the procedure of performing the odd extension of the boundary value function $f(x)$ is justified.

Next, in order to solve for $u_2(x, y)$, we first derive the modified Schwarz integral formula for the Dirichlet problem in the right half-plane $x > 0$. In the new derivation procedure, we choose the corresponding closed contour to be the infinitely large semi-circle in the right half-plane plus the vertical diameter. The chosen image point outside the closed contour is $-x + iy$, which is the image of $x + iy$ with the y-axis as the mirror. The modified Schwarz integral formula for the domain $x > 0$ becomes

$$u(x, y) = \frac{1}{2\pi i} \int_\infty^{-\infty} \left[\frac{1}{it - (x+iy)} - \frac{1}{it - (-x+iy)} \right] u(0, t)\, i\, dt$$

$$= \frac{1}{\pi} \int_{-\infty}^\infty \frac{x}{(t-y)^2 + x^2} u(0, t)\, dt, \quad x > 0.$$

We need to cope with the zero boundary condition $u_2(x, 0) = 0$ along the positive x-axis. By following the same procedure as above, we perform the odd extension of $g(y)$ over the interval $(-\infty, 0)$ by defining

$$g(y) = -g(-y), \quad -\infty < y < 0.$$

By the modified Schwarz integral formula derived above, the solution for $u_2(x, y)$ is found to be

$$u_2(x, y) = \frac{1}{\pi} \int_0^\infty \frac{x}{(t-y)^2 + x^2} g(t)\, dt - \frac{1}{\pi} \int_{-\infty}^0 \frac{x}{(t-y)^2 + x^2} g(-t)\, dt \quad \text{(ii)}$$

$$= \frac{x}{\pi} \int_0^\infty \left[\frac{1}{(t-y)^2 + x^2} - \frac{1}{(t+y)^2 + x^2} \right] g(t)\, dt, \quad x > 0 \text{ and } y > 0.$$

Note that the boundary condition $u_2(x, 0) = 0$ is satisfied by the above solution. Finally, the solution for $u(x, y)$ is given by the sum of $u_1(x, y)$ and $u_2(x, y)$ as given in eq. (i) and eq. (ii), respectively.

7.1.3 Neumann problems

We would like to find a solution $u(x, y)$ that is harmonic in a bounded domain $\mathcal{D}$ and satisfies the prescribed boundary values taken by $\frac{\partial u}{\partial n}$ along the boundary $\partial \mathcal{D}$ of the domain. For simplicity, the domain $\mathcal{D}$ is assumed to be simply connected. In order that a solution exists for the Neumann problem, the boundary values taken by $\frac{\partial u}{\partial n}$ must satisfy the *compatibility condition*

$$\oint_{\partial \mathcal{D}} \frac{\partial u}{\partial n} \, ds = 0, \tag{7.1.15}$$

where s is the arc length along the boundary. The proof for the necessity of the compatibility condition is constructed as follows:

$$
\begin{aligned}
0 &= \iint_{\mathcal{D}} \nabla^2 u \, dxdy && \text{since } u \text{ is harmonic} \\
&= \iint_{\mathcal{D}} \nabla \cdot \nabla u \, dxdy && \\
&= \oint_{\partial \mathcal{D}} \frac{\partial u}{\partial n} \, ds && \text{by virtue of Green's theorem.}
\end{aligned}
$$

The following physical interpretation may be helpful to appreciate why the compatibility condition is necessary. Suppose $u(x, y)$ is the solution to the steady state temperature distribution inside $\mathcal{D}$; then the normal temperature gradient $\frac{\partial u}{\partial n}$ on $\partial \mathcal{D}$ is proportional to the heat flux across the boundary. In order that steady state temperature prevails, simple physical intuition dictates that the net heat flux across the whole boundary must be zero. This is precisely the compatibility condition imposed on $\frac{\partial u}{\partial n}$ along the boundary (assuming constant conductivity of the material).

The solution to a Neumann problem is *unique to within an additive constant* since the addition of a constant to u does not alter the value of $\frac{\partial u}{\partial n}$.

Dirichlet problems and Neumann problems are closely related, as we now show. Let $v(x, y)$ denote a harmonic conjugate to the harmonic solution $u(x, y)$ of the given Neumann problem with prescribed normal derivative

$$\frac{\partial u}{\partial n} = g(s) \quad \text{along } \partial \mathcal{D}. \tag{7.1.16}$$

By the Cauchy–Riemann relations, we deduce that (see Example 2.6.1)

$$\frac{\partial v}{\partial s} = \frac{\partial u}{\partial n} = g(s) \quad \text{along } \partial \mathcal{D}. \tag{7.1.17}$$

To obtain the Dirichlet boundary condition for $v(x, y)$ along $\partial \mathcal{D}$, we integrate the above equation with respect to s to obtain

$$v = G(s) = G_0 + \int_0^s g(s)\, ds \quad \text{on } \partial \mathcal{D}, \tag{7.1.18}$$

where G_0 is an arbitrary constant. Now, $v(x, y)$ is harmonic with $G(s)$ as the corresponding Dirichlet boundary condition along ∂D.

The final goal in this subsection is to find the integral representation of the function $u(r, \theta)$ that is harmonic inside the circular domain $|z| < R$ with normal derivative $\frac{\partial u}{\partial n}$ along the boundary $|z| = R$, where

$$\left.\frac{\partial u}{\partial n}\right|_{|z|=R} = g(s), \quad s = Re^{i\theta}. \tag{7.1.19}$$

According to eq. (7.1.18), we compute $G(s)$, $s = Re^{i\theta}$, by

$$G(Re^{i\theta}) = G_0 + \int_0^\theta g(Re^{i\phi})i Re^{i\phi}\, d\phi, \tag{7.1.20}$$

where $G_0 = G(Re^{i\theta})|_{\theta=0}$. We expect

$$G(Re^{i\theta})|_{\theta=0} = G(Re^{i\theta})|_{\theta=2\pi},$$

and this condition would implicitly require

$$\int_0^{2\pi} g(Re^{i\phi})i Re^{i\phi}\, d\phi = 0.$$

This is precisely the compatibility condition on $g(s)$ as stated in eq. (7.1.15).

As an intermediate step, we solve the Dirichlet problem with the prescribed Dirichlet boundary condition $G(s)$ along the boundary $|z| = R$. Let $v(r, \theta)$ denote the solution to this Dirichlet problem in the domain $|z| < R$. The integral representation of $v(r, \theta)$ is given by the Poisson integral formula (7.1.5), which takes the form

$$v(r, \theta) = \frac{1}{2\pi} \int_0^{2\pi} P(R, r, \phi - \theta) G(Re^{i\phi})\, d\phi. \tag{7.1.21}$$

To find $u(r, \theta)$, it suffices to find a harmonic conjugate to $-v(r, \theta)$.

Next we find the harmonic conjugate of the Poisson kernel $P(R, r, \phi - \theta)$. Since $P(R, r, \phi - \theta) = \text{Re}\left(\frac{\zeta+z}{\zeta-z}\right)$, where $\zeta = Re^{i\phi}$ and $z = re^{i\theta}$, its harmonic conjugate is

$$Q(R, r, \phi - \theta) = \text{Im}\left(\frac{\zeta+z}{\zeta-z}\right) = -\frac{2Rr\sin(\phi-\theta)}{R^2 - 2Rr\cos(\phi-\theta)+r^2}.$$

A harmonic conjugate to $v(r, \theta)$ can be obtained readily by replacing the kernel $P(R, r, \phi - \theta)$ by $Q(R, r, \phi - \theta)$. Since $-u(r, \theta)$ is a harmonic conjugate to $v(r, \theta)$, the integral representation of $u(r, \theta)$ is given by

$$u(r, \theta) = U_0 + \frac{1}{2\pi} \int_0^{2\pi} \frac{2Rr \sin(\phi - \theta)}{R^2 - 2Rr \cos(\phi - \theta) + r^2} G(Re^{i\phi}) \, d\phi. \quad (7.1.22)$$

An additive constant U_0 is added here since the solution to a Neumann problem is unique to within an additive constant. By setting $r = 0$ in eq. (7.1.22), we see that U_0 equals the value of u at the origin.

Now, $G(Re^{i\theta})$ is unique to within an additive constant. The representation of $G(Re^{i\theta})$ in eq. (7.1.20) contains the arbitrary constant G_0. However, since $\int_0^{2\pi} Q(R, r, \phi - \theta) \, d\phi = 0$ (see Problem 7.3), the value of the integral in eq. (7.1.22) is not affected by any choice of the value of the arbitrary constant G_0. For simplicity, we may take $G_0 = 0$, and correspondingly, $G(Re^{i\theta})|_{\theta=2\pi} = 0$.

It is more desirable to express the solution in terms of $g(s)$. Integrating the integral in eq. (7.1.22) by parts and observing

$$\frac{\partial}{\partial \phi} \ln(R^2 - 2Rr \cos(\phi - \theta) + r^2) = \frac{2Rr \sin(\phi - \theta)}{R^2 - 2Rr \cos(\phi - \theta) + r^2}$$

and $G(Re^{i\theta})|_{\theta=0} = G(Re^{i\theta})|_{\theta=2\pi} = 0$, we obtain

$$u(r, \theta) = U_0 - \frac{R}{2\pi} \int_0^{2\pi} \ln(R^2 - 2Rr \cos(\phi - \theta) + r^2) g(Re^{i\phi}) \, d\phi, \quad r < R.$$
$$(7.1.23)$$

This is the integral representation of the harmonic function $u(r, \theta)$, which satisfies the Neumann boundary condition as specified in eq. (7.1.19).

7.2 The Laplace transform and its inversion

Given a function $f(t)$ defined for $t \geq 0$, and a complex parameter $s = \sigma + i\tau$, the Laplace transform of $f(t)$ is defined by

$$F(s) = \mathcal{L}\{f(t)\} = \int_0^\infty e^{-st} f(t) \, dt, \quad (7.2.1)$$

provided that the integral exists.

To guarantee the existence of the integral over any finite interval $[0, b]$, it suffices to assume that the integrand is piecewise continuous over $[0, b]$, where b takes any finite value. A possibility that leads to the non-existence of the Laplace transform is that the integrand $e^{-st}f(t)$ may diverge over large t. Suppose there exist constants M and a such that

$$|f(t)| \leqslant Me^{at} \quad t > 0. \qquad (7.2.2)$$

then

$$|f(t)e^{-st}| \leq Me^{at}e^{-\sigma t} = Me^{(a-\sigma)t}.$$

If s is chosen such that Re $s = \sigma > a$, then the integral is absolutely convergent for Re $s > a$. A function $f(t)$ that satisfies condition (7.2.2) is said to be of *exponential order* (with growth exponent a).

In summary, any piecewise continuous function of exponential order has a Laplace transform. These conditions are sufficient but not necessary. For example, the function $f(t) = \frac{1}{\sqrt{t}}$ has an infinite singularity at $t = 0$ and so it is not piecewise continuous, but its Laplace transform exists (see Example 7.2.2).

It can be shown that the Laplace transform is analytic in the domain Re $s > a$ in the s-plane. Also, the linearity property is observed, where

$$\mathcal{L}\{\alpha f(t) + \beta g(t)\} = \alpha\mathcal{L}\{f(t)\} + \beta\mathcal{L}\{g(t)\},$$

provided that the individual Laplace transforms exist.

Some basic transform formulas

The Laplace transforms of some elementary functions are shown below:

$$\mathcal{L}\{e^{at}\} = \int_0^\infty e^{-st}e^{at}dt = \frac{1}{s-a}, \quad \text{Re } s > a. \qquad (7.2.3a)$$

$$\mathcal{L}\{t^n\} = \int_0^\infty e^{-st}t^n dt$$

$$= \frac{n}{s}\int_0^\infty e^{-st}t^{n-1}dt = \cdots = \frac{n!}{s^{n+1}}, \quad \text{Re } s > 0. \qquad (7.2.3b)$$

$$\mathcal{L}\{\sin wt\} = \text{Im}\int_0^\infty e^{-st}e^{iwt}\,dt$$

$$= \text{Im}\frac{1}{s-iw} = \frac{w}{s^2+w^2}, \quad \text{Re } s > 0. \qquad (7.2.3c)$$

$$\mathcal{L}\{\cosh \beta t\} = \int_0^\infty e^{-st}\frac{e^{\beta t}+e^{-\beta t}}{2}\,dt$$

$$= \frac{1}{2}\left(\frac{1}{s-\beta}+\frac{1}{s+\beta}\right) = \frac{s}{s^2-\beta^2}, \quad s > |\beta|. \qquad (7.2.3d)$$

Step function

Recall that the Heaviside step function $H(x)$ is defined by

$$H(t) = \begin{cases} 1 & t > 0 \\ 0 & t < 0 \end{cases}.$$ (7.2.4a)

The Laplace transform of $H(t - \tau)$, $\tau > 0$, is found to be

$$\mathcal{L}\{H(t - \tau)\} = \int_\tau^\infty e^{-st}\, dt = \frac{e^{-s\tau}}{s}.$$ (7.2.4b)

Shifting

Consider the shifting of the function $f(t)$ as defined by

$$f_\tau(t) = \begin{cases} f(t - \tau) & t \geq \tau \\ 0 & t < \tau \end{cases}, \quad \tau > 0;$$

then its Laplace transform is found to be

$$\mathcal{L}\{f_\tau(t)\} = \int_0^\infty e^{-st} f_\tau(t)\, dt$$

$$= \int_\tau^\infty e^{-st} f(t - \tau)\, dt = e^{-s\tau} \int_\tau^\infty e^{-s(t-\tau)} f(t - \tau)\, d(t - \tau)$$

$$= e^{-s\tau} \int_0^\infty e^{-st'} f(t')\, dt' = e^{-s\tau} \mathcal{L}\{f(t)\}.$$ (7.2.5)

Dirac function

The Laplace transform of the Dirac function $\delta(t - t_0)$, $t_0 > 0$, is found to be

$$\mathcal{L}\{\delta(t - t_0)\} = \int_0^\infty e^{-st} \delta(t - t_0)\, dt = e^{-st_0}.$$ (7.2.6)

Derivatives

Suppose $f(t)$, $f'(t)$, $\ldots$, $f^{(n)}(t)$ are piecewise continuous and of exponential order with growth exponent a; then

$$\mathcal{L}\{f'(t)\} = \int_0^\infty e^{-st} f'(t)\, dt = [e^{-st} f(t)]_0^\infty + s \int_0^\infty e^{-st} f(t)\, dt.$$

For Re $s > a$, we have $\lim_{t \to \infty} e^{-st} f(t) = 0$ and $\lim_{t \to 0} e^{-st} f(t) = f(0)$; so

$$\mathcal{L}\{f'(t)\} = s\mathcal{L}\{f(t)\} - f(0).$$ (7.2.7a)

Deductively,

$$\mathcal{L}\{f''(t)\} = s\mathcal{L}\{f'(t)\} - f'(0) = s^2 \mathcal{L}\{f(t)\} - sf(0) - f'(0);$$

and, in general,

$$\mathcal{L}\{f^{(n)}(t)\} = s^n \mathcal{L}\{f(t)\} - s^{n-1} f(0) - s^{n-2} f'(0) - \cdots - f^{(n-1)}(0).$$

$$(7.2.7b)$$

Convolution property

Suppose $\mathcal{L}\{f(t)\}$ and $\mathcal{L}\{g(t)\}$ exist for Re $s > a$, and recall that the convolution of $f(t)$ and $g(t)$ is given by

$$(f * g)(t) = \int_0^t f(\tau)g(t - \tau) \, d\tau.$$

The Laplace transform of the convolution is given by

$$\mathcal{L}\{(f * g)(t)\} = \int_0^\infty e^{-st} \int_0^t f(\tau)g(t - \tau) \, d\tau \, dt$$

$$= \int_0^\infty f(\tau) \int_\tau^\infty e^{-st} g(t - \tau) \, dt \, d\tau,$$

where the last integral is obtained by changing the order of integration. By setting $t - \tau = t'$, we obtain

$$\mathcal{L}\{(f * g)(t)\} = \int_0^\infty e^{-s\tau} f(\tau) \, d\tau \int_0^\infty e^{-st'} g(t') \, dt'$$

$$= \mathcal{L}\{f(t)\}\mathcal{L}\{g(t)\}. \qquad (7.2.8)$$

Example 7.2.1 Solve each of the following equations using the Laplace transform method:

(a) $\displaystyle\int_0^x f(u)(x - u) \, du + f(x) = \sin 2x, \quad x > 0;$

(b) $\begin{cases} \dfrac{d^2y}{dt^2} + 2\dfrac{dy}{dt} + 2y = \delta(t - 1) + f(t) \\[2mm] y(0) = y'(0) = 1 \end{cases}$

Solution

(a) This is an integral equation with the unknown function $f(x)$. The integral takes the form of the convolution of $f(x)$ and x. Taking the Laplace transform on both sides of the equation and applying the convolution formula (7.2.8), we obtain

$$\frac{\mathcal{L}\{f(x)\}}{s^2} + \mathcal{L}\{f(x)\} = \frac{2}{s^2 + 4}.$$

Solving for $\mathcal{L}\{f(x)\}$, we obtain

$$\mathcal{L}\{f(x)\} = \frac{2s^2}{s^4 + 5s^2 + 4} = \frac{4}{3}\frac{2}{s^2 + 4} - \frac{2}{3}\frac{1}{s^2 + 1}.$$

The inversion of the above Laplace transform is easily seen to be

$$f(x) = \frac{4}{3}\sin 2x - \frac{2}{3}\sin x, \quad x > 0.$$

(b) Let $Y(s)$ and $F(s)$ denote the Laplace transforms of $y(t)$ and $f(t)$, respectively. By taking the Laplace transform of the equation, we obtain

$$s^2Y(s) - sy(0) - y'(0) + 2[sY(s) - y(0)] + 2Y(s) = e^{-s} + F(s).$$

Applying the given initial conditions and solving for $Y(s)$, we have

$$Y(s) = \frac{e^{-s}}{(s+1)^2 + 1} + \frac{(s+1) + 2}{(s+1)^2 + 1} + \frac{F(s)}{(s+1)^2 + 1}.$$

The Laplace transform inversions of the individual terms are

$$\mathcal{L}^{-1}\left\{\frac{e^{-s}}{(s+1)^2 + 1}\right\} = H(t-1)e^{-(t-1)}\sin(t-1),$$

$$\mathcal{L}^{-1}\left\{\frac{(s+1) + 2}{(s+1)^2 + 1}\right\} = e^{-t}(\cos t + 2\sin t),$$

$$\mathcal{L}^{-1}\left\{\frac{F(s)}{(s+1)^2 + 1}\right\} = \int_0^t f(t)e^{-(t-\tau)}\sin(t-\tau)\,d\tau.$$

The solution is then given by

$$y(t) = H(t-1)e^{-(t-1)}\sin(t-1) + e^{-t}(\cos t + 2\sin t)$$
$$+ \int_0^t f(t)e^{-(t-\tau)}\sin(t-\tau)\,d\tau.$$

Note that $y(t)$ is continuous but $y'(t)$ is discontinuous at $t = 1$, due to the impulsive effect caused by the term $\delta(t-1)$ in the differential equation.

7.2.1 Bromwich integrals

The inversion of the Laplace transform of a function can be effected by integrating the corresponding Bromwich integral, the details of which are stated in the following theorem.

Theorem 7.2.1 *Assume $f(t)$ to be a piecewise continuous function of exponential order defined for $t \geq 0$, and that there exists a real constant $a > 0$ such* that the real integral $\int_0^\infty e^{at}|f(t)|\,dt$ exists. Let $F(s) = \mathcal{L}\{f(t)\}$ denote the Laplace transform of $f(t)$. The inverse Laplace transform of $F(s)$ is given by

$$f(t) = \frac{1}{2\pi i} \lim_{R\to\infty} \int_{a-iR}^{a+iR} e^{st} F(s)\,ds, \quad t > 0. \tag{7.2.9}$$

Proof We apply the Fourier transform inversion formula to the function $e^{-at} f(t)$, assuming $f(t) = 0$ for $t < 0$ [see eq. (6.4.5)], to obtain

$$e^{-at} f(t) = \frac{1}{2\pi} \lim_{R\to\infty} \int_{-R}^{R} e^{-iut} \left[\int_0^\infty e^{iut'} e^{-at'} f(t')\,dt' \right] du.$$

The above result is reformulated as

$$f(t) = \frac{1}{2\pi} \lim_{R\to\infty} \int_{-R}^{R} e^{(a-iu)t} \left[\int_0^\infty e^{-(a-iu)t'} f(t')\,dt' \right] du, \quad t > 0, \tag{7.2.10}$$

where the inner integral is recognized as the Laplace transform of $f(t)$ if $a - iu$ becomes the transform variable s. Changing the variable of integration from u to $s = a - iu$, the above formula becomes

$$f(t) = \frac{1}{2\pi i} \lim_{R\to\infty} \int_{a-iR}^{a+iR} e^{st} F(s)\,ds, \quad t > 0, \tag{7.2.11}$$

where the integration path is the infinite vertical line $\mathrm{Re}\, s = a$. Recall that the constant $a > 0$ is chosen to be sufficiently large to ensure the existence of $\int_0^\infty e^{-at}|f(t)|\,dt$. The vertical line $\mathrm{Re}\, s = a$ is called the *Bromwich line*, and subsequently, the inversion integral in eq. (7.2.11) is called the *Bromwich integral*.

Remark

   (i) At a point with a finite jump, $f(t)$ is assigned the average of its left-hand and right-hand limits.
   (ii) Suppose $F(s)$ is analytic except for a finite number of poles in the half-plane to the left of the vertical line $\mathrm{Re}\, s = a$. We construct a positively oriented closed contour C_B which consists of the vertical Bromwich line ℓ_B from $a - iR$ to $a + iR$, the upper horizontal line ℓ_u from $a + iR$ to iR, the left semi-circle C_R: $z = Re^{i\theta}$, $\frac{\pi}{2} \le \theta \le \frac{3\pi}{2}$, and the lower horizontal line ℓ_d from $-iR$ to $a - iR$ (see Figure 7.6).

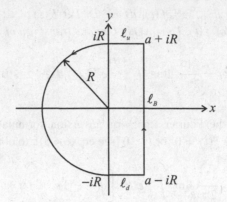

Figure 7.6. A closed contour C_B is constructed which consists of the vertical Bromwich line ℓ_B, the horizontal lines ℓ_u and ℓ_d, and the left semi-circle C_R.

Suppose $F(s)$ tends uniformly to zero as $R \to \infty$ when s assumes values on $\ell_u \cup C_R \cup \ell_d$. It can be shown that

$$f(t) = \frac{1}{2\pi i} \lim_{R \to \infty} \int_{a-iR}^{a+iR} e^{st} F(s)\, ds$$

$$= \frac{1}{2\pi i} \lim_{R \to \infty} \oint_{C_B} e^{st} F(s)\, ds, \qquad (7.2.12)$$

since the contour integrals along the line segments ℓ_u and ℓ_d and the left semi-circle C_R all tend to zero as $R \to \infty$ (hints about proving these claims are given in Problem 7.20). Now the Laplace transform inversion formula becomes a contour integral with a closed contour. As $R \to \infty$, the closed contour C_B contains all the poles of $F(s)$ in the half-plane to the left of ℓ_B. By the Residue Theorem, it then follows that

$$f(t) = \sum_n \text{Res}(e^{st} F(s), s_n), \qquad (7.2.13)$$

where the points s_n denote the poles of $F(s)$ that lie on the left side of ℓ_B, and the summation of residue values is taken over all these poles.

Example 7.2.2   Find the inverse Laplace transform of

(a) $F(s) = \dfrac{s}{(s+1)^3}$;   (b) $F(s) = \dfrac{1}{\sqrt{s}}$;   (c) $F(s) = \dfrac{e^{-\lambda \sqrt{s}}}{s}$, $\lambda > 0$.

Solution

(a) Note that $F(s) = \frac{s}{(s+1)^3}$ has one pole of order 3 at $s = -1$. One may choose the Bromwich line to be Re $s = a$, $a > -1$. By observing that the pole $s = -1$ lies to the left of the Bromwich line and $\frac{s}{(s+1)^3}$ tends to zero uniformly as $|s| \to \infty$, we then apply eq. (7.2.13) to obtain

$$ f(t) = \text{Res}\left(e^{st} \frac{s}{(s+1)^3}, s = -1 \right) $$

The most effective method for evaluating the above residue is to find the coefficient of $\frac{1}{s+1}$ in the Laurent series expansion of $\frac{e^{st}s}{(s+1)^3}$ valid in a deleted neighborhood of $s = -1$. We obtain

$$ e^{st} \frac{s}{(s+1)^3} $$

$$ = e^{(s+1)t} e^{-t} \frac{(s+1) - 1}{(s+1)^3} $$

$$ = \left[1 + (s+1)t + \frac{(s+1)^2 t^2}{2!} + \cdots \right]\left[\frac{1}{(s+1)^2} - \frac{1}{(s+1)^3} \right] e^{-t} $$

$$ = \left[-\frac{1}{(s+1)^3} + \frac{1-t}{(s+1)^2} + \frac{t - \frac{t^2}{2}}{s+1} + \cdots \right] e^{-t}; $$

so

$$ f(t) = \text{coefficient of } \frac{1}{s+1} \text{ in the above Laurent expansion} $$

$$ = \left(t - \frac{t^2}{2} \right) e^{-t}, \quad t > 0. $$

Remark There is a more direct method of finding $\mathcal{L}^{-1}\left\{ \frac{s}{(s+1)^3} \right\}$. By making use of the inversion formulas

$$ \mathcal{L}\{e^{at} f(t)\} = \int_0^\infty e^{-st} e^{at} f(t)\, dt = F(s - a), \quad F(s) = \mathcal{L}\{f(t)\}, $$

and

$$ \mathcal{L}\{t^n\} = \frac{n!}{s^{n+1}}, \quad n \text{ is a non-negative integer}, $$

we deduce that

$$ \mathcal{L}^{-1}\left\{ \frac{s}{(s+1)^3} \right\} = \mathcal{L}^{-1}\left\{ \frac{1}{(s+1)^2} - \frac{1}{(s+1)^3} \right\} = \left(t - \frac{t^2}{2} \right) e^{-t}. $$

(b) We choose the branch cut of $F(s) = \frac{1}{\sqrt{s}}$ to be along the negative real axis. The Bromwich line can be any vertical line in the right half-plane. The closed contour with the Bromwich line as the vertical side has to be

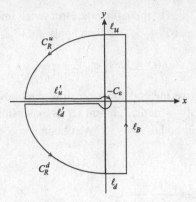

Figure 7.7. To avoid crossing the branch cut along the negative real axis, the closed contour C_B is modified to include the extra excursion to the branch point $s = 0$ along the upper and lower sides of the branch cut.

modified to include the extra excursion to the branch point $s = 0$ along the upper and lower sides of the branch cut, plus an infinitesimal circle C_ϵ looped in the clockwise sense around the branch point $s = 0$ (see Figure 7.7). The rationale for such a construction is that the closed contour C_B should not cross the branch cut along the negative real axis. Note that $F(s) = \frac{1}{\sqrt{s}}$ tends uniformly to zero as $|s| \to \infty$. Following analogous arguments that establish eq. (7.2.12), it is seen that the contributions along the outer quarter circles and the line segments joining the outer quarter circles with the Bromwich line are zero (see Problem 7.20). The remaining contributions to the contour integral around the closed contour C_B come from the Bromwich line ℓ_B, the upper line ℓ'_u and the lower line ℓ'_d along the branch cut and the small circle C_ϵ (looped in the negative sense). Note that the contour integration along ℓ_B gives $f(t)$. On the other hand, since the closed contour C_B contains no pole of $\frac{e^{st}}{\sqrt{s}}$, the contour integral around C_B is zero. We then have

$$
0 = \frac{1}{2\pi i} \oint_{C_B} \frac{e^{st}}{\sqrt{s}} \, ds
$$

$$
= f(t) + \frac{1}{2\pi i} \int_{\ell'_u} \frac{e^{st}}{\sqrt{s}} \, ds
$$

$$
+ \frac{1}{2\pi i} \int_{\ell'_d} \frac{e^{st}}{\sqrt{s}} \, ds - \frac{1}{2\pi i} \oint_{C_\epsilon} \frac{e^{st}}{\sqrt{s}} \, ds.
$$

Along the upper side of the branch cut ℓ'_u, $s = e^{i\pi}\xi$, where ξ runs from R to ϵ, and along the lower side of the branch cut ℓ'_d, $s = e^{-i\pi}\xi$, where ξ runs from ϵ to R. On the infinitesimal circle C_ϵ, we have $s = \epsilon e^{i\theta}$, $-\pi < \theta \le \pi$. Taking the limits $R \to \infty$ and $\epsilon \to 0$, we obtain

$$f(t) = \frac{1}{2\pi i}\int_0^\infty e^{-\xi t}\frac{1}{i\sqrt{\xi}}(-d\xi) - \frac{1}{2\pi i}\int_0^\infty e^{-\xi t}\frac{1}{i\sqrt{\xi}}(-d\xi)$$
$$+ \frac{1}{2\pi i}\lim_{\epsilon\to 0}\int_{-\pi}^{\pi} e^{t\epsilon e^{i\theta}}\frac{1}{\sqrt{\epsilon}e^{i\theta/2}}i\epsilon e^{i\theta}\,d\theta.$$

The modulus of the last integral is $O(\sqrt{\epsilon})$ and so it becomes zero as $\epsilon \to 0$. The above expression can be simplified as

$$f(t) = \frac{1}{\pi}\int_0^\infty \frac{e^{-\xi t}}{\sqrt{\xi}}\,d\xi$$
$$= \frac{2}{\pi\sqrt{t}}\int_0^\infty e^{-y^2}\,dy, \quad y^2 = \xi t$$
$$= \frac{1}{\sqrt{\pi t}}.$$

(c) With the appearance of $\sqrt{s}$ in the exponent of $F(s) = \frac{e^{-\lambda\sqrt{s}}}{s}$, the closed contour is chosen to be the same as that shown in Figure 7.7. Since $F(s) = \frac{e^{-\lambda\sqrt{s}}}{s}$ also tends uniformly to zero as $|s| \to \infty$, the contributions to the contour integral come only from the Bromwich line, the upper and lower line segments along the branch cut and the infinitesimal circle around the branch point $s = 0$. Again, there is no pole enclosed within the closed contour C_B. Following a similar approach to part (b), we obtain

$$f(t) = \frac{1}{2\pi i}\int_0^\infty e^{-\xi t}\frac{e^{-\lambda\sqrt{\xi}i}}{-\xi}(-d\xi) - \frac{1}{2\pi i}\int_0^\infty e^{-\xi t}\frac{e^{-(-\lambda\sqrt{\xi}i)}}{-\xi}(-d\xi)$$
$$+ \frac{1}{2\pi i}\lim_{\epsilon\to 0}\int_{-\pi}^{\pi} e^{t\epsilon e^{i\theta}}\frac{e^{-\lambda\sqrt{\epsilon}e^{i\theta/2}}}{\epsilon e^{i\theta}}i\epsilon e^{i\theta}\,d\theta.$$

The last integral becomes 1 as $\epsilon \to 0$. The above expression can be simplified as

$$f(t) = 1 - \frac{1}{\pi}\int_0^\infty e^{-\xi t}\frac{\sin\lambda\sqrt{\xi}}{\xi}\,d\xi$$
$$= 1 - \frac{2}{\pi}\int_0^\infty e^{-tu^2}\left(\int_0^\lambda \cos\alpha u\,d\alpha\right)du, \quad u = \sqrt{\xi}$$
$$= 1 - \frac{2}{\pi}\int_0^\lambda\left(\int_0^\infty e^{-tu^2}\cos\alpha u\,du\right)d\alpha.$$

It is shown in Example 4.2.4 that

$$\int_0^\infty e^{-tu^2} \cos \alpha u \, du = \frac{1}{2}\sqrt{\frac{\pi}{t}} e^{-\alpha^2/4t},$$

so we obtain

$$f(t) = 1 - \frac{1}{\sqrt{\pi t}} \int_0^\lambda e^{-\alpha^2/4t} \, d\alpha.$$

The above integral can be expressed in terms of the complementary error function erfc(x) defined by

$$\text{erfc}(x) = 1 - \frac{2}{\sqrt{\pi}} \int_0^x e^{-\eta^2} d\eta.$$

By taking $\eta = \dfrac{\alpha}{2\sqrt{t}}$ and $x = \dfrac{\lambda}{2\sqrt{t}}$, we finally obtain

$$f(t) = \text{erfc}\left(\frac{\lambda}{2\sqrt{t}}\right).$$

Remark   By differentiating the Laplace transform formula

$$\mathcal{L}\left\{\text{erfc}\left(\frac{\lambda}{2\sqrt{t}}\right)\right\} = \frac{e^{-\lambda\sqrt{s}}}{s}$$

with respect to the parameter λ repeatedly, we obtain the following related Laplace transform formulas:

$$\mathcal{L}\left\{\frac{1}{\sqrt{\pi t}} e^{-\lambda^2/4t}\right\} = \frac{e^{-\lambda\sqrt{s}}}{\sqrt{s}},$$

$$\mathcal{L}\left\{\frac{\lambda}{2\sqrt{\pi t^3}} e^{-\lambda^2/4t}\right\} = e^{-\lambda\sqrt{s}}.$$

These formulas will be used in solving the heat conduction problems discussed in the next section.

7.3 Initial-boundary value problems

The Laplace transform method is known to be an effective tool for solving initial-boundary value problems arising from mathematical physics. Two classical initial-boundary value problems are examined in this section: heat conduction and wave propagation. Here we show how the complex variables techniques are applied in the analytic inversion of the Laplace transform to obtain the solutions. Though the governing equations for heat conduction and wave propagation are derived in many texts on partial differential equations, for the purpose of making this text self-contained, we spare a few paragraphs to examine how these governing equations arise from the modeling of the underlying physical phenomena.

7.3.1 Heat conduction

The heat conduction problem is closely related to the steady state temperature distribution problems discussed in the earlier chapters. Consider the same control volume inside a two-dimensional conducting body as shown in Figure 2.7. Let c and ρ denote the specific heat of the material and density of the conducting body, respectively. The material parameters are assumed to be constant. The heat content $H(t)$ contained in the control volume at time t is given by

$$H(t) = c\rho T(x, y, t)\Delta x \Delta y,$$

where the temperature $T(x, y, t)$ is a function of x, y and t. The rate of change of the heat content is given by

$$\frac{dH}{dt} = c\rho \frac{\partial T}{\partial t} \Delta x \Delta y.$$

The net accumulation of heat energy per unit time inside the control volume through conduction along its four sides is given by

$$K \Delta x \Delta y \left(\frac{\partial^2 T}{\partial x^2} + \frac{\partial^2 T}{\partial y^2} \right),$$

where K is the thermal conductivity of the material (see Section 2.6.2). By the law of conservation of energy, the rate of change of heat energy in the control volume should equal the heat flux across its boundary through conduction. From the above two expressions, we then obtain

$$\frac{\partial T}{\partial t} = \frac{K}{c\rho} \left(\frac{\partial^2 T}{\partial x^2} + \frac{\partial^2 T}{\partial y^2} \right). \tag{7.3.1}$$

This is the *two-dimensional heat conduction equation*. When steady state conditions prevail, the temporal rate of change of temperature $\frac{\partial T}{\partial t}$ becomes zero. The above equation then reduces to

$$\frac{\partial^2 T}{\partial x^2} + \frac{\partial^2 T}{\partial y^2} = 0. \tag{7.3.2}$$

This is the familiar Laplace equation that governs steady state temperature distribution. The one-dimensional version of the heat conduction equation is given by

$$\frac{\partial T}{\partial t} = a^2 \frac{\partial^2 T}{\partial x^2}, \quad a^2 = \frac{K}{c\rho}, \tag{7.3.3}$$

where $T = T(x, t)$. The above equation governs the heat conduction process inside a thin longitudinal rod.

Heat conduction in an infinite rod

Consider the heat conduction problem in an infinitely long rod with initial condition $T(x, 0) = f(x)$, $-\infty < x < \infty$. The boundary conditions at the infinite ends are not specified except that the temperature values are assumed to remain bounded as $|x| \to \infty$. The set of governing equations for the initial value problem can be stated as

$$\begin{cases} \dfrac{\partial T}{\partial t} = a^2 \dfrac{\partial^2 T}{\partial x^2}, \quad -\infty < x < \infty, \quad t > 0 \\[2mm] T(x, 0) = f(x) \end{cases} \tag{7.3.4}$$

First we try to find the fundamental solution $F(x, t)$ that satisfies

$$\begin{cases} \dfrac{\partial F}{\partial t} = a^2 \dfrac{\partial^2 F}{\partial x^2}, \quad -\infty < x < \infty \\[2mm] F(x, 0) = \delta(x) \end{cases} \tag{7.3.5}$$

where $\delta(x)$ is the Dirac function. By the defining property of the Dirac function, the initial condition $f(x)$ can be expressed as

$$f(x) = \int_{-\infty}^{\infty} f(\xi)\delta(x - \xi)\, d\xi. \tag{7.3.6}$$

The above relation can be interpreted as the decomposition of $f(x)$ into impulses distributed over $(-\infty, \infty)$ with magnitude $f(\xi)$ at the position ξ. The solution to the heat conduction problem with initial condition $f(\xi)\delta(x - \xi)$ is recognized to be $f(\xi)F(x - \xi, t)$. Since the heat equation is linear, the principle of superposition of solutions applies, so the solution to eq. (7.3.4) can be obtained by integrating $f(\xi)F(x - \xi, t)$ with respect to ξ from $-\infty$ to ∞. Hence, the solution to eq. (7.3.4) can be represented as

$$T(x, t) = \int_{-\infty}^{\infty} f(\xi)F(x - \xi, t)\, d\xi. \tag{7.3.7}$$

To solve for $F(x, t)$, we take the Laplace transform with respect to t on both sides of eq. (7.3.5). Let $\hat{F}(x, s)$ denote the Laplace transform of $F(x, t)$. We obtain the following ordinary differential equation for $\hat{F}(x, s)$:

$$a^2 \frac{d^2 \hat{F}}{dx^2}(x, s) - s\hat{F}(x, s) = -\delta(x), \qquad -\infty < x < \infty. \tag{7.3.8}$$

The solution to the above equation takes the form

$$\hat{F}(x, s) = \begin{cases} A_+(s)e^{-\sqrt{s}x/a} + B_+(s)e^{\sqrt{s}x/a} & x > 0 \\ A_-(s)e^{-\sqrt{s}x/a} + B_-(s)e^{\sqrt{s}x/a} & x < 0 \end{cases}.$$

To ensure that the solution remains bounded as $|x| \to \infty$, we must set $B_+(s) = A_-(s) = 0$. To determine $A_+(s)$ and $B_-(s)$, we observe that $\hat{F}(x, s)$ is continuous across $x = 0$ and there is a jump of $\frac{\partial \hat{F}}{\partial x}$ across $x = 0$ due to the term $\delta(x)$ in eq. (7.3.8) [see also Example 7.2.1, part (b)]. The continuity of $\hat{F}(x, s)$ across $x = 0$ gives the following condition on $\hat{F}$:

$$\hat{F}(0^+, s) - \hat{F}(0^-, s) = 0.$$

By integrating eq. (7.3.8) with respect to x from $x = 0^-$ to $x = 0^+$, we obtain another condition on $\hat{F}$:

$$\frac{\partial \hat{F}}{\partial x}(0^+, s) - \frac{\partial \hat{F}}{\partial x}(0^-, s) = -\frac{1}{a^2}.$$

Using these two conditions, the solutions to $A_+(s)$ and $B_-(s)$ are found to be

$$A_+(s) = B_-(s) = \frac{1}{2a\sqrt{s}};$$

so

$$\hat{F}(x, s) = \frac{1}{2a\sqrt{s}}e^{-\sqrt{s}|x|/a}. \tag{7.3.9}$$

The inverse of the above Laplace transform is found to be (see Example 7.2.2)

$$F(x, t) = \frac{1}{\sqrt{4\pi a^2 t}}e^{-x^2/4a^2 t}.$$

Finally, by eq. (7.3.7), the solution to eq. (7.3.4) is found to be

$$T(x, t) = \frac{1}{\sqrt{4\pi a^2 t}}\int_{-\infty}^{\infty} f(\xi)e^{-(x-\xi)^2/4a^2 t}\, d\xi. \tag{7.3.10}$$

Heat conduction in a semi-infinite rod

Next we consider the heat conduction problem in a semi-infinite rod $x \geq 0$. For simplicity, we assume the initial temperature distribution to be zero. At the finite end $x = 0$, the temperature is maintained at $T(0, t) = g(t)$. The problem can then be formulated as

$$\begin{cases} \dfrac{\partial T}{\partial t} = a^2 \dfrac{\partial^2 T}{\partial x^2}, & x > 0, \ t > 0 \\[3mm] T(x,0) = 0 \text{ and } T(0,t) = g(t) \end{cases} \tag{7.3.11}$$

We follow a similar technique of taking the Laplace transform with respect to t. Let $\hat{T}(x,s)$ and $G(s)$ denote the Laplace transforms of $T(x,t)$ and $g(t)$, respectively. We obtain the following ordinary differential equation for $\hat{T}(x,s)$:

$$\begin{cases} s\hat{T}(x,s) = a^2 \dfrac{\partial^2 \hat{T}}{\partial x^2}(x,s), & x > 0 \\[3mm] \hat{T}(0,s) = G(s) \end{cases}$$

Following a similar procedure as before, the term $e^{\sqrt{s}x/a}$ in the general solution is discarded since the solution remains bounded as $x \to \infty$. The solution to the above equation is then found to be

$$\hat{T}(x,s) = G(s)e^{-\sqrt{s}x/a}, \quad x > 0.$$

Using the Laplace transform inversion formula derived in Example 7.2.2

$$\mathcal{L}^{-1}\{e^{-\sqrt{s}x/a}\} = \frac{x}{2\sqrt{\pi a^2} t^{3/2}} e^{-x^2/4a^2 t},$$

and the convolution formula (7.2.8), the solution to eq. (7.3.11) is found to be

$$T(x,t) = \frac{x}{\sqrt{4\pi a^2}} \int_0^t g(\tau) \frac{1}{(t-\tau)^{3/2}} e^{-x^2/4a^2(t-\tau)} \, d\tau. \tag{7.3.12}$$

Example 7.3.1   Consider the heat conduction problem within a rod of unit length. Suppose the two ends are kept at zero temperature, and the initial temperature is kept at the constant value $T(x,0) = 1$, $0 < x < 1$; find the temperature along the rod at any time $t > 0$.

Solution

The heat conduction problem can be stated as

$$\begin{cases} \dfrac{\partial T}{\partial t} = \dfrac{\partial^2 T}{\partial x^2}, & 0 < x < 1, \ t > 0 \\[3mm] T(x,0) = 1, & T(0,t) = T(1,t) = 0 \end{cases},$$

where the coefficient a^2 has been taken to be unity for simplicity. The differential equation governing the Laplace transform $\hat{T}(x, s)$ is given by

$$\begin{cases} \dfrac{\partial^2 \hat{T}}{\partial x^2}(x, s) - s\hat{T}(x, s) = -1 \\[3mm] \hat{T}(0, s) = \hat{T}(1, s) = 0 \end{cases}$$

The solution to the above differential equation is found to be

$$\hat{T}(x, s) = \frac{1}{s(e^{\sqrt{s}} - e^{-\sqrt{s}})}[e^{\sqrt{s}} - e^{-\sqrt{s}} + (e^{-\sqrt{s}} - 1)e^{\sqrt{s}x} - (e^{\sqrt{s}} - 1)e^{-\sqrt{s}x}].$$

The direct analytic valuation of the Bromwich inversion integral for $\hat{T}(x, s)$ is intractable. Alternatively, we expand the transform function in series for large s and perform the inverse Laplace transforms term by term. It can be shown that the "large s" approximation in $\hat{T}(x, s)$ corresponds to the "small t" approximation in $T(x, t)$ (see Problem 7.21). We expand $\hat{T}(x, s)$ in the "large s" approximation as follows:

$$\hat{T}(x, s) = \frac{1}{s} + \frac{e^{-\sqrt{s}}}{s}(1 + e^{-2\sqrt{s}} + e^{-4\sqrt{s}} + \cdots)$$

$$(e^{-\sqrt{s}(1-x)} - e^{\sqrt{s}(1-x)} - e^{\sqrt{s}x} + e^{-\sqrt{s}x})$$

$$= \frac{1}{s} + \frac{1}{s}\sum_{n=1}^{\infty}(-1)^n e^{-\sqrt{s}(n-x)} - \sum_{n=0}^{\infty}(-1)^n e^{-\sqrt{s}(n+x)}.$$

We apply the inversion formula (see Example 7.2.2)

$$\mathcal{L}^{-1}\left\{\frac{1}{s}e^{-\lambda\sqrt{s}}\right\} = \operatorname{erfc}\left(\frac{\lambda}{2\sqrt{t}}\right),$$

so the termwise inversion of the above series gives the following solution to the temperature distribution:

$$T(x, t) = 1 + \sum_{n=1}^{\infty}(-1)^n \operatorname{erfc}\left(\frac{n-x}{2\sqrt{t}}\right) - \sum_{n=0}^{\infty}(-1)^n \operatorname{erfc}\left(\frac{n+x}{2\sqrt{t}}\right).$$

7.3.2 *Longitudinal oscillations of an elastic thin rod*

Let us consider the phenomenon of wave propagation in a thin elastic rod under longitudinal oscillations. The longitudinal movement of a segment of the rod leads to contraction or elongation of its neighboring portions, which in turn affect the other parts of the rod in sequence. Subsequently, the longitudinal oscillations are propagated to the whole rod at a certain speed of propagation.

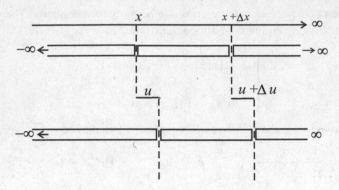

Figure 7.8. Longitudinal oscillations of a differential segment of length Δx along an infinite rod.

Let $u(x, t)$ denote the displacement of the rod at the position x and time t. Consider the differential segment $[x, x + \Delta x]$ of length Δx along the rod. The displacements of the rod at positions x and $x + \Delta x$ are u and $u + \frac{\partial u}{\partial x}\Delta x$, respectively (see Figure 7.8). The magnitude of the strain at position x is given by

$$\frac{u + \frac{\partial u}{\partial x}\Delta x - u}{\Delta x} = \frac{\partial u}{\partial x}.$$

Assuming that the material of the rod observes the law of linear elasticity, the stress at position x is given by $E\frac{\partial u}{\partial x}(x, t)$, where E is the Young modulus of the material. Let A denote the uniform cross-sectional area of the rod. The forces at the left and right ends of the segment $[x, x + \Delta x]$ due to the elastic stresses are $EA\frac{\partial u}{\partial x}(x, t)$ and $EA\frac{\partial u}{\partial x}(x + \Delta x, t)$, respectively. The mass of the differential segment is $\rho A\Delta x$, where ρ is the mass per unit length of the rod.

Using Newton's second law of motion, the equation that governs the dynamics of the longitudinal oscillations can be expressed as

$$\rho A\Delta x \frac{\partial^2 u}{\partial t^2}(x, t) = EA\frac{\partial u}{\partial x}(x + \Delta x, t) - EA\frac{\partial u}{\partial x}(x, t).$$

Taking the limit $\Delta x \to 0$, we obtain

$$\frac{\partial^2 u}{\partial t^2} = c^2 \frac{\partial^2 u}{\partial x^2}, \quad c = \sqrt{\frac{E}{\rho}} > 0. \tag{7.3.13}$$

This is the *wave propagation equation* that governs the longitudinal oscillations of a thin elastic rod. The parameter c gives the speed of propagation of longitudinal oscillations along the rod [see eq. (7.3.15)].

Wave propagation along a semi-infinite elastic rod

Since the wave equation involves the second-order time derivative, the full prescription of the set of initial conditions requires the specification of the initial displacement and velocity. Consider the wave propagation along a semi-infinite elastic rod where the time-dependent movement at $x = 0$ is specified to be $f(t)$, and the initial displacement and velocity are zero. Also, the displacement vanishes at the far end, $x \to \infty$. The longitudinal oscillation problem can be formulated as

$$\begin{cases} \dfrac{\partial^2 u}{\partial t^2} = c^2 \dfrac{\partial^2 y}{\partial x^2}, \quad 0 < x < \infty, \ t > 0 \\[2mm] u(0, t) = f(t), \ u(x, 0) = \dfrac{\partial u}{\partial t}(x, 0) = 0 \end{cases} \tag{7.3.14}$$

Let $U(x, s)$ and $F(s)$ denote the Laplace transform of $u(x, t)$ and $f(t)$ with respect to time, respectively. The governing differential equation for $U(x, s)$ is given by

$$\begin{cases} s^2 U(x, s) = c^2 \dfrac{\partial^2 U}{\partial x^2}(x, s) \\[2mm] U(0, s) = F(s) \end{cases}$$

Solving the above equation, we obtain

$$U(x, s) = A(s)e^{-sx/c} + B(s)e^{sx/c},$$

where the arbitrary functions $A(s)$ and $B(s)$ are determined by the boundary conditions. Since $u(x, t) \to 0$ as $x \to \infty$, we have $B(s) = 0$. Incorporating the boundary condition $U(0, s) = F(s)$, $U(x, s)$ is found to be

$$U(x, s) = F(s)e^{-sx/c}.$$

Inversion of the above Laplace transform gives [see eq. (7.2.5)]

$$u(x, t) = H\left(t - \frac{x}{c}\right) f\left(t - \frac{x}{c}\right). \tag{7.3.15}$$

The waveform represented by the solution remains unchanged if $t - \frac{x}{c}$ is kept constant, indicating that the waveform propagates at the speed c.

Example 7.3.2   An elastic rod of length L is fixed at one end and the other end is subject to the periodic force $F \sin \omega t$. The initial displacement and velocity of the rod are zero. Determine the subsequent motion of the elastic rod under longitudinal oscillations.

Solution   The present problem can be formulated as

$$\begin{cases} \dfrac{\partial^2 u}{\partial t^2} = c^2 \dfrac{\partial^2 y}{\partial x^2}, \quad 0 < x < L, \, t > 0 \\[3mm] u(x, 0) = \dfrac{\partial u}{\partial t}(x, 0) = 0 \\[3mm] u(0, t) = 0 \text{ and } \dfrac{\partial u}{\partial x}(L, t) = \dfrac{F}{E} \sin \omega t \end{cases},$$

where E is the Young modulus of the rod material, $c = \sqrt{\dfrac{E}{\rho}}$, and ρ is the linear density of the rod. Let $U(x, s)$ denote the Laplace transform of $u(x, t)$ with respect to t. By taking the Laplace transform of the governing equation with respect to t and incorporating the initial conditions, we obtain

$$s^2 U(x, s) = c^2 \frac{\partial^2 U}{\partial x^2}(x, s),$$

subject to the boundary conditions

$$U(0, s) = 0 \quad \text{and} \quad \frac{\partial U}{\partial x}(L, s) = \frac{F}{E} \frac{\omega}{s^2 + \omega^2}.$$

The solution is found to be

$$U(x, s) = \frac{F}{E} \frac{c\omega}{s(s^2 + \omega^2)} \frac{\sinh \frac{sx}{c}}{\cosh \frac{sL}{c}}.$$

The poles of $e^{st} U(x, s)$ are

$$s = 0, \quad \pm \omega i \quad \text{and} \quad \pm \frac{c}{L}\left(k - \frac{1}{2}\right) \pi i, \quad k = 1, 2, 3, \ldots.$$

It can be checked easily that $U(x, s)$ tends uniformly to zero as $R \to \infty$ when s assumes values on $\ell_u \cup C_R \cup \ell_d$ (see Figure 7.6). By the Laplace transform inversion formula (7.2.13), the solution to the displacement function is given by

$$u(x, t) = \sum_i \text{Res}\left(e^{st} \frac{F}{E} \frac{c\omega}{s(s^2 + \omega^2)} \frac{\sinh \frac{sx}{c}}{\cosh \frac{sL}{c}}, s_i \right),$$

where s_i are the poles of $e^{st} U(x, s)$. The poles are all simple and so their residues are given by

$$\text{Res}(e^{st} U(x, s), s_i) = \frac{e^{st} F c\omega \sinh \frac{sx}{c}}{\frac{d}{ds}\left[Es(s^2 + \omega^2) \cosh \frac{sL}{c} \right]}\Bigg|_{s=s_i}.$$

The respective residues are found to be:

(i) $\text{Res}(e^{st}U(x, s), 0) = 0;$

(ii) $\text{Res}(e^{st}U(x, s), \pm i\omega) = \frac{Fc\omega}{E}\left(\mp_n^{\pm i\omega t} \frac{i \sin \frac{\omega x}{c}}{2\omega^2 \cos \frac{\omega L}{c}}\right),$

(iii) $\text{Res}\left(e^{st}U(x, s), \pm\frac{c}{L}\left(k - \frac{1}{2}\right)\pi i\right)$

$$= \frac{Fc\omega}{E}\left\{\pm\frac{8i(-1)^k L^2 \sin \frac{(2k-1)\pi x}{L}}{(2k-1)\pi[4L^2\omega^2 - (2k-1)^2 c^2\pi^2]}e^{\pm i(2k-1)c\pi t/(2L)}\right\}.$$

Therefore, the displacement of the elastic rod is given by

$$u(x, t) = \frac{F}{E}\frac{c}{\omega}\frac{\sin \frac{\omega x}{c}}{\cos \frac{\omega L}{c}}\sin \omega t$$

$$+ \frac{16}{\pi}\frac{FL^2}{E}c\omega\sum_{k=1}^{\infty}\frac{(-1)^{k-1}\sin \frac{(2k-1)\pi x}{2L}}{(2k-1)[4L^2\omega^2 - (2k-1)^2 c^2\pi^2]}\sin \frac{(2k-1)c\pi t}{2L}.$$

Example 7.3.3 Use the Fourier transform method to solve the following wave equation with the non-homogeneous term $f(x, t)$:

$$\begin{cases}\dfrac{\partial^2 u}{\partial t^2} - c^2\dfrac{\partial^2 u}{\partial x^2} = f(x, t), \quad -\infty < x < \infty \text{ and } t > 0 \\[3mm] u(x, 0) = 0 \text{ and } \dfrac{\partial u}{\partial t}(x, 0) = 0.\end{cases}$$

Solution First, we take the Fourier transform of the equation and the initial conditions with respect to x. Let the Fourier transform of $u(x, t)$ and $f(x, t)$ be denoted by

$$U(t, \omega) = \int_{-\infty}^{\infty} e^{i\omega x}u(x, t)\, dx \quad \text{and} \quad F(t, \omega) = \int_{-\infty}^{\infty} e^{i\omega x}f(x, t)\, dx,$$

respectively. By observing

$$\mathcal{F}\left\{\frac{\partial^2 u}{\partial x^2}(x, t)\right\} = (-i\omega)^2\mathcal{F}\{u(x, t)\},$$

the resulting equation for $U(t, \omega)$ becomes

$$\begin{cases}\dfrac{\partial^2 U}{\partial t^2}(t, \omega) + c^2\omega^2 U(t, \omega) = F(t, \omega) \\[3mm] U(0, \omega) = 0 \text{ and } \dfrac{\partial U}{\partial t}(0, \omega) = 0\end{cases}$$

The solution to the above differential equation for $U(t, \omega)$ can be obtained by using the Laplace transform method and applying the Laplace inversion formula for the convolution of two functions [see Example 7.2.1, part (b)]. Let $\widehat{U}(s, \omega)$ and $\widehat{F}(s, \omega)$ denote the Laplace transforms of $U(t, \omega)$ and $F(t, \omega)$, respectively. Taking the Laplace transform of the above equation, we obtain

$$\widehat{U}(s, \omega) = \frac{\widehat{F}(s, \omega)}{s^2 + c^2\omega^2} = \frac{\widehat{F}(s, \omega)}{2c\omega i}\left(\frac{1}{s - c\omega i} - \frac{1}{s + c\omega i}\right).$$

The inversion of the above Laplace transform gives

$$U(t, \omega) = \frac{1}{2c\omega i}\int_0^t F(\tau, \omega)[e^{ic\omega(t-\tau)} - e^{-ic\omega(t-\tau)}]\,d\tau.$$

Lastly, by taking the inversion of the Fourier transform $U(t, \omega)$, the solution $u(x, t)$ can be expressed as

$$u(x, t) = \frac{1}{2c}\int_0^t \left[\frac{1}{2\pi}\int_{-\infty}^{\infty}\frac{F(\tau, \omega)}{i\omega}e^{ic\omega(t-\tau)}e^{-i\omega x}\,d\omega\right]d\tau$$
$$- \frac{1}{2c}\int_0^t \left[\frac{1}{2\pi}\int_{-\infty}^{\infty}\frac{F(\tau, \omega)}{i\omega}e^{-ic\omega(t-\tau)}e^{-i\omega x}\,d\omega\right]d\tau.$$

Recall the Fourier transform rules

$$\mathcal{F}\{u'(x)\} = -i\omega\mathcal{F}\{u(x)\} \quad \text{and} \quad \mathcal{F}\{u(x - a)\} = e^{ia\omega}\mathcal{F}\{u(x)\};$$

correspondingly, we define

$$g(x, \tau) = \int_{x_0}^x f(\xi, \tau)\,d\xi, \quad \text{where } x_0 \text{ is some constant.}$$

We then see that

$$\mathcal{F}\{g(x \pm c(t - \tau), \tau)\} = -\frac{F(\tau, \omega)}{i\omega}e^{\mp ic\omega(t-\tau)},$$

so

$$u(x, t) = \frac{1}{2c}\int_0^t g(x + c(t - \tau), \tau)\,d\tau - \frac{1}{2c}\int_0^t g(x - c(t - \tau), \tau)\,d\tau$$
$$= \frac{1}{2c}\int_0^t \int_{x-c(t-\tau)}^{x+c(t-\tau)} f(\xi, \tau)\,d\xi\,d\tau.$$

7.4 Problems

7.1. Consider the steady state temperature distribution inside the domain $|z| < 1$. Given that the boundary temperature value $T(1, \theta)$ along the circumference of the unit circle is prescribed by

$$T(1, \theta) = \begin{cases} 1 & 0 < \theta < \frac{\pi}{2} \\ 0 & \text{otherwise} \end{cases},$$

show that the temperature $T(r, \theta)$ inside the unit circle is given by

$$T(r, \theta) = \frac{1}{\pi} \tan^{-1} \frac{1 - r^2}{r^2 - 2r(\cos\theta + \sin\theta) + 1}, \quad r < 1,$$

where the inverse tangent function assumes values in $[0, \pi]$.

7.2. Let $u(r, \theta)$ and $U(r, \theta)$ be harmonic in the respective domains that are interior and exterior to the circle $|z| = R$, and let both functions satisfy the same boundary condition on $|z| = R$. Show that $u(r, \theta)$ and $U(r, \theta)$ are related by

$$U(r, \theta) = u\left(\frac{R^2}{r}, \theta - 2\pi\right).$$

Hence show that the integral representation of $U(r, \theta)$ is given by

$$U(r, \theta) = -\frac{1}{2\pi} \int_0^{2\pi} \frac{r^2 - R^2}{r^2 - 2rR\cos(\phi - \theta) + R^2} U(R, \phi)\, d\phi, \quad r > R.$$

7.3. Let $u(r, \theta)$ be the harmonic function in the domain $|z| < R$ that satisfies the boundary value $f(\theta)$, $0 \le \theta < 2\pi$, on the circumference of the circle. Let $v(r, \theta)$ be the harmonic conjugate of $u(r, \theta)$ satisfying the condition $v(0, \theta) = 0$. Show that $v(r, \theta)$ is given by

$$v(r, \theta) = -\frac{1}{2\pi} \int_0^{2\pi} \frac{2Rr\sin(\phi - \theta)}{R^2 - 2Rr\cos(\phi - \theta) + r^2} f(\phi)\, d\phi, \quad r < R.$$

Further, show that

$$\int_0^{2\pi} \frac{2Rr\sin(\phi - \theta)}{R^2 - 2Rr\cos(\phi - \theta) + r^2}\, d\phi = 0.$$

Hint: Show that

$$-\frac{2Rr\sin(\phi - \theta)}{R^2 - 2Rr\cos(\phi - \theta) + r^2} = \operatorname{Im}\left(\frac{\zeta + z}{\zeta - z}\right)$$

$$= 2 \sum_{n=1}^{\infty} \left(\frac{r}{R}\right)^n \sin n(\phi - \theta),$$

where $\zeta = Re^{i\phi}$ and $z = re^{i\theta}$.

7.4. Let $f(z) = u(z) + iv(z)$ be analytic on and inside the circle $|z| = R$. Show that

$$f(z) = \frac{1}{2\pi i} \oint_{|\zeta|=R} \frac{\zeta + z}{\zeta - z} \frac{u(\zeta)}{\zeta} d\zeta + iv(0), \quad |z| < R.$$

7.5. Consider the Dirichlet problem of finding the function that is harmonic inside the unit disk $|z| < 1$. The boundary values are prescribed as 1 on the upper semi-circle and -1 on the lower semi-circle of the boundary. Show that the solution to the Dirichlet problem is given by

$$u(z) = \frac{2}{\pi} \text{Arg} \frac{1+z}{1-z}.$$

From the form of the solution, explain why the value of $u(z)$, when z approaches 1 or -1, depends on the direction of approach to these two points.

Hint: The value of $\text{Arg}(z - 1) = \tan^{-1} \frac{y}{x-1}$, $z = x + iy$, depends on the angle of inclination of the line segment joining (x, y) and $(1, 0)$.

7.6. Show that the Poisson kernel $P(R, r, \phi - \theta)$ can be expressed as

$$\frac{R^2 - r^2}{|Re^{i\phi} - re^{i\theta}|^2}.$$

Explain why

$$R - r \leq |Re^{i\phi} - re^{i\theta}| \leq R + r,$$

and hence deduce

$$\frac{R - r}{R + r} \leq P(R, r, \phi - \theta) \leq \frac{R + r}{R - r}.$$

Suppose u is a positive harmonic function inside the circle $|z| = R$; use the above inequalities to show that

$$\frac{R - r}{R + r} u(0) \leq u(re^{i\theta}) \leq \frac{R + r}{R - r} u(0), \quad r < R \text{ and } -\pi < \theta \leq \pi.$$

7.7. Suppose $u(r, \theta)$ is harmonic inside the disk $|z| < R$ and assumes the following boundary values on $r = R$:

$$u(R, \theta) = \begin{cases} 1 & \theta_1 < \theta < \theta_2 \\ 0 & \text{otherwise} \end{cases}.$$

Let P_1 and P_2 denote the two points $Re^{i\theta_1}$ and $Re^{i\theta_2}$ on the boundary $|z| = R$. Take any arbitrary point z inside the disk and denote it by P. Draw the two chords PP_1 and PP_2 which intersect the circle at P_1' and P_2' respectively. Let s denote the arc length of $P_1'P_2'$ (see the figure). Show that the value of $u(r, \theta)$ at the point P is given by

$$u(r, \theta) = \frac{S}{2\pi R}.$$

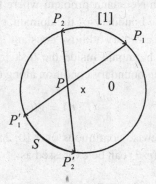

7.8. Use the Schwarz integral formula to show that

$$u(x, y) = \frac{U}{\pi}\mathrm{Arg}\,\frac{z - x_2}{z - x_1}, \qquad z = x + iy,$$

is harmonic in the upper half-plane $\mathrm{Im}\,z > 0$, and satisfies the boundary condition

$$u(x, 0) = \begin{cases} U & x_1 < x < x_2 \\ 0 & \text{otherwise} \end{cases}.$$

Remark: Compare this with a similar result in eq. (3.3.14).

7.9. Let $f(z) = u(x, y) + iv(x, y)$, $z = x + iy$, be analytic in the closed upper half-plane $\mathrm{Im}\,z \geq 0$. Show that

$$v(x, y) = \frac{1}{\pi}\int_{-\infty}^{\infty}\frac{x - t}{(t - x)^2 + y^2}\,u(t, 0)\,dt, \qquad \mathrm{Im}\,z > 0.$$

7.10. Let u be harmonic inside a domain $\mathcal{D}$ and assume the zero value along the boundary $\partial\mathcal{D}$ of the domain; that is,

$$\nabla^2 u = 0 \text{ in } \mathcal{D} \quad \text{and} \quad u = 0 \text{ on } \partial\mathcal{D}.$$

Using Green's theorem in the form

$$\oint_{\partial D} u \left(\frac{\partial u}{\partial y} dx - \frac{\partial u}{\partial x} dy \right)$$
$$= -\iint_D \left[u \left(\frac{\partial^2 u}{\partial x^2} + \frac{\partial^2 u}{\partial y^2} \right) + \left(\frac{\partial u}{\partial x} \right)^2 + \left(\frac{\partial u}{\partial y} \right)^2 \right] dx dy,$$

show that $u = 0$ throughout the domain $\mathcal{D}$.

7.11. Following a similar approach as used in Problem 7.10, given that a solution exists for a Neumann problem where the normal derivative is prescribed along the boundary of the domain, show that the solution is unique to within an arbitrary additive constant.

7.12. Suppose $u(r, \theta)$ is harmonic inside the disk $|z| < R$ and satisfies the prescribed Neumann boundary condition along the boundary

$$\frac{\partial u}{\partial r}(R, \theta) = g(\theta),$$

where $g(\theta)$ is piecewise continuous over $[0, 2\pi]$. Show that the series representation of $u(r, \theta)$ can be expressed as

$$u(r, \theta) = \frac{a_0}{2} + \sum_{n=1}^{\infty} \left(\frac{r}{R} \right)^n (a_n \cos n\theta + b_n \sin n\theta),$$

where a_0 is arbitrary, and the coefficients a_n and b_n are given by

$$a_n = \frac{R}{n\pi} \int_0^{2\pi} g(\phi) \cos n\phi \, d\phi, \quad n = 1, 2, \dots,$$

$$b_n = \frac{R}{n\pi} \int_0^{2\pi} g(\phi) \sin n\phi \, d\phi, \quad n = 1, 2, \dots.$$

7.13. Find the function $u(x, y)$ that is harmonic in the upper half-plane and satisfies the following Neumann boundary condition along the x-axis:

$$\frac{\partial u}{\partial y}(x, 0) = \begin{cases} 0 & \text{if } |x| > 1 \\ K & \text{if } |x| < 1 \end{cases}.$$

7.14. Find the function that is harmonic inside the half-disk $|z| < R$ and Im $z > 0$, and satisfies the following mixed Dirichlet–Neumann boundary conditions:

$$\begin{cases} u(R, \theta) = f(\theta), \quad 0 < \theta < \pi \\ \dfrac{\partial u}{\partial \theta}(R, 0) = 0 \quad \text{and} \quad \dfrac{\partial u}{\partial \theta}(R, \pi) = 0 \end{cases}.$$

7.15. Show that

$$u(x, y) = \frac{1}{2\pi} \int_0^\infty \ln \frac{(t-x)^2 + y^2}{(t+x)^2 + y^2} g(t)\, dt, \quad x > 0 \text{ and } y > 0,$$

is harmonic inside the domain {z: Re z > 0 and Im z > 0} and satisfies
the boundary conditions

$$\begin{cases} u(0, y) = 0, & y > 0 \\[2mm] \lim_{y \to 0} \dfrac{\partial u}{\partial y}(x, y) = g(x), & x > 0 \end{cases}$$

7.16. Show that

$$\mathcal{L}\{t^n f(t)\} = (-1)^n \frac{d^n}{ds^n} \mathcal{L}\{F(s)\}, \quad \text{where } n \text{ is any positive integer.}$$

Let $Y(s)$ denote the Laplace transform of $y(t)$. In each of the following
ordinary differential equations, find the corresponding governing equation for $Y(s)$:

(a) $\dfrac{d^2 y}{dt^2} - 2t\dfrac{dy}{dt} + \lambda y = 0, \quad y(0) = y_0 \text{ and } y'(0) = y_0';$

(b) $t\dfrac{d^2 y}{dt^2} + (1 - t)\dfrac{dy}{dt} + \lambda y = 0, \quad y(0) = 0 \text{ and } y'(0) = 0.$

7.17. Let $F(s)$ denote the Laplace transform of $f(t)$, $t > 0$, and define

$$F_T(s) = \int_0^T e^{-st} f(t)\, dt.$$

Show that

$$\mathcal{L}\{f(t + T)\} = e^{sT}[F(s) - F_T(s)].$$

Illustrate the result with the function $f(t) = \cos(t + \theta)$.

7.18. Suppose $\mathcal{L}\left\{\dfrac{f(t)}{t}\right\} = G(s)$ and $\mathcal{L}\{f(t)\} = F(s)$; show that

$$G(s) = \int_s^\infty F(z)\, dz.$$

Use the formula to find

(a) $\mathcal{L}\left\{\dfrac{\sin t}{t}\right\};$ (b) $\mathcal{L}\left\{\dfrac{\cos t}{t}\right\}.$

7.19. Show that

$$\mathcal{L}\left\{\int_0^t f(\xi)\, d\xi\right\} = \frac{1}{s}\mathcal{L}\{f(t)\}.$$

Use the formula to find

(a) $\mathcal{L}\left\{\int_t^\infty \dfrac{\sin u}{u}\, du\right\}$;   (b) $\mathcal{L}\left\{\int_t^\infty \dfrac{\cos u}{u}\, du\right\}$.

7.20. Suppose $F(s)$ tends uniformly to zero as $R \to \infty$ when s assumes values on $\ell_u \cup C_R \cup \ell_d$ (see Figure 7.6); that is, $|F(s)| \le K_R$ on $\ell_u \cup C_R \cup \ell_d$, where $K_R \to 0$ as $R \to \infty$. Show that

(a) $\lim\limits_{R\to\infty} \displaystyle\int_{\ell_u} e^{st} F(s)\, ds = 0$;   (b) $\lim\limits_{R\to\infty} \displaystyle\int_{C_R} e^{st} F(s)\, ds = 0$.

Hint: For part (a), $s = x + Ri$ on ℓ_u and so

$$|e^{st}| = |e^{t(x+Ri)}| = e^{xt} \le e^{at}, \quad \text{since } 0 < x < a.$$

For part (b), $s = Re^{i\theta}$ on C_R, $\frac{\pi}{2} \le \theta \le \frac{3\pi}{2}$. Consider

$$\left|\int_{C_R} e^{st} F(s)\, ds\right| \le 2RK_R \int_{\frac{\pi}{2}}^{\pi} e^{tR\cos\theta}\, d\theta$$

$$= 2RK_R \int_0^{\frac{\pi}{2}} e^{-tR\sin\theta}\, d\theta,$$

then apply the Jordan lemma (see Subsection 6.4.2).

7.21. Use the Laplace transform formula

$$\mathcal{L}\{f'(t)\} = sF(s) - f(0^+),$$

to show

$$\lim_{s\to\infty} sF(s) = \lim_{t\to 0^+} f(t).$$

Verify the result with the following functions:

(a) $f(t) = \sin t$;   (b) $f(t) = \dfrac{\sin t}{t}$.

7.22. Find the inversion of the following Laplace transform functions:

(a) $F(s) = \mathrm{Log}\, \dfrac{s+1}{s}$;   (b) $F(s) = \dfrac{1}{(s^2 + a^2)^{1/2}}$, $a > 0$.

Hint: For part (a), consider the closed contour shown below:

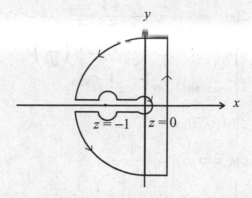

For part (b), the function $F(s) = \frac{1}{(s^2+a^2)^{1/2}}$ has branch points at $z = \pm ai$. Suppose we choose the branch cut to be the line segment joining $z = -ai$ and $z = ai$; modify the above closed contour so that it includes an excursion along the branch cut with indentations around the two branch points.

7.23. Solve the following integral equation for $y(t)$:

$$\phi(t) = \int_0^t \frac{y(\tau)}{(t-\tau)^{1/2}} \, d\tau.$$

Hint: Use the convolution formula and apply the relation

$$s\mathcal{L}\{\phi(t)\} = \int_0^\infty e^{-st} \phi'(t) \, dt.$$

7.24. Consider the differential equation

$$t^2 y''(t) + t y'(t) + t^2 y(t) = 0$$

with initial conditions: $y(0) = 1$ and $y'(0) = 0$. Let $Y(s)$ denote the Laplace transform of $y(t)$. Show that

$$\frac{dF(s)}{ds} = -\frac{s}{s^2+1} F(s).$$

By expanding $F(s)$ in negative powers of s, show that the solution is given by

$$y(t) = \sum_{n=0}^\infty \frac{(-1)^n t^{2n}}{(2^n n!)^2}.$$

7.25. Prove the following Laplace transform formulas:

(a) $\mathcal{L}\{e^{-\lambda t}\cos\omega t\} = \dfrac{s+\lambda}{(s+\lambda)^2+\omega^2}$;

(b) $\mathcal{L}\left\{\dfrac{1}{\sqrt{\pi t}}e^{-2a\sqrt{t}}\right\} = \dfrac{1}{\sqrt{s}}e^{a^2/s}\,\mathrm{erfc}\left(\dfrac{a}{\sqrt{s}}\right)$;

(c) $\mathcal{L}\left\{\dfrac{1}{\sqrt{\pi a}}\sin 2\sqrt{at}\right\} = \dfrac{1}{s\sqrt{s}}e^{-a/s}$;

(d) $\mathcal{L}\{\mathrm{erf}(\sqrt{at})\} = \dfrac{\sqrt{a}}{s\sqrt{s+a}}$;

(e) $\mathcal{L}\{e^t\,\mathrm{erfc}(\sqrt{t})\} = \dfrac{1}{s+\sqrt{s}}$;

(f) $\mathcal{L}\left\{\dfrac{1}{\sqrt{\pi t}} - e^t\,\mathrm{erfc}(\sqrt{t})\right\} = \dfrac{1}{1+\sqrt{s}}$;

(g) $\mathcal{L}\left\{\dfrac{1}{\sqrt{\pi t}}\sin\dfrac{1}{2t}\right\} = \dfrac{1}{\sqrt{s}}e^{-\sqrt{s}}\sin\sqrt{s}$;

(h) $\mathcal{L}\left\{\dfrac{1}{\sqrt{\pi t}}\cos\dfrac{1}{2t}\right\} = \dfrac{1}{\sqrt{s}}e^{-\sqrt{s}}\cos\sqrt{s}$.

7.26. Solve the heat conduction equation

$$\frac{\partial T}{\partial t} = \frac{\partial^2 T}{\partial x^2}, \quad 0 < x < 1 \text{ and } t > 0,$$

with boundary conditions $T(0, t) = T(1, t) = 2$, and initial condition $T(x, 0) = 2 + \sin\pi x$.

7.27. Solve the heat conduction problem in a semi-infinite rod with insulated end as modeled by

$$\begin{cases} \dfrac{\partial T}{\partial t} = a^2\dfrac{\partial^2 T}{\partial x^2} & x > 0,\ t > 0 \\[2mm] \dfrac{\partial T}{\partial x}(0, t) = 0 & t > 0 \\[2mm] T(x, 0) = f(x) & x > 0 \end{cases}$$

7.28. Suppose the temperature on the earth's surface (assumed to be an infinite flat surface) has seasonal variations. Let the x-axis be along the direction normal to the earth's surface. We would like to find the seasonal variations of the temperature in the free space above the earth's surface. For simplicity, we assume the temperature function T depends on time t and spatial coordinate x only. The governing equation for $T(x, t)$ can be described by

$$\frac{\partial T}{\partial t} = a^2 \frac{\partial^2 T}{\partial x^2}, \quad x > 0 \text{ and } t > 0.$$

Let the seasonal temperature variations on the earth's surface be modeled by the boundary condition $T(0, t) = \sin \omega t$, and the initial temperature is assumed to be zero. Find the solution for $T(x, t)$.

7.29. Consider the heat conduction problem in a semi-infinite rod subject to the heat flux $q(t)$ incident at the free end $x = 0$; that is,

$$-k \frac{\partial T}{\partial x}\bigg|_{x=0} = q(t),$$

where k is the conductivity of the material. For simplicity, we assume that the heat conduction equation takes the form

$$\frac{\partial T}{\partial t} = \frac{\partial^2 T}{\partial x^2}, \quad x > 0, \ t > 0,$$

and the initial temperature is $T(x, 0) = 0, \ x > 0$. Solve the problem under the following two special cases:

(a) $q(t) = q_0$; (b) $q(t) = q_0 \sin \omega t$.

Hint: The temperature is obtained by evaluating the Bromwich integral

$$T(x, t) = \frac{1}{2\pi i k} \oint_{C_B} e^{st - \sqrt{s}x} \frac{Q(s)}{\sqrt{s}} \, ds,$$

where $Q(s) = \mathcal{L}\{q(t)\}$. In part (a) where $q(t) = q_0$, it is necessary to consider the derivative

$$\frac{\partial T}{\partial x} = -\frac{q_0}{k} \frac{1}{2\pi i} \oint_{C_B} \frac{e^{st - \sqrt{s}x}}{s} \, ds$$

so that the contour integral around the infinitesimal circle $|z| = \epsilon$ (see Figure 7.7) becomes zero.

7.30. Use the Laplace transform method to solve the heat conduction problem within a finite rod of length L. The governing equation is given by

$$\frac{\partial T}{\partial t} = a^2 \frac{\partial^2 T}{\partial x^2}, \quad 0 < x < L, \ t > 0,$$

subject to the boundary conditions $T(0, t) = T_1$ and $T(L, t) = T_2$, where T_1 and T_2 are constants, and initial condition $T(x, 0) = 0, \ 0 < x < L$.

Hint:

$$\mathcal{L}^{-1}\left\{\frac{\sinh x\sqrt{s}}{s\sinh L\sqrt{s}}\right\} = \frac{x}{L} + \frac{2}{\pi}\sum_{n=1}^{\infty}\frac{(-1)^n}{n}e^{-n^2\pi^2 t/L^2}\sin\frac{n\pi x}{L}.$$

7.31. The heat conduction problem in an infinite rod with the non-homogeneous term $f(x,t)$ and initial condition $\phi(x)$ is posed as

$$\begin{cases}\dfrac{\partial T}{\partial t} = a^2\dfrac{\partial^2 T}{\partial x^2} + f(x,t), & -\infty < x < \infty,\ t > 0 \\[2ex] T(x,0) = \phi(x)\end{cases}$$

Show that the solution is given by

$$\begin{aligned}u(x,t) = {}& \frac{1}{\sqrt{4\pi a^2 t}}\int_{-\infty}^{\infty}\phi(\xi)e^{-\frac{(x-\xi)^2}{4a^2 t}}\,d\xi \\ & + \frac{1}{2a\sqrt{\pi}}\int_{0}^{t}\int_{-\infty}^{\infty}\frac{f(\xi,\tau)}{\sqrt{t-\tau}}e^{-\frac{(x-\xi)^2}{4a^2(t-\tau)}}\,d\xi\,d\tau.\end{aligned}$$

7.32. Use the Fourier transform method to solve the following wave equation:

$$\begin{cases}\dfrac{\partial^2 u}{\partial t^2} = \dfrac{\partial^2 u}{\partial x^2} + t\sin x, & -\infty < x < \infty \text{ and } t > 0 \\[2ex] u(x,0) = 0 \text{ and } \dfrac{\partial u}{\partial t}(x,0) = \sin x.\end{cases}$$

7.33. Use the Laplace transform method to solve the following wave equation:

$$\begin{cases}\dfrac{\partial^2 u}{\partial t^2} = \dfrac{\partial^2 u}{\partial x^2} + \sin\pi x, & 0 < x < 1 \text{ and } t > 0 \\[2ex] u(x,0) = 0 \text{ and } \dfrac{\partial u}{\partial t}(x,0) = 0 \\[2ex] u(0,t) = u(1,t) = 0\end{cases}$$

7.34. Consider the wave propagation along an infinite elastic rod with initial displacement and velocity defined by $f(x)$ and $g(x)$, respectively, and suppose the displacement and velocity die off at the far ends. Let $u(x,t)$ denote the displacement function. The wave propagation problem is formulated as follows:

$$
\begin{cases}
\dfrac{\partial^2 u}{\partial t^2} = c^2 \dfrac{\partial^2 u}{\partial x^2}, & -\infty < x < \infty, \ t > 0 \\[2mm]
u(x,0) = f(x) \ \text{and} \ \dfrac{\partial u}{\partial t}(x,0) = g(x)
\end{cases}
$$

By taking the Laplace transform of the wave equation with respect to t, show that the governing equation for $U(x,s) = \mathcal{L}\{u(x,t)\}$ is given by

$$
s^2 U(x,s) - sf(x) - g(x) - c^2 \frac{\partial^2 u}{\partial x^2}(x,s) = 0.
$$

Solve the above equation and explain the steps that lead to

$$
U(x,s) = \frac{1}{2c} \int_x^\infty \frac{e^{-s(\xi-x)/c}}{s} g(\xi)\,d\xi + \frac{1}{2c} \int_{-\infty}^x \frac{e^{-s(x-\xi)/c}}{s} g(\xi)\,d\xi
$$

$$
+ \frac{1}{2c} \int_x^\infty \frac{e^{-s(\xi-x)/c}}{s} sf(\xi)\,d\xi + \frac{1}{2c} \int_{-\infty}^x \frac{e^{-s(x-\xi)/c}}{s} sf(\xi)\,d\xi.
$$

By performing the inversion of the Laplace transform $U(x,s)$, show that

$$
u(x,t) = \frac{1}{2c} \int_{x-ct}^{x+ct} g(\xi)\,d\xi + \frac{1}{2}[f(x+ct) + f(x-ct)].
$$

The result is called the *D'Alembert formula*.

Hint: Using the Laplace inversion formula

$$
\mathcal{L}^{-1}\left\{ \frac{e^{-s(\xi-x)/c}}{s} \right\} = H\left(t - \frac{\xi-x}{c}\right),
$$

show that

$$
\mathcal{L}^{-1}\left\{ \frac{1}{2c} \int_x^\infty \frac{e^{-s(\xi-x)/c}}{s} g(\xi)\,d\xi \right\} = \frac{1}{2c} \int_x^{x+ct} g(\xi)\,d\xi.
$$

7.35. Use the Laplace transform method to solve

$$
x\frac{\partial u}{\partial t} + \frac{\partial u}{\partial x} = x, \quad x > 0, \ t > 0,
$$

subject to the auxiliary conditions $u(x,0) = 0$ for $x > 0$ and $u(0,t) = 0$ for $t > 0$.

8

Conformal Mappings and Applications

A complex function $w = f(z)$ can be regarded as a mapping from its domain in the z-plane to its range in the w-plane. In this chapter, we go beyond the previous chapters by analyzing in greater depth the geometric properties associated with mappings represented by complex functions. First, we examine the linkage between the analyticity of a complex function and the conformality of a mapping. A mapping is said to be conformal at a point if it preserves the angle of intersection between a pair of smooth arcs through that point. The invariance of the Laplace equation under a conformal mapping is also established. This invariance property allows us to use conformal mappings to solve various types of physical problem, like steady state temperature distribution, electrostatics and fluid flows, where problems with complicated configurations can be transformed into those with simple geometries.

First, we introduce various techniques for effecting the mappings of regions. Two special classes of transformation, the bilinear transformations and the Schwarz–Christoffel transformations, are discussed fully. A bilinear transformation maps the class of circles and lines to the same class, and it is conformal at every point except at its pole. The Schwarz–Christoffel transformations take half-planes onto polygonal regions. These polygonal regions can be unbounded with one or more of their vertices at infinity. We also consider the class of hodograph transformations, where the roles of the dependent and independent variables are reversed.

8.1 Conformal mappings

Let f be a complex function that is analytic at a point z_0 and $f'(z_0) \neq 0$. Let γ denote a smooth arc through z_0, whose parametric form is given by

$$z(t) = x(t) + iy(t), \quad a \leq t \leq b, \tag{8.1.1}$$

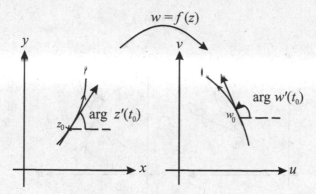

Figure 8.1. The curve γ through z_0 is carried under the mapping $w = f(z)$ to the curve Γ through $w_0 = f(z_0)$.

and $z(t_0) = z_0$. Recall that the arc represented by $z(t)$ is smooth if $x(t)$ and $y(t)$ are both continuously differentiable and $x'(t)$ and $y'(t)$ do not both vanish at the same value of t. The image of γ under the mapping $w = f(z)$, denoted by Γ, can be represented by

$$w(t) = f(z(t)), \quad a \le t \le b. \tag{8.1.2}$$

We would like to examine the change in the direction of a curve through the point z_0 under the mapping $w = f(z)$. The chain rule of differentiation gives

$$w'(t) = f'(z(t))z'(t), \quad a \le t \le b,$$

in particular,

$$w'(t_0) = f'(z_0)z'(t_0).$$

Given that $z'(t_0) \ne 0$ (corresponding to γ being a smooth curve) and $f'(z_0) \ne 0$, it then follows that $w'(t_0)$ is nonzero. Hence we have

$$\arg w'(t_0) = \arg f'(z_0) + \arg z'(t_0). \tag{8.1.3}$$

Here, $\arg z'(t_0)$ and $\arg w'(t_0)$ give the angles of inclination of the tangent vectors to the curve γ at z_0 and the image curve Γ at $w_0 = f(z_0)$, respectively (see Figure 8.1). By virtue of eq. (8.1.3), the tangent vector to γ at z_0 is rotated in the anticlockwise sense through the angle $\arg f'(z_0)$ under the mapping $w = f(z)$. The magnitude and sense of this rotation are dependent only on the mapping $f(z)$, but independent of the curve γ.

Let γ_1 and γ_2 be a pair of smooth arcs intersecting at z_0, and let α be the angle between the tangent vectors to the two arcs at z_0. Since the two tangent vectors are rotated in the same sense and by the same magnitude under the

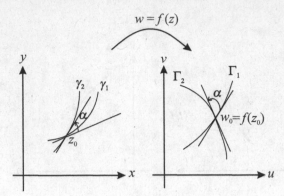

Figure 8.2. The mapping $w = f(z)$ is conformal at z_0 since it preserves the angle of intersection between a pair of smooth arcs through the point.

mapping $w = f(z)$, the angle between the tangent vectors to the image curves Γ_1 and Γ_2 at $w_0 = f(z_0)$ is also α (see Figure 8.2). This property of angle preservation under a mapping is formally defined as follows:

Definition 8.1.1 A mapping that preserves the magnitude and sense of the angle between any two smooth arcs passing through a given point is said to be *conformal* at that point.

The relation between analyticity and the above angle-preserving property of a complex function is stated in the following theorem.

Theorem 8.1.1 *An analytic function f is conformal at every point z_0 for which $f'(z_0) \neq 0$.*

Proof In the z-plane, we draw two smooth arcs γ_1 and γ_2 through z_0. The smooth arcs are parametrized by

$$z_1(t) = x_1(t) + iy_1(t), \quad a_1 \leq t \leq b_1,$$
$$z_2(s) = x_2(s) + iy_2(s), \quad a_2 \leq s \leq b_2,$$

with $z_1(t_0) = z_2(s_0) = z_0$. The included angle α, measured in the anti-clockwise sense, from the tangent vector $z_1'(t_0)$ to the tangent vector $z_2'(s_0)$ is given by (see Figure 8.2)

$$\alpha = \arg z_2'(s_0) - \arg z_1'(t_0).$$

Sometimes, special care may be needed to adjust appropriately the argument of one of the tangent vectors by adding or subtracting 2π. The image curves Γ_1

and Γ_2 under the mapping $w = f(z)$ are represented by $w_1(t) = f(z_1(t))$ and $w_2(s) = f(z_2(s))$, respectively, and they intersect at $w_0 = f(z_0)$ (see Figure 8.2). The angle, measured in the anticlockwise sense, from the tangent vector $w_1'(t_0)$ to the tangent vector $w_2'(s_0)$ is given by $\arg w_2'(s_0) - \arg w_1'(t_0)$. By virtue of eq. (8.1.3), this included angle between the image curves is the same as α since

$$\arg w_2'(s_0) - \arg w_1'(t_0) = [\arg f'(z_0) + \arg z_2'(s_0)] - [\arg f'(z_0) + \arg z_1'(t_0)]$$

$$= \arg z_2'(s_0) - \arg z_1'(t_0) = \alpha. \tag{8.1.4}$$

Note that analyticity of f at z_0 and non-vanishing of $f'(z_0)$ are required to establish eq. (8.1.4). The mapping is seen to preserve the magnitude and sense of the angle between any two smooth arcs passing through z_0, and so the function f is conformal at z_0.

Scale factor

We have seen that $\arg f'(z_0)$ gives the angle of rotation of the tangent vector to the curve γ at z_0 under the mapping $w = f(z)$. Can we attach any geometric meaning to the modulus $|f'(z_0)|$? Let $z - z_0 = re^{i\theta}$ and $w - w_0 = \rho e^{i\phi}$, where z and w are points on the curve γ and the image curve Γ, respectively. Let Δs be the arc length between z_0 and z along γ, and $\Delta \sigma$ be the arc length between w_0 and w along Γ. We see that as z tends to z_0, w tends to w_0. Consider the ratio

$$\frac{w - w_0}{z - z_0} = \frac{f(z) - f(z_0)}{z - z_0} = \frac{\rho e^{i\phi}}{re^{i\theta}} = \frac{\Delta\sigma}{\Delta s}\frac{\rho}{\Delta\sigma}\frac{\Delta s}{r}e^{i(\phi-\theta)},$$

and observe that

$$\lim_{w \to w_0} \frac{\rho}{\Delta\sigma} = 1 \quad \text{and} \quad \lim_{z \to z_0} \frac{\Delta s}{r} = 1.$$

We obtain

$$|f'(z_0)| = \left|\lim_{z \to z_0} \frac{f(z) - f(z_0)}{z - z_0}\right| = \lim_{z \to z_0} \frac{\Delta\sigma}{\Delta s} = \lim_{z \to z_0} \frac{|f(z) - f(z_0)|}{|z - z_0|}. \tag{8.1.5}$$

As $z \to z_0$, $|f(z) - f(z_0)| \approx |f'(z_0)||z - z_0|$, so the quantity $|f'(z_0)|$ is visualized as the *scale factor* of the conformal transformation $f(z)$ at z_0 (see also Example 8.1.2).

There is a close link between the *one-to-one property* of a mapping and non-vanishing of the derivative value of the mapping function. We state the following result without proof: if f is analytic at z_0 and $f'(z_0) \neq 0$, then there exists a neighborhood of z_0 such that f is one-to-one inside that neighborhood.

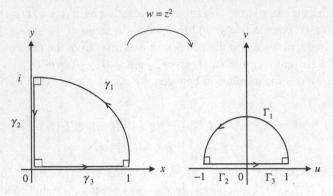

Figure 8.3. The mapping $w = z^2$ takes the quarter circle $|z| \leq 1$ and $0 \leq \text{Arg } z \leq \frac{\pi}{2}$ in the z-plane onto the upper semi-circle $|w| \leq 1$ and $0 \leq \text{Arg } w \leq \pi$ in the w-plane.

Example 8.1.1 Discuss the conformal property of $w = f(z) = z^2$ inside the domain

$$\mathcal{D} = \left\{ z : 0 < |z| < 1, \ 0 < \text{Arg } z < \frac{\pi}{2} \right\}.$$

Solution The analytic function $w = f(z) = z^2$ is seen to be conformal at all points inside $\mathcal{D}$ since f' vanishes only at $z = 0$ (which does not lie inside $\mathcal{D}$). The image of $\mathcal{D}$ in the w-plane under the given mapping is the upper semi-circle

$$\mathcal{D}' = \{ w : 0 < |w| < 1, \ 0 < \text{Arg } w < \pi \}.$$

Let $\gamma = \gamma_1 \cup \gamma_2 \cup \gamma_3$ be the boundary of $\mathcal{D}$, where γ_1 is the arc of the unit circle, γ_2 is the line segment along the imaginary axis, and γ_3 is the line segment along the real axis. The corresponding image curves of γ_1, γ_2 and γ_3 are denoted by Γ_1, Γ_2 and Γ_3, respectively (see Figure 8.3).

The curves γ_1 and γ_2 intersect orthogonally at $z = i$, as do the image curves Γ_1 and Γ_2 at $w = -1$. The same phenomenon holds for the curves γ_3 and γ_1 at $z = 1$ and the image curves Γ_3 and Γ_1 at $w = 1$. These observations agree with the conformal property of $w = z^2$ at $z = i$ and $z = 1$.

However, though the curves γ_2 and γ_3 intersect orthogonally at $z = 0$, the angle of intersection of the image curves Γ_2 and Γ_3 at $w = 0$ is zero. This is not surprising since the mapping $w = z^2$ fails to be conformal at $z = 0$. Actually, the magnification of the included angle at a critical point of the mapping function depends on the order of the critical point (see discussion below).

Mapping behavior around a critical point

Suppose z_0 is a critical point of an analytic function f, that is, $f'(z_0) = 0$. In this case, f fails to be conformal at z_0. Let m be the lowest order such that $f^{(m)}(z_0) \neq 0$, that is, $f'(z_0) = \cdots = f^{(m-1)}(z_0) = 0$. What would be the mapping property of $w = f(z)$ around $z = z_0$?

The Taylor series for $f(z)$ at z_0 becomes

$$w = f(z) = f(z_0) + \frac{f^{(m)}(z_0)}{m!}(z - z_0)^m + \frac{f^{(m+1)}(z_0)}{(m+1)!}(z - z_0)^{m+1} + \cdots .$$

At points close to z_0, we may neglect the higher-order terms and obtain the approximation

$$w - w_0 \approx \frac{f^{(m)}(z_0)}{m!}(z - z_0)^m, \quad w_0 = f(z_0). \tag{8.1.6}$$

The arguments and moduli of $w - w_0$ and $z - z_0$ are then approximately related by

$$\arg(w - w_0) \approx \arg f^{(m)}(z_0) + m \arg(z - z_0) \tag{8.1.7a}$$

and

$$|w - w_0| \approx \left| \frac{f^{(m)}(z_0)}{m!} \right| |z - z_0|^m. \tag{8.1.7b}$$

Suppose we write $z - z_0 = re^{i\theta}$ and $w - w_0 = \rho e^{i\phi}$, where z and w are points on the curve γ and the image curve Γ, respectively. From eq. (8.1.7a), we deduce that

$$\phi \approx \arg f^{(m)}(z_0) + m\theta. \tag{8.1.8}$$

As $z \to z_0$, θ and ϕ tend to the angle of inclination of the tangent vector to the curve γ at z_0 and the image curve Γ at w_0, respectively. Let α be the angle between the two smooth arcs γ_1 and γ_2 intersecting at z_0; then $\alpha = \theta_2 - \theta_1$, where θ_1 and θ_2 are the respective angles of inclination of the tangent vectors to γ_1 and γ_2 at z_0. The corresponding angle between the tangent vectors to the image arcs Γ_1 and Γ_2 intersecting at w_0 is given by

$$[\arg f^{(m)}(z_0) + m\theta_2] - [\arg f^{(m)}(z_0) + m\theta_1] = m(\theta_2 - \theta_1) = m\alpha. \tag{8.1.9}$$

That is, the included angle between the image arcs in the w-plane is m times that between the smooth arcs in the z-plane.

Referring to the mapping function $w = z^2$ considered in Example 8.1.1, $z = 0$ is a critical point with $m = 2$. Consider the pair of curves $z = t, t \geq 0$, and $z = is, s \geq 0$, where they intersect at $z = 0$ in the z-plane. The image curves under the mapping $w = z^2$ in the z-plane become $w = t^2, t \geq 0$, and $w = -s^2, s \geq 0$. These image curves intersect at $w = 0$. The included angle between the pair of curves in the z-plane is $\frac{\pi}{2}$ while that between the image curves in the w-plane is π. This is expected since $m = 2$, so the included angle is magnified by a factor of 2 under the mapping.

Example 8.1.2 Suppose the conformal transformation $w = f(z)$ maps a domain $\mathcal{D}$ in the z-plane onto the domain Δ in the w-plane. Show that the area of Δ is given by

$$A = \int\int_{\mathcal{D}} |f'(z)|^2 \, dx \, dy.$$

Solution The area of the domain Δ in the w-plane is given by

$$A = \int\int_{\Delta} du \, dv = \int\int_{\mathcal{D}} \frac{\partial(u, v)}{\partial(x, y)} \, dx \, dy, \quad w = u + iv, \quad z = x + iy,$$

where $\frac{\partial(u,v)}{\partial(x,y)}$ is the Jacobian of the transformation. One can establish

$$\frac{\partial(u, v)}{\partial(x, y)} = \frac{\partial u}{\partial x}\frac{\partial v}{\partial y} - \frac{\partial v}{\partial x}\frac{\partial u}{\partial y} = \left(\frac{\partial u}{\partial x}\right)^2 + \left(\frac{\partial u}{\partial y}\right)^2,$$

by virtue of the Cauchy–Riemann relations. On the other hand, we have

$$f'(z) = \frac{\partial u}{\partial x} + i\frac{\partial v}{\partial x} = \frac{\partial u}{\partial x} - i\frac{\partial u}{\partial y},$$

so

$$|f'(z)|^2 = \left(\frac{\partial u}{\partial x}\right)^2 + \left(\frac{\partial u}{\partial y}\right)^2;$$

thus the area formula is established. The result is not surprising since $|f'(z)|$ gives the local scale factor of the conformal transformation at z [see eq. (8.1.5)].

8.1.1 Invariance of the Laplace equation

In the previous chapters, we have obtained various forms of the solution to the Laplace equation where the domain of the problem has a simple configuration, like the unit circle and the upper half-plane. However, most real life physical problems involve configurations with complicated geometries. Suppose a conformal mapping can be found that maps a complicated domain onto a simple domain; can the solution to the Laplace equation in the mapped domain be carried over to the original domain? In other words, given that $\phi(x, y)$ is harmonic in a domain $\mathcal{D}$ in the z-plane, $z = x + iy$, and that $w = u + iv = f(z)$

is a conformal mapping in $\mathcal{D}$ which takes the domain $\mathcal{D}$ onto the domain Δ in the w-plane, does $\phi(u, v)$ remain harmonic in Δ?

Fortunately, the answer to the above question is 'yes'. This property is termed the invariance of the Laplace equation, where

$$\frac{\partial^2 \phi}{\partial x^2} + \frac{\partial^2 \phi}{\partial y^2} = |f'(z)|^2 \left(\frac{\partial^2 \phi}{\partial u^2} + \frac{\partial^2 \phi}{\partial v^2} \right). \tag{8.1.10}$$

Note that $\phi(x, y)$ is transformed into the function $\phi(x(u, v), y(u, v))$ by the transformation $w = u + iv = f(z)$, $z = x + iy$. Suppose $f(z)$ is conformal in $\mathcal{D}$ so that $f'(z) \neq 0$ in $\mathcal{D}$; eq. (8.1.10) dictates that

$$\frac{\partial^2 \phi}{\partial x^2} + \frac{\partial^2 \phi}{\partial y^2} = 0 \text{ in } \mathcal{D}$$

if and only if

$$\frac{\partial^2 \phi}{\partial u^2} + \frac{\partial^2 \phi}{\partial v^2} = 0 \text{ in } \Delta.$$

The mathematical steps required to show eq. (8.1.10) amount to straightforward differentiation. First, we observe that

$$\frac{\partial \phi}{\partial x} = \frac{\partial \phi}{\partial u}\frac{\partial u}{\partial x} + \frac{\partial \phi}{\partial v}\frac{\partial v}{\partial x} \quad \text{and} \quad \frac{\partial \phi}{\partial y} = \frac{\partial \phi}{\partial u}\frac{\partial u}{\partial y} + \frac{\partial \phi}{\partial v}\frac{\partial v}{\partial y},$$

so the second derivatives are found to be

$$\frac{\partial^2 \phi}{\partial x^2} = \frac{\partial \phi}{\partial u}\frac{\partial^2 u}{\partial x^2} + \left(\frac{\partial^2 \phi}{\partial u^2}\frac{\partial u}{\partial x} + \frac{\partial^2 \phi}{\partial v \partial u}\frac{\partial v}{\partial x} \right)\frac{\partial u}{\partial x} + \frac{\partial \phi}{\partial v}\frac{\partial^2 v}{\partial x^2}$$
$$+ \left(\frac{\partial^2 \phi}{\partial u \partial v}\frac{\partial u}{\partial x} + \frac{\partial^2 \phi}{\partial v^2}\frac{\partial v}{\partial x} \right)\frac{\partial v}{\partial x},$$

$$\frac{\partial^2 \phi}{\partial y^2} = \frac{\partial \phi}{\partial u}\frac{\partial^2 u}{\partial y^2} + \left(\frac{\partial^2 \phi}{\partial u^2}\frac{\partial u}{\partial y} + \frac{\partial^2 \phi}{\partial v \partial u}\frac{\partial v}{\partial y} \right)\frac{\partial u}{\partial y} + \frac{\partial \phi}{\partial v}\frac{\partial^2 v}{\partial y^2}$$
$$+ \left(\frac{\partial^2 \phi}{\partial u \partial v}\frac{\partial u}{\partial y} + \frac{\partial^2 \phi}{\partial v^2}\frac{\partial v}{\partial y} \right)\frac{\partial v}{\partial y}.$$

Adding the above two equations together, we obtain

$$\frac{\partial^2 \phi}{\partial x^2} + \frac{\partial^2 \phi}{\partial y^2} = \frac{\partial \phi}{\partial u}\left(\frac{\partial^2 u}{\partial x^2} + \frac{\partial^2 u}{\partial y^2} \right) + \frac{\partial^2 \phi}{\partial u^2}\left[\left(\frac{\partial u}{\partial x} \right)^2 + \left(\frac{\partial u}{\partial y} \right)^2 \right]$$
$$+ 2\frac{\partial^2 \phi}{\partial v \partial u}\left(\frac{\partial u}{\partial x}\frac{\partial v}{\partial x} + \frac{\partial u}{\partial y}\frac{\partial v}{\partial y} \right) + \frac{\partial \phi}{\partial v}\left(\frac{\partial^2 v}{\partial x^2} + \frac{\partial^2 v}{\partial y^2} \right)$$
$$+ \frac{\partial^2 \phi}{\partial v^2}\left[\left(\frac{\partial v}{\partial x} \right)^2 + \left(\frac{\partial v}{\partial y} \right)^2 \right].$$

Since $w = u + iv = f(z)$ is analytic, both u and v are harmonic and so the first and fourth terms in the above equation are zero. Further, since u and v satisfy the Cauchy–Riemann relations, the third term also vanishes. Hence, the above equation reduces to

$$\frac{\partial^2 \phi}{\partial x^2} + \frac{\partial^2 \phi}{\partial y^2} = \left[\left(\frac{\partial u}{\partial x} \right)^2 + \left(\frac{\partial u}{\partial y} \right)^2 \right] \frac{\partial^2 \phi}{\partial u^2} + \left[\left(\frac{\partial v}{\partial x} \right)^2 + \left(\frac{\partial v}{\partial y} \right)^2 \right] \frac{\partial^2 \phi}{\partial v^2}$$

$$= |f'(z)|^2 \left(\frac{\partial^2 \phi}{\partial u^2} + \frac{\partial^2 \phi}{\partial v^2} \right),$$

and thus eq. (8.1.10) is established.

Given the invariance property of the Laplace equation, a wide range of physical problems with complicated configurations that are governed by the Laplace equation can be solved by finding an appropriate conformal mapping that maps the given domain onto either the unit circle or the upper half-plane. Such a solution approach is illustrated in the following examples, which include applications in electrostatics, steady state temperature distribution and fluid flows.

Example 8.1.3 Find the electric potential inside the semi-infinite strip $\{z = x + iy : -\frac{\pi}{2} < x < \frac{\pi}{2}, \ y > 0\}$, where the potential value along the bottom boundary is kept constant at ϕ_0, and the potential values along the vertical boundaries are zero.

Solution Let $\phi(x, y)$ denote the electric potential inside the strip. The electric potential function is known to satisfy the Laplace equation (see Subsection 4.4.2). It was shown in Subsection 3.2.1 that the mapping function $w = \sin z$ carries the above strip onto the upper half-plane conformally, with the base $\left[-\frac{\pi}{2}, \frac{\pi}{2} \right]$ of the strip going to the segment $[-1, 1]$ and the vertical sides to the horizontal rays $(-\infty, -1]$ and $[1, \infty)$ (see Figure 8.4).

In the transformed plane, by virtue of the invariance property of the Laplace equation, the electric potential $\phi(u, v)$ remains harmonic. The corresponding boundary potential values along the u-axis are

$$\phi(u, 0) = \begin{cases} \phi_0 & -1 < u < 1 \\ 0 & u > 1 \text{ or } u < -1 \end{cases}.$$

Using the techniques developed in Subsection 3.3.2, the solution to the electric potential in the w-plane is readily found to be

$$\phi(u, v) = \frac{\phi_0}{\pi} \text{Arg} \frac{w - 1}{w + 1}, \quad w = u + iv.$$

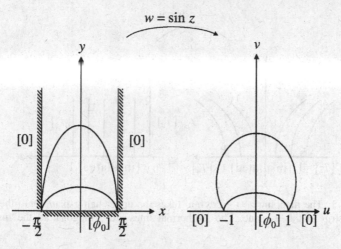

Figure 8.4. The mapping $w = \sin z$ carries the semi-infinite strip onto the upper half-plane. The curves inside the strip are the equipotential lines.

Referring back to the z-plane, the electric potential inside the strip is given by

$$\phi(x, y) = \frac{\phi_0}{\pi} \text{Arg} \frac{\sin z - 1}{\sin z + 1} = \frac{\phi_0}{\pi} \tan^{-1} \frac{2 \text{Im}(\sin z)}{|\sin^2 z| - 1}$$

$$= \frac{\phi_0}{\pi} \tan^{-1} \frac{2 \cos x \sinh y}{\sinh^2 y - \cos^2 x} = \frac{\phi_0}{\pi} \tan^{-1} \frac{2 \cos x / \sinh y}{1 - (\cos x / \sinh y)^2}$$

$$= \frac{2\alpha}{\pi} \phi_0,$$

where $\tan \alpha = \frac{\cos x}{\sinh y}$ and $\tan^2 \alpha = \frac{2 \tan \alpha}{1 - \tan^2 \alpha}$. Since $\cos x$ assumes value within $[0, 1]$ when $x \in \left[-\frac{\pi}{2}, \frac{\pi}{2}\right]$ and $\sinh y$ assumes a positive value for $y > 0$, α can assume any value within the semi-infinite strip. The equipotential lines inside the strip correspond to the level curves $\frac{\cos x}{\sinh y} = k$, for some positive constant k.

Example 8.1.4 Find the steady state temperature $T(x, y)$ in the upper half-plane $y > 0$ if the boundary temperature along the x-axis is kept at

$$T(x, 0) = \begin{cases} T_0 & -\infty < x < -1 \\ T_1 & 1 < x < \infty \end{cases},$$

and the wall is insulated within $|x| < 1$; that is,

$$\frac{\partial T}{\partial y}(x, 0) = 0, \quad |x| < 1.$$

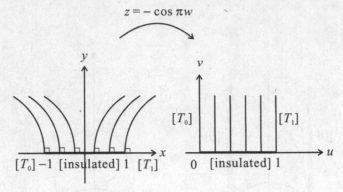

Figure 8.5. The mapping $z = -\cos \pi w$ takes the upper half z-plane onto the semi-infinite strip in the w-plane. The isothermal lines are orthogonal to the insulated boundaries.

This is a mixed Dirichlet–Neumann problem where values of the dependent variable are prescribed on some part of the boundary and derivative values are prescribed on the remaining part of the boundary.

Solution   In the last example, we facilitate the solution by finding a mapping that takes the given strip onto the upper half-plane. Though the configuration in the present problem is already the upper half-plane, the solution procedure is complicated by the presence of the Neumann condition along a part of the boundary. Interestingly, we solve this problem by considering a mapping that takes the upper half-plane onto a strip in the transformed plane, just the reverse of the usual procedure. This is because the Neumann boundary condition can be treated easier under the configuration of a strip.

We choose the mapping $z = -\cos \pi w$ which takes the upper half z-plane onto the semi-infinite strip $\{w = u + iv : 0 < u < 1 \text{ and } 0 < v < \infty\}$ (see Figure 8.5).

In the transformed w-plane, the boundary conditions are:

(i) $T = T_0$ along the left vertical edge $u = 0$, $v > 0$;
(ii) $\frac{\partial T}{\partial v} = 0$ along the bottom end $v = 0$, $0 < u < 1$;
(iii) $T = T_1$ along the right vertical edge $u = 1$, $v > 0$.

Since the normal derivative of the temperature at points along an insulated boundary must be zero, the isothermal lines must be orthogonal to an insulated surface. In the w-plane, the isothermal lines within the strip must be parallel to the v-axis. One then naturally deduces that the temperature $T(u, v)$ in the

w-plane is given by

$$T(u, v) = T_0 + (T_1 - T_0)u.$$

To verify that this is the solution, we check that $T(u, v)$ is harmonic inside the strip and all boundary conditions along the boundaries of the strip are satisfied. By the uniqueness property of solutions to the Laplace equation, this is the solution to the mixed Dirichlet–Neumann problem.

The next step is to express u in terms of x and y. Noting that

$$z = -\cos\pi(u + iv) = -\cos\pi u \cosh\pi v + i\sin\pi u \sinh\pi v,$$

we have

$$(\cosh\pi v - \cos\pi u)^2 = \cosh^2\pi v - 2\cosh\pi v \cos\pi u + \cos^2\pi u$$
$$= x^2 + 2x + 1 + y^2.$$

By taking the positive square root on both sides, we have

$$\cosh\pi v - \cos\pi u = \sqrt{(x + 1)^2 + y^2}.$$

Similarly, we obtain

$$\cosh\pi v + \cos\pi u = \sqrt{(x - 1)^2 + y^2}.$$

In the original z-plane, the temperature is then given by

$$T(x, y) = T_0 + \frac{T_1 - T_0}{\pi}\cos^{-1}\left(\frac{1}{2}\left(\sqrt{(x - 1)^2 + y^2} - \sqrt{(x + 1)^2 + y^2}\right)\right).$$

The isothermal lines in the z-plane are given by the level curves:

$$\sqrt{(x - 1)^2 + y^2} - \sqrt{(x + 1)^2 + y^2} = k,$$

where k is some value lying between T_0 and T_1. These isothermal lines can be shown to be circles that cut orthogonally the insulated boundary along the segment $[-1, 1]$ on the x-axis. The temperature value along the boundary $y = 0$ is found to be

$$T(x, 0) = T_0 + \frac{T_1 - T_0}{\pi}\cos^{-1}\left(\frac{\sqrt{(x - 1)^2} - \sqrt{(x + 1)^2}}{2}\right)$$

$$= \begin{cases} T_0 & \text{if } x < -1 \\ T_0 + \dfrac{T_1 - T_0}{\pi}\cos^{-1}(-x) & \text{if } -1 \le x \le 1 \\ T_1 & \text{if } x > 1 \end{cases}.$$

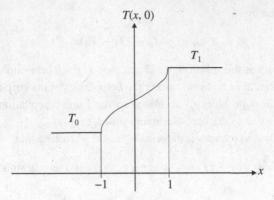

Figure 8.6. The boundary temperature $T(x, 0)$ takes constant value outside $[-1, 1]$ and assumes the value of an inverse cosine function within $[-1, 1]$.

The plot of $T(x, 0)$ against x is shown in Figure 8.6. Interestingly, the temperature assumes the value of an inverse cosine function along the insulated surface.

Example 8.1.5   Find the complex potential for the potential flow around a flat plate of length $2a$ that is oriented to be perpendicular to the incoming uniform flow of velocity U.

Solution   The configuration of a uniform flow past a perpendicular obstacle is shown schematically in Figure 8.7. By symmetry, the x-axis is a streamline dividing the flow field into two equal halves. It suffices to seek the solution to the flow field in the upper half-plane.

We try to find a mapping that carries the upper half z-plane (excluding the x-axis) minus the segment from $z = 0$ to $z = ai$ along the imaginary axis to the upper half w-plane. First, we take $w_1 = z^2$ which maps the above domain onto the whole w_1-plane minus the segment $[-a^2, \infty)$ along the real axis. Next, we take the translation $w_2 = w_1 + a^2$ so that the mapped domain is the whole w_2-plane minus the segment $[0, \infty)$ along the real axis. Finally, we choose $w = \sqrt{w_2} = \sqrt{z^2 + a^2}$ (the branch which takes $z = \sqrt{2}ai$ to $w = ai$). This mapping carries the given domain in the z-plane to the upper half w-plane. The sequence of mappings is shown in Figure 8.8.

In the w-plane, the flow configuration becomes uniform and rectilinear. The speed of the flow stays at the same value U since

$$\lim_{z \to \infty} \frac{w}{z} = \lim_{z \to \infty} \frac{\sqrt{z^2 + a^2}}{z} = 1.$$

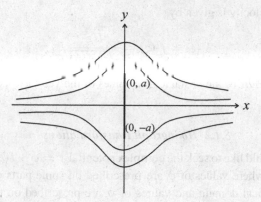

Figure 8.7. The configuration of a uniform flow past a perpendicular obstacle of length $2a$.

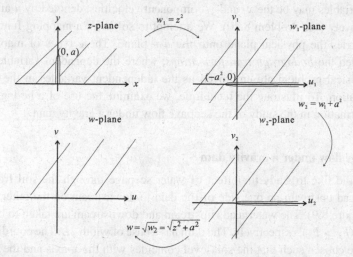

Figure 8.8. The sequence of mappings that takes the upper half-plane minus the vertical segment onto the whole upper half-plane.

The complex potential of the flow in the w-plane is given by

$$f(w) = Uw.$$

Referring back to the z-plane, the complex potential then becomes

$$f(z) = U\sqrt{z^2 + a^2}.$$

The complex velocity is given by

$$V(z) = f'(z) = \frac{Uz}{\sqrt{z^2 + a^2}}.$$

At the tips $z = \pm ai$ of the obstacle, the values of the velocity become infinite.

8.1.2 Hodograph transformations

Suppose we would like to seek the complex potential $f = \phi + i\psi$ of a potential flow problem, where values of ϕ are prescribed on some parts of the boundary of the physical domain and values of ψ are prescribed on the remaining parts. The solution approach then amounts to finding a mapping function that carries the physical plane onto the ϕ-ψ plane. In this case, the desired complex potential $f = \phi + i\psi$ can be viewed as a mapping. In other cases, the dependent variables may be the x- and y-component velocities, denoted by u and v, respectively (see Problem 8.20). We attempt to solve for a mapping function that carries the physical plane onto the u-v plane. These types of mapping are called the *hodograph transformations*, where the dependent variables in the physical problem are employed as the independent variables in the new formulation. To illustrate the technique, we examine the use of a hodograph transformation in the study of the seepage flow under a gravity dam.

Seepage flow under a gravity dam

We would like to study the effect of water seepage through the soil from a waterhead upstream on one side of the dam to downstream on the other side (see Figure 8.9). The waterheads upstream and downstream are taken to be h_1 and h_2 $(h_1 > h_2)$, respectively. The dam has a base of width $2L$. The coordinate axes are chosen such that the soil level coincides with the x-axis and the base of the dam lies between $-L \le x \le L$.

A rather ideal form of seepage flow through soil and flow configuration is assumed here. The soil underneath the water and the dam is homogeneous and has an infinite depth so that the domain of the seepage is the whole lower half-plane. The velocity of the flow is so slow that viscosity is negligible, and so potential flow can be assumed for the seepage. The dam is assumed to be completely impermeable.

The flow problem is considered solved if the complex potential of the seepage is known. The complex potential is of the form

$$w = f(z) = \phi(x, y) + i\psi(x, y), \quad z = x + iy, \tag{8.1.11}$$

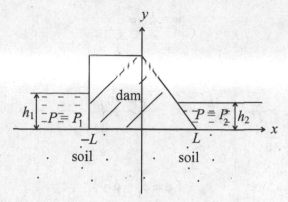

Figure 8.9. Seepage flow through the soil from the waterhead upstream on one side of the dam to downstream on the other side.

where ϕ and ψ are the velocity potential and stream function, respectively. We adopt Darcy's law for fluid flow inside a porous medium, which states that the velocity of fluid flow through fine sands and soils is directly proportional to the pressure gradient. That is

$$|\nabla P| \propto |\nabla \phi|, \tag{8.1.12}$$

where P is the pressure. With reference to the material properties associated with flow in a porous medium, Darcy's law becomes

$$P = -\frac{\eta}{K}\phi, \tag{8.1.13}$$

where η is the coefficient of viscosity of water and K is the coefficient of permeability of the soil. Here, both coefficients are assumed to be constant. Using this empirical law of seepage flow, we manage to relate the waterheads with the velocity potential.

The boundary conditions along the x-axis for the seepage are prescribed as follows. The impervious wall is a streamline so that $\psi(x, 0)$ is constant for $-L \leq x \leq L$. Without loss of generality, we take

$$\psi(x, 0) = 0, \quad -L \leq x \leq L. \tag{8.1.14}$$

Along the soil surface to the left side of the dam, $x \leq -L$ and $y = 0$, the pressure head is given by $P(x, 0) = \rho g h_1$, where ρ is the density of water and g is the acceleration due to gravity. From eq. (8.1.13), we then have

$$\phi(x, 0) = -\frac{K}{\eta}\rho g h_1 = \phi_1, \quad x \leq -L. \tag{8.1.15a}$$

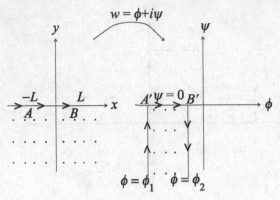

Figure 8.10. The boundary values assumed by ϕ and ψ along the soil surface dictate the mapping that carries the lower half x-y plane onto a semi-infinite strip in the ϕ-ψ plane.

Similarly, the boundary condition along the soil surface to the right side of the dam, $x \geq L$ and $y = 0$, is given by

$$\phi(x, 0) = -\frac{K}{\eta}\rho g h_2 = \phi_2, \quad x \geq L. \tag{8.1.15b}$$

Both ϕ_1 and ϕ_2 are negative quantities. Now, the hodograph transformation $w = f(z) = \phi(x, y) + i\psi(x, y)$, $z = x + iy$, maps the lower half z-plane onto the semi-infinite strip $\phi_1 \leq \phi \leq \phi_2$ and $\psi \leq 0$ in the w-plane (see Figure 8.10). The stream function values in the lower half z-plane are negative since $\psi(x, 0) = 0$ along a part of the real axis and the seepage runs from the left side to the right side of the dam, implying $\frac{\partial \psi}{\partial y} > 0$.

Recall that the mapping $\zeta = \cos^{-1}(\frac{z}{L})$ maps the lower half z-plane onto the strip $\{\zeta : 0 \leq \operatorname{Re} \zeta \leq \pi \text{ and } \operatorname{Im} \zeta \geq 0\}$ in the ζ-plane. Hence, the required transformation which effects the mapping shown in Figure 8.10 is deduced to be

$$w = f(z) = \frac{\phi_1 - \phi_2}{\pi}\cos^{-1}\frac{z}{L} + \phi_2. \tag{8.1.16}$$

It is not possible to express ϕ and ψ explicitly as functions of x and y. Rather, the reverse is possible, where x and y can be expressed in terms of ϕ and ψ. From eq. (8.1.16), we can obtain

$$\frac{x}{L} = \cos\frac{\pi(\phi - \phi_2)}{\phi_1 - \phi_2}\cosh\frac{\pi\psi}{\phi_1 - \phi_2}$$

and

$$\frac{y}{L} = -\sin\frac{\pi(\phi - \phi_2)}{\phi_1 - \phi_2}\sinh\frac{\pi\psi}{\phi_1 - \phi_2}.$$

Once the complex potential $f(z)$ is known, the properties of the seepage flow field can be analyzed. For example, the streamline $\psi = \psi_0 \ (< 0)$ is the lower half-portion of the ellipse $\frac{x^2}{a^2} + \frac{y^2}{b^2} = 1$, where

$$a = L\cosh\frac{\pi\psi_0}{(\phi_1 - \phi_2)} \quad \text{and} \quad b = L\sinh\frac{\pi\psi_0}{(\phi_1 - \phi_2)}.$$

The equipotential line $\phi = \phi_0 \ (\phi_1 \le \phi_0 \le \phi_2)$ is the lower half-portion of the hyperbola $\frac{x^2}{A^2} - \frac{y^2}{B^2} = 1$, where

$$A = L\frac{\cos\pi(\phi_0 - \phi_2)}{\phi_1 - \phi_2} \quad \text{and} \quad B = L\frac{\sin\pi(\phi_0 - \phi_2)}{\phi_1 - \phi_2}.$$

A civil engineer would be interested in finding the upward lift exerted by the seepage on the base of the dam. It turns out that this upward lift is given by the product of the width of the base and the average pressure due to the waterheads upstream and downstream. To prove the claim, we consider the pressure $P(x, 0)$ along the base of the dam, $-L \le x \le L$. From Darcy's law, $P(x, 0) = -\frac{\eta}{K}\phi(x, 0)$. Further, since $\psi(x, 0) = 0$, the complex potential $f(x)$ is real when $z = x$ and correspondingly, $\phi(x, 0) = f(x)$. Using eq. (8.1.16), we obtain

$$P(x, 0) = -\frac{\eta}{K}\left(\frac{\phi_1 - \phi_2}{\pi}\cos^{-1}\frac{x}{L} + \phi_2\right)$$

$$= \frac{P_1 - P_2}{\pi}\cos^{-1}\frac{x}{L} + P_2. \qquad (8.1.17)$$

The total uplift exerted on the base of the dam is given by

$$\int_{-L}^{L} P(x, 0)\,dx = \int_{-L}^{L}\frac{P_1 - P_2}{\pi}\cos^{-1}\frac{x}{L} + P_2\,dx = \frac{P_1 + P_2}{2}(2L). \qquad (8.1.18)$$

8.2 Bilinear transformations

The transformation defined by

$$w = f(z) = \frac{az + b}{cz + d}, \qquad (8.2.1)$$

where a, b, c and d are complex numbers and $ad - bc \ne 0$ is called a bilinear transformation (or linear fractional transformation). It is so named because it takes the form of the ratio of two linear functions. The bilinear transformation can be decomposed as

$$w = f(z) = \frac{a}{c} + \frac{bc - ad}{c} \frac{1}{cz + d}, \quad c \neq 0. \tag{8.2.2}$$

Clearly, the restriction $ad - bc \neq 0$ is essential in order to ensure that $f(z)$ is not a constant function. When $c = 0$, $f(z)$ becomes a linear transformation.

One-to-one mapping

The bilinear transformation is a one-to-one mapping of the extended plane onto itself. In other words, the bilinear transformation maps distinct points onto distinct images. This means

$$f(z_1) = f(z_2) \quad \text{if and only if} \quad z_1 = z_2.$$

The "if" part is obvious. To show the "only if" part, we assume

$$\frac{az_1 + b}{cz_1 + d} = f(z_1) = f(z_2) = \frac{az_2 + b}{cz_2 + d}.$$

After some manipulation, we obtain

$$(ad - bc)z_1 = (ad - bc)z_2,$$

which implies $z_1 = z_2$ since $ad - bc \neq 0$.

Since a bilinear transformation is one-to-one, its inverse always exists. The inverse transformation is obtained by solving eq. (8.2.1) for z. This gives

$$z = \frac{-dw + b}{cw - a}. \tag{8.2.3}$$

When $c \neq 0$, $w = f(z)$ has a simple pole at $-\frac{d}{c}$ so that $f\left(-\frac{d}{c}\right) = \infty$. Similarly, the inverse transformation has a simple pole at $\frac{a}{c}$ so that $f(\infty) = \frac{a}{c}$. When $c = 0$, we have $f(\infty) = \infty$.

Consider the derivative of f in the form

$$f'(z) = \frac{ad - bc}{(cz + d)^2}, \tag{8.2.4}$$

which is well defined everywhere except at the pole $-\frac{d}{c}$. Also, $f'(z)$ never assumes the zero value in the finite complex plane, provided that $ad - bc \neq 0$. Hence, a bilinear transformation is conformal at every point in the finite complex plane except at its pole.

Triples to triples

Apparently, we have four coefficients in the bilinear transformation, but only three of them are independent. There exists a unique bilinear transformation that maps three distinct points z_1, z_2, z_3 in the z-plane onto three distinct points w_1, w_2, w_3 in the w-plane. First, we assume that the six points are all finite. Since z_i is mapped to w_j, $j = 1, 2, 3$, it follows that

$$w_j = \frac{az_j + b}{cz_j + d}, \quad j = 1, 2, 3.$$

Using the relations

$$w - w_j = \frac{(ad - bc)(z - z_j)}{(cz + d)(cz_j + d)}, \quad j = 1, 2,$$

and

$$w_3 - w_j = \frac{(ad - bc)(z_3 - z_j)}{(cz_3 + d)(cz_j + d)}, \quad j = 1, 2,$$

we obtain the following formula for the required bilinear transformation:

$$\frac{w - w_1}{w - w_2} \bigg/ \frac{w_3 - w_1}{w_3 - w_2} = \frac{z - z_1}{z - z_2} \bigg/ \frac{z_3 - z_1}{z_3 - z_2}. \tag{8.2.5}$$

What happens when some of these points are not finite? For example, when $z_1 \to \infty$, the right-hand side of eq. (8.2.5) is then replaced by

$$\lim_{z_1 \to \infty} \frac{z - z_1}{z - z_2} \bigg/ \frac{z_3 - z_1}{z_3 - z_2} = \frac{z_3 - z_2}{z - z_2}.$$

This technique can be applied to other limiting cases, like $z_2 \to \infty$, $w_1 \to \infty$, etc., to find the corresponding reduced form of the bilinear transformation formula.

Example 8.2.1 Find the bilinear transformation that carries the points $-1, \infty, i$ onto the points
(a) i, 1, $1 + i$; (b) ∞, i, 1.

Solution Write $z_1 = -1$, $z_2 = \infty$ and $z_3 = i$.

(a) Here, $w_1 = i$, $w_2 = 1$ and $w_3 = 1 + i$. Taking the limit $z_2 \to \infty$, eq. (8.2.5) becomes

$$\frac{w - i}{w - 1} \bigg/ \frac{1 + i - i}{1 + i - 1} = \frac{z + 1}{i + 1}.$$

Rearranging the terms, we obtain

$$w = \frac{z + 2 + i}{z + 2 - i}.$$

(b) Now, $w_1 = \infty$, $w_2 = i$ and $w_3 = 1$. Taking the limits $z_2 \to \infty$ and $w_1 \to \infty$ in eq. (8.2.5), we obtain

$$\frac{1-i}{w-i} = \frac{z+1}{i+1} \quad \text{or} \quad w = \frac{iz+2+i}{z+1}.$$

8.2.1 Circle-preserving property

The decomposition of $f(z)$ in eq. (8.2.2) reveals that a bilinear transformation can be expressed as a composition of three successive transformations, namely,

(i) linear transformation:

$$w_1 = cz + d; \tag{8.2.6a}$$

(ii) inversion:

$$w_2 = \frac{1}{w_1}; \tag{8.2.6b}$$

(iii) linear transformation:

$$w = f(z) = \frac{a}{c} + \frac{bc - ad}{c} w_2. \tag{8.2.6c}$$

Expressing c in polar form, where $c = re^{i\theta}$, the linear transformation defined in eq. (8.2.6a) can be decomposed into (i) rotation: $z_1 = e^{i\theta}z$; (ii) magnification: $z_2 = rz_1$; and (iii) translation: $w_1 = z_2 + d$. Combining these transformations, a bilinear transformation can then be visualized as the composition of a sequence of translation, rotation, magnification and inversion.

The four types of transformation – translation, rotation, magnification and inversion – all share the *circle-preserving property*. They map the class of circles and lines to the same class. Here, we treat a straight line as a circle with infinite radius. It is quite obvious that translation, rotation and magnification do preserve circles. A few steps are required to show that the inversion transformation also shares the circle-preserving property. Note that the general form of a circle in the complex plane can be expressed as

$$Az\bar{z} + \bar{E}z + E\bar{z} + D = 0, \quad A \text{ and } D \text{ are real}, \tag{8.2.7}$$

where $A \neq 0$ and $|E|^2 - AD > 0$. When $A = 0$, the above equation represents a straight line. Consider the inversion transformation $w = \frac{1}{z}$. Substitute $z = \frac{1}{w}$ into eq. (8.2.7) to obtain

$$Dw\bar{w} + \bar{E}\bar{w} + Ew + A = 0. \tag{8.2.8}$$

This is an equation of a circle if $D \neq 0$ or a straight line if $D = 0$.

Under what condition is a circle mapped to a straight line by the bilinear transformation defined in eq. (8.2.1)? Since a straight line passes through the point of infinity and $f\left(-\frac{d}{c}\right) = \infty$, we deduce that a circle or a straight line that passes through the pole ~~~~~~ onto a straight line.

Example 8.2.2 A bilinear transformation is defined by

$$w = f(z) = \frac{2iz - 2}{2z - i}.$$

(a) Determine the invariant points of the transformation.
(b) Find the point ξ for which the equation $f(z) = \xi$ has no solution for z in the finite complex plane.
(c) Show that the imaginary axis is mapped onto itself.
(d) Determine the image of the disc $|z| < 1$.

Solution

(a) The invariant points are found by solving

$$z = f(z) = \frac{2iz - 2}{2z - i} \Leftrightarrow 2z^2 - 3iz + 2 = 0.$$

This gives $z = -\frac{i}{2}$ and $z = 2i$.

(b) The point ξ is given by

$$\xi = \lim_{z \to \infty} f(z) = i.$$

In other words, the equation $f(z) = i$ is satisfied only by $z = \infty$.

(c) Along the imaginary axis, $z = it$, where t is real. The image point of $z = it$ is

$$f(it) = \frac{2i(it) - 2}{2it - i} = \frac{1 + t}{t - \frac{1}{2}},$$

which is purely imaginary. Hence, the imaginary axis is mapped onto itself.

(d) We rearrange $w = \frac{2iz-2}{2z-i}$ to obtain $z = \frac{iw-2}{2w-2i}$. Now,

$$|z| < 1 \Leftrightarrow \left|\frac{iw - 2}{2w - 2i}\right| < 1 \Leftrightarrow w\overline{w} - 2i\overline{w} + 2iw > 0 \Leftrightarrow |w - 2i| > 2.$$

The mapped region in the w-plane is the exterior of the circle centered at $2i$ and with radius 2.

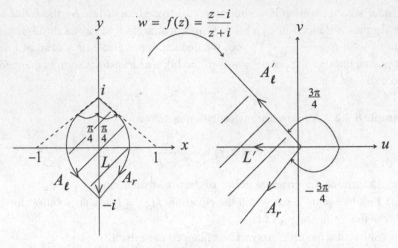

Figure 8.11. The lens is mapped to the infinite wedge bounded within the lines Arg $w = \frac{3\pi}{4}$ and Arg $w = -\frac{3\pi}{4}$.

Example 8.2.3 Consider the two circles whose centers are at $z = -1$ and $z = 1$, both with radius $\sqrt{2}$. They intersect at $z = -i$ and $z = i$, and their overlapping region is in the shape of a lens. Find the image of this lens under the bilinear transformation

$$w = f(z) = \frac{z - i}{z + i}.$$

Solution Since $z = -i$ is a pole of the bilinear transformation, any circle that passes through $z = -i$ is mapped onto a straight line. Hence, the two given circles are mapped onto two straight lines. These two straight lines also pass through $w = 0$ since both circles pass through $z = i$ and $f(i) = 0$.

Next, we try to find the inclination of these two mapped straight lines in the w-plane. Let A_r (A_ℓ) be the right (left) edge of the lens corresponding to the circle with center at $z = -1$ ($z = 1$), and L be the line segment along the imaginary axis $z = iy$, $-1 < y < 1$ (see Figure 8.11). The image of L is the negative real axis in the w-plane since $f(iy) = \frac{y-1}{y+1} < 0$. Suppose we move along the two arcs A_r and A_ℓ and the vertical line L from $z = i$ to $z = -i$; the image points move along the mapped straight lines A'_r, A'_ℓ and L', respectively, from $w = 0$ to $w = \infty$ (see the direction arrows shown in Figure 8.11).

Since the bilinear transformation is conformal at $z = i$, the angle included between A_ℓ and L is preserved under the transformation. It is straightforward to show that the line joining $z = -1$ and $z = i$ is tangent to the arc A_ℓ at $z = i$. With regard to the directed curves shown in Figure 8.11, the arc A_ℓ

is turned $\frac{\pi}{4}$ radians in the anticlockwise sense at $z = i$ to the line L. By the conformal property of the bilinear transformation, the image line A'_ℓ should correspondingly turn the same angle in the same sense to the image line L'. Hence, we deduce that A'_ℓ is the line Arg $w = \frac{3\pi}{4}$. By comparing the same argument, or alternatively using the principle of symmetry, one deduces that the arc A_r is mapped onto the line A'_r : Arg $w = -\frac{3\pi}{4}$.

The images of the boundary curves of the lens are now known. The final step is to determine whether the interior of the lens is mapped onto the infinite wedge bounded within the lines Arg $w = \frac{3\pi}{4}$ and Arg $w = -\frac{3\pi}{4}$, or to the region exterior to the wedge. A simple rule of thumb is that the left (right) region of a directed curve is mapped onto the left (right) region of the directed image curve. Note that the line segment L is left of the directed curve A_ℓ. By the conformal property of the bilinear transformation, the image line segment L' should also lie to the left of the directed image curve A'_ℓ. By connectivity, one can conclude that the left region is mapped onto the left region. Hence, in this case, the lens is mapped onto the infinite wedge bounded within A'_ℓ and A'_r (see Figure 8.11).

8.2.2 Symmetry-preserving property

In the preceding subsection, we explored the circle-preserving property of bilinear transformations. This subsection further examines the symmetry-preserving property of bilinear transformations. First, we present the definition of symmetric points of a circle.

Definition 8.2.1 Given the circle C: $|z - \alpha| = R$ in the z-plane, two points z_1 and z_2 are said to be *symmetric* with respect to the circle C if z_1 and z_2 satisfy

$$\text{Arg}(z_1 - \alpha) = \text{Arg}(z_2 - \alpha), \tag{8.2.9a}$$
$$|z_1 - \alpha|\,|z_2 - \alpha| = R^2. \tag{8.2.9b}$$

By convention, the center α of the circle C and the complex infinity ∞ are symmetric with respect to C.

Geometrically, by virtue of relation (8.2.9a), the two symmetric points z_1 and z_2 and the center α of the circle are on the same ray through α, and their distances to the center satisfy the relation (8.2.9b). Given a point z_1, its symmetric point with respect to the circle C can be found using the same method of construction shown in Figure 1.4 (that figure shows the special case where $\alpha = 0$). When z_1 is on the circle, the symmetric point is simply itself. The two relations (8.2.9a,b) can be combined into the following single equation:

$$(z_1 - \alpha)\overline{(z_2 - \alpha)} = R^2. \tag{8.2.10}$$

Representation of the equation of a circle in terms of symmetric points

Consider the circle $C : |z - \alpha| = R$, and let z_1 and z_2 be a pair of symmetric points with respect to C. We write $z_1 = \alpha + d_1 e^{i\phi}$, where $d_1 = |z_1 - \alpha|$ and $\phi = \operatorname{Arg}(z_1 - \alpha)$. Since z_1, z_2 and α lie on the same ray, by virtue of relation (8.2.9b), we obtain $z_2 = \alpha + d_2 e^{i\phi}$, where $d_2 = |z_2 - \alpha| = \frac{R^2}{d_1}$. Suppose z is a point on the circle C, and write $z = \alpha + R e^{i\theta}$. We then have

$$\left| \frac{z - z_1}{z - z_2} \right| = \left| \frac{R e^{i\theta} - d_1 e^{i\phi}}{R e^{i\theta} - d_2 e^{i\phi}} \right| = \frac{d_1}{R} \left| \frac{R e^{i\theta} - d_1 e^{i\phi}}{d_1 e^{i\theta} - R e^{i\phi}} \right|.$$

By symmetry, the two moduli $|R e^{i\theta} - d_1 e^{i\phi}|$ and $|d_1 e^{i\theta} - R e^{i\theta}|$ are equal so that

$$\left| \frac{z - z_1}{z - z_2} \right| = \frac{d_1}{R} = \frac{R}{d_2}.$$

This gives an alternative representation of the equation of the circle C in terms of symmetric points z_1 and z_2.

Conversely, given an equation of the form

$$\left| \frac{z - z_1}{z - z_2} \right| = k, \quad z_1 \neq z_2 \text{ and } k \text{ is real and non-negative}, \tag{8.2.11}$$

it can be shown that this equation represents a circle with z_1 and z_2 as a pair of symmetric points. We wish to find the center and radius of the circle in terms of z_1, z_2 and k. Using the relation $|z - z_1| = k|z - z_2|$, it is straightforward to show that

$$|(z - z_1) - k^2(z - z_2)| = k|(z - z_1) - (z - z_2)|.$$

Upon rearranging, we obtain

$$\left| z - \frac{z_1 - k^2 z_2}{1 - k^2} \right| = \frac{k|z_1 - z_2|}{|1 - k^2|}.$$

The above equation is in the form $|z - \alpha| = R$, where

$$\alpha = \frac{z_1 - k^2 z_2}{1 - k^2} \quad \text{and} \quad R = \frac{k|z_1 - z_2|}{|1 - k^2|}. \tag{8.2.12}$$

Further, since we have

$$z_1 - \alpha = \frac{k^2}{1 - k^2}(z_2 - z_1) \quad \text{and} \quad z_2 - \alpha = \frac{1}{1 - k^2}(z_2 - z_1),$$

it follows that z_1 and z_2 satisfy the relation

$$z_2 - \alpha = \frac{R^2}{\overline{z_1 - \alpha}}. \qquad (8.2.13)$$

Therefore, the two points z_1 and z_2 are symmetric with respect to the circle whose center and radius are given in eq. (8.2.12).

Consider the two special cases $k = 0$ and $k = 1$. When $k = 0$, the circle reduces to a point since the radius becomes zero. When $k = 1$, the radius becomes infinite and so the circle becomes a straight line. In fact, the equation $|z - z_1| = |z - z_2|$ represents the perpendicular bisector of the line joining z_1 and z_2. The two symmetric points are mirror images of each other with the perpendicular bisector as the mirror.

Apollonius' family of circles

The above result shows that any circle can be represented in the form

$$\left| \frac{z - z_1}{z - z_2} \right| = k, \quad k > 0, \qquad (8.2.14)$$

with z_1 and z_2 as its symmetric points. When $k \neq 1$, the center and radius of the circle are given by eq. (8.2.12). The family of circles with the same pair of symmetric points z_1 and z_2 is called the *Apollonius family of circles*. Alternatively, given any two circles C_1 and C_2, we can find the pair of points z_1 and z_2 that are symmetric with respect to both circles. As a result, C_1 and C_2 belong to the same Apollonius family. Another interesting property that can be deduced immediately is that the bilinear transformation

$$w = \frac{z - z_1}{z - z_2} \qquad (8.2.15)$$

maps the Apollonius family of circles defined by eq. (8.2.14) onto the family of concentric circles $|w| = k$. These properties can be used together to find a bilinear transformation that maps a given pair of circles onto two concentric circles. The details of the technique are shown in the following two examples.

Example 8.2.4 Find the bilinear transformation that maps the region bounded within the circles $|z| = 1$ and $|z - 1| = \frac{5}{2}$ onto the annular region $1 < |w| < r$, for some r to be determined, and the unit circle $|z| = 1$ in the z-plane onto the unit circle $|w| = 1$ in the w-plane.

Solution First, we find the pair of points z_1 and z_2 which are symmetric points of both circles in the z-plane. Since the centers of these two circles are on the real axis, this pair of symmetric points z_1 and z_2 must also lie on the real axis.

By substituting $\alpha = 0$, $R = 1$ and $\alpha = 1$, $R = \frac{5}{2}$ successively into eq. (8.2.9b), we deduce that z_1 and z_2 satisfy

$$z_1 z_2 = 1 \quad \text{and} \quad (z_1 - 1)(z_2 - 1) = \left(\frac{5}{2}\right)^2.$$

The solutions to the above simultaneous equations give $z_1 = -\frac{1}{4}$ and $z_2 = -4$. By applying eq. (8.2.11), we deduce that the two given circles in the z-plane belong to the following Apollonius family of circles:

$$\left|\frac{z + \frac{1}{4}}{z + 4}\right| = k, \quad \text{for some real positive number } k.$$

The center of the unit circle $|z| = 1$ is $\alpha = 0$. Using eq. (8.2.12), the corresponding value of k for the above unit circle can be found by solving

$$0 = \alpha = \frac{z_1 - k^2 z_2}{1 - k^2} = \frac{-\frac{1}{4} + 4k^2}{1 - k^2},$$

which gives $k = \frac{1}{4}$. Therefore, an alternative representation of the unit circle is $|\frac{z + \frac{1}{4}}{z + 4}| = \frac{1}{4}$. Following a similar procedure, the other circle $|z - 1| = \frac{5}{2}$ can be represented as $|\frac{z + \frac{1}{4}}{z + 4}| = \frac{1}{2}$.

Since $|z| = 1$ is equivalent to $|\frac{z + \frac{1}{4}}{z + 4}| = \frac{1}{4}$, for any z on the unit circle $|z| = 1$, we have

$$\left|4 e^{i\theta} \frac{z + \frac{1}{4}}{z + 4}\right| = 1, \quad \text{for any real } \theta.$$

We deduce that the circle $|z| = 1$ is mapped onto the circle $|w| = 1$ by the bilinear transformation

$$w = 4 e^{i\theta} \frac{z + \frac{1}{4}}{z + 4}, \quad \theta \text{ is real.}$$

The determination of the required bilinear transformation is unique up to the multiplicative constant $e^{i\theta}$. Likewise, the other circle $|z - 1| = \frac{5}{2}$, which has the alternative representation $2\left|\frac{z + \frac{1}{4}}{z + 4}\right| = 1$, is mapped onto the circle $|w| = 2$ by the above bilinear transformation. The region bounded within the two circles $|z| = 1$ and $|z - 1| = \frac{5}{2}$ is then mapped conformally onto the annular region $1 < |w| < 2$ (see Figure 8.12).

Example 8.2.5 A circular pipe of radius a lies below the earth's surface at a depth h ($h > a$). Assuming that the surface of the buried pipe is kept at constant

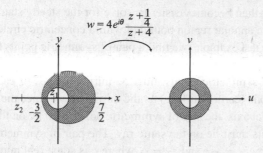

Figure 8.12. The region bounded within the circles $|z| = 1$ and $|z - 1| = \frac{5}{2}$ is mapped conformally onto the annular region $1 < |w| < 2$ by the bilinear transformation $w = 4e^{i\theta}\frac{z+\frac{1}{4}}{z+4}$.

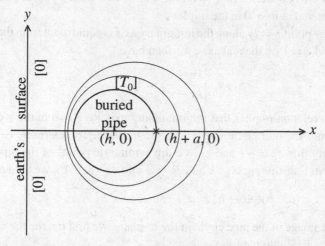

Figure 8.13. The isothermal lines lie below the earth's surface (right half-plane) surrounding the pipe and form a system of coaxial circles along the x-axis.

temperature T_0 and the temperature of the earth's surface is zero, find the steady state temperature distribution below the earth's surface surrounding the pipe.

Solution For convenience, we assign the earth's surface to be the vertical imaginary axis and the center of the buried pipe to be placed at the point $(h, 0)$. Both the pipe and the earth's surface can be recognized as circles in the z-plane (see Figure 8.13). We seek a bilinear transformation that maps the pipe and the earth's surface onto concentric circles in the w-plane. Under the new configuration, it then becomes easier to solve for the steady state temperature distribution in an annular region bounded within concentric circles.

Similar to the last example, we find a pair of symmetric points for the buried pipe and the earth's surface in the z-plane. Since the earth's surface is a straight line, the pair of symmetric points must be mirror images of each other with the imaginary axis (earth's surface) as the mirror. Also, since the center of the pipe is on the real axis, the pair of symmetric points also lie on the real axis as these three points must lie on the same ray. The pair of symmetric points can be represented by $z_1 = -c$ and $z_2 = c$, where c is some real number. Without loss of generality, we may assume $c > 0$. The bilinear transformation

$$w = f(z) = \frac{z+c}{z-c}$$

then maps the two circles $\text{Re } z = 0$ and $|z - h| = a$ onto a pair of concentric circles centered at $w = 0$ in the w-plane.

Since any point $z = iy$ along the imaginary axis is equidistant from the points $z = -c$ and $z = c$ on the real axis, we then have

$$\left| \frac{iy+c}{iy-c} \right| = 1.$$

The above relation implies that the imaginary axis $\text{Re } z = 0$ in the z-plane is mapped onto the unit circle $|w| = 1$ in the w-plane. To determine c, we use the property that $z_1 = -c$ and $z_2 = c$ are symmetric points of the pipe circle $|z - h| = a$. Substituting $\alpha = h$ and $R = a$ into eq. (8.2.9b), we obtain

$$(c - h)(-c - h) = a^2, \quad \text{so} \quad c = \sqrt{h^2 - a^2} > 0.$$

To find the image of the pipe circle in the w-plane, we first rewrite the equation of the pipe circle into the form

$$\left| \frac{z+c}{z-c} \right| = k.$$

From eq. (8.2.12), k is governed by

$$h = \frac{-c - k^2 c}{1 - k^2}$$

so that

$$k = \sqrt{\frac{h+c}{h-c}} > 1.$$

Hence, the pipe circle is mapped onto the circle $|w| = \sqrt{\frac{h+c}{h-c}}$ in the w-plane.

At this stage, we have found the bilinear transformation

$$w = \frac{z+c}{z-c}, \quad c = \sqrt{h^2 - a^2} > 0,$$

which maps the region outside the buried pipe and below the earth's surface in the z-plane onto the annular region $1 < |w| < k$, $k = \sqrt{\frac{h+c}{h-c}} > 1$, in the w-plane.

To find the steady state temperature distribution within the annular region in the w-plane, we observe that the function

$$T(w) = \frac{T_0}{\ln k} \, \text{Re}(\text{Log } w)$$

is harmonic inside $1 < |w| < k$. Also, $T(w)$ becomes 0 on the circle $|w| = 1$ and equals T_0 on the circle $|w| = k$. Hence, this is *the* solution to the present temperature distribution problem. Referring back to the z-plane, the solution to the temperature distribution outside the pipe and below the earth's surface is given by

$$T(z) = \frac{T_0}{\ln \sqrt{\frac{h+c}{h-c}}} \ln \left| \frac{z+c}{z-c} \right| = \frac{T_0}{\ln \frac{h+\sqrt{h^2-a^2}}{h-\sqrt{h^2-a^2}}} \ln \frac{(x + \sqrt{h^2 - a^2})^2 + y^2}{(x - \sqrt{h^2 - a^2})^2 + y^2}.$$

The isothermal lines are given by

$$\frac{(x + \sqrt{h^2 - a^2})^2 + y^2}{(x - \sqrt{h^2 - a^2})^2 + y^2} = \mu, \quad \mu \text{ is some real constant.}$$

They lie below the earth's surface surrounding the pipe and form a system of coaxial circles along the x-axis (see Figure 8.13). One isothermal line $T(x, y) = T_0$ is the buried pipe and another isothermal line $T(x, y) = 0$ is the earth's surface. The other isothermal lines assume temperature values that lie between 0 and T_0.

Mapping of symmetric points by a bilinear transformation

Consider the bilinear transformation

$$w = f(z) = \frac{az + b}{cz + d}, \tag{8.2.16}$$

which maps the circle C_z in the z-plane onto the circle C_w in the w-plane. Suppose z_1 and z_2 are a pair of symmetric points of C_z, $z_1 \neq z_2$, and $w_1 = f(z_1)$ and $w_2 = f(z_2)$ are image points of z_1 and z_2, respectively. Are w_1 and w_2 a pair of symmetric points of C_w? With the property of conformality of a bilinear transformation in mind, the answer to the above question is "yes". This is called the *symmetry-preserving property* of bilinear transformations. The validity of this property can be shown using either the geometric or algebraic approach.

The algebraic proof is relatively straightforward. A circle C_z that has z_1 and z_2 as a pair of symmetric points can be represented by

$$\left| \frac{z - z_1}{z - z_2} \right| = k, \quad k > 0. \tag{8.2.17}$$

The inverse of the bilinear transformation (8.2.16) is found to be

$$z = \frac{-dw + b}{cw - a}. \tag{8.2.18}$$

Substituting eq. (8.2.18) into eq. (8.2.17), we obtain

$$\left| \frac{(cz_1 + d)w - (az_1 + b)}{(cz_2 + d)w - (az_2 + b)} \right| = k. \tag{8.2.19}$$

Suppose both z_1 and z_2 differ from $-\frac{d}{c}$ (the pole of the bilinear transformation); the above equation can be rewritten as

$$\left| \frac{w - w_1}{w - w_2} \right| = k \left| \frac{cz_2 + d}{cz_1 + d} \right| = k', \quad k' > 0,$$

indicating that w_1 and w_2 are a pair of symmetric points of C_w.

What happens when $z_1 = -\frac{d}{c}$? We then have $w_1 = f\left(-\frac{d}{c}\right) = \infty$. By observing $cz_1 + d = 0$, $cz_2 + d \neq 0$, and $az_1 + b \neq 0$, eq. (8.2.19) becomes

$$|w - w_2| = \frac{1}{k} \left| \frac{az_1 + b}{cz_2 + d} \right|,$$

which reveals that C_w is a circle with w_2 as the center. Recall that the center of a circle and the complex infinity are symmetric with respect to the circle. Hence, w_1 and w_2 are symmetric with respect to C_w. A similar argument can be applied when $z_2 = -\frac{d}{c}$.

The alternative proof using the geometric approach employs the conformal property of bilinear transformations. First, we establish a basic geometric property about symmetric points: any circle passing through a pair of symmetric points z_1 and z_2 of C_z always cuts C_z orthogonally (see Figure 8.14).

To prove the claim, we draw a circle γ that passes through the pair of symmetric points z_1 and z_2 of C_z. Through the center of C_z, a tangent is drawn that touches γ at the point z'. Let α and R denote the center and radius of C_z, respectively. The three points α, z_1 and z_2 are collinear. Here, $|z' - \alpha|$ is the length of the tangent to the circle γ through the external point α. We recall the well-known result in elementary geometry that $|z' - \alpha|$ satisfies

$$|z' - \alpha|^2 = |z' - z_1|\,|z' - z_2|,$$

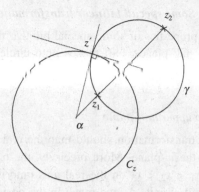

Figure 8.14. The circle γ passes through the symmetric points z_1 and z_2 of C_z. A tangent is drawn from α (the center of C_z) to γ that touches γ at the point z'. The line segment joining α and z' can be shown to be the radius of C_z.

where the three points α, z_1 and z_2 all lie on the same straight line, z_1 and z_2 are points on the circle γ. On the other hand, since z_1 and z_2 are a pair of symmetric points of C_z, we know from eq. (8.2.9b) that $|z' - z_1|\,|z' - z_2| = R^2$. We then obtain the result $|z' - \alpha| = R$ so that the line joining z' and α is a radius of C_z; thus γ and C_z cut each other orthogonally at z'.

Reversing the argument, one can show the converse of the above result: if every circle passing through the two points z_1 and z_2 intersects C_z orthogonally, then z_1 and z_2 are a pair of symmetric points of C_z.

The proof of the symmetry-preserving property follows by using arguments that combine the above geometric property about symmetric points of a circle, the conformal property and the circle-preserving property of bilinear transformations. Let Γ be the image of a circle γ that passes through the pair of symmetric points z_1 and z_2 of the circle C_z. Since a bilinear transformation is circle-preserving, the image curve Γ is a circle that passes through w_1 and w_2, which are the image points of z_1 and z_2, respectively. The question is whether w_1 and w_2 are a pair of symmetric points of C_w, the image circle of C_z under the bilinear transformation. Since γ and C_z cut each other orthogonally, we argue that the respective image circles Γ and C_w also cut each other orthogonally, by virtue of the conformality of the bilinear transformation. If this is true for any image circle Γ that passes through w_1 and w_2, then w_1 and w_2 must be a pair of symmetric points of C_w. Hence, we can conclude that bilinear transformations are symmetry-preserving. As shown in the examples below, this symmetry-preserving property is a versatile technique which can be employed to find the appropriate bilinear transformation that maps one given region onto another region.

8.2.3 *Some special bilinear transformations*

We now examine the properties of some special bilinear transformations that map half-planes onto half-planes, half-planes onto circles, and circles onto circles.

Upper half-plane onto upper half-plane

The required bilinear transformation should map the real axis in the z-plane onto the real axis in the w-plane. More precisely, the transformation maps three distinct points $x_1 < x_2 < x_3$ on the real axis onto three distinct points $w_1 < w_2 < w_3$ on the real axis. As we move in the positive direction along the two real axes, the left-hand region (upper half z-plane) is mapped onto the left-hand region (upper half w-plane). The angle of turning under the bilinear transformation at x_i, $i = 1, 2, 3$, is zero, that is,

$$\text{Arg } f'(x_i) = 0 \quad \text{so that} \quad f'(x_i) > 0, \quad i = 1, 2, 3. \quad (8.2.20)$$

The coefficients in the bilinear transformation can be determined (up to a multiplicative constant) by the three conditions

$$w_i = f(x_i) = \frac{ax_i + b}{cx_i + d}, \quad i = 1, 2, 3.$$

Since x_1, x_2, x_3 and w_1, w_2, w_3 are all real, these coefficients must be real. Further, we have

$$f'(x_i) = \frac{ad - bc}{(cx_i + d)^2}, \quad i = 1, 2, 3,$$

which are known to be real and positive by virtue of eq. (8.2.20). We then deduce that the coefficients must observe the condition $ad - bc > 0$.

Conversely, consider a bilinear transformation

$$w = f(z) = \frac{az + b}{cz + d},$$

where the coefficients a, b, c and d are real, and $ad - bc > 0$. With real coefficients, the transformation then maps the real axis onto the real axis. In addition, we observe that

$$f'(x) = \frac{ad - bc}{(cx + d)^2} > 0$$

for all x since $ad - bc > 0$, thus Arg $f'(x) = 0$. Hence, the positive direction along the real axis remains the same direction along the image real axis under the bilinear transformation. The upper half z-plane is then mapped onto the upper half w-plane.

Remark Supposing instead the sign of $ad - bc$ is negative, the bilinear transformation with real coefficients maps the upper (lower) half-plane onto the lower (upper) half-plane.

Example 8.2.6 Find the bilinear transformation that maps the upper half z-plane onto the upper half w-plane, the point $z = 0$ to $w = 0$, and the point $z = i$ to $w = 1 + i$.

Solution Since the required bilinear transformation maps the real axis onto the real axis, it should take the form

$$w = f(z) = \frac{z + b}{cz + d},$$

where the coefficients b, c and d are real. Since $f(0) = 0$, we obtain $b = 0$. Further, from $f(i) = 1 + i$, we have

$$1 + i = \frac{i}{ci + d} \quad \text{or} \quad (d - c) + i(d + c) = i.$$

Equating the real and imaginary parts, we obtain the simultaneous equations

$$d - c = 0 \quad \text{and} \quad d + c = 1.$$

Solving the above equations, the bilinear transformation is found to be

$$w = f(z) = \frac{2z}{z + 1}.$$

Upper half-plane onto unit circle

We would like to determine the general form of bilinear transformations that map the upper half-plane Im $z > 0$ onto the unit circle $|w| < 1$.

Since Im $z > 0$ is mapped to $|w| < 1$, there exists a particular point α, Im $\alpha > 0$, that is mapped onto $w = 0$. The symmetric point of α with respect to the real axis is $\bar{\alpha}$. By virtue of the symmetry-preserving property, the point $\bar{\alpha}$ will be mapped to the symmetric point of $w = 0$ with respect to the unit circle $|w| = 1$. Since $w = 0$ is the center of the unit circle, its symmetric point is $w = \infty$. Let the bilinear transformation be represented by

$$w = f(z) = \frac{az + b}{cz + d}.$$

From the deduced properties $f(\alpha) = 0$ and $f(\overline{\alpha}) = \infty$, we obtain $a\alpha + b = 0$ and $c\overline{\alpha} + d = 0$, thus giving $\alpha = -b/a$ and $\overline{\alpha} = -d/c$. The bilinear transformation can be rewritten as

$$w = f(z) = \frac{a\left(z + \frac{b}{a}\right)}{c\left(z + \frac{d}{c}\right)} = \frac{a}{c}\frac{z - \alpha}{z - \overline{\alpha}}.$$

Recall that the real axis $\operatorname{Im} z = 0$ is mapped onto the circle $|w| = 1$. By setting $z = x$ in the above equation, we obtain

$$\left|\frac{a}{c}\frac{x - \alpha}{x - \overline{\alpha}}\right| = \left|\frac{a}{c}\right|\left|\frac{x - \alpha}{x - \overline{\alpha}}\right| = 1.$$

Since $|x - \alpha| = |x - \overline{\alpha}|$, we obtain $|a/c| = 1$ and so $a/c = e^{i\theta}$ for some real θ. As a summary, the required bilinear transformation takes the general form

$$w = f(z) = e^{i\theta}\frac{z - \alpha}{z - \overline{\alpha}}, \qquad \theta \text{ is any real value.} \qquad (8.2.21)$$

Example 8.2.7   Find the function $u(x, y)$ that is harmonic in the upper half-plane and assumes the boundary value

$$u(x, 0) = \frac{x}{x^2 + 1}$$

along the real axis.

Solution   The obvious choice

$$u(z) = \operatorname{Re}\frac{z}{z^2 + 1}, \qquad z = x + iy,$$

fails since the function has a singularity $z = i$ in the upper half-plane. One may try to solve the problem by transforming the domain of the problem from the upper half-plane to the unit disk. Suppose we choose the bilinear transformation

$$w = \frac{i - z}{i + z}$$

that maps the upper half-plane $\operatorname{Im} z \geq 0$ onto the unit disk $|w| \leq 1$. For any point on $|w| = 1$, we write $w = e^{i\phi}$; and for any point on $\operatorname{Im} z = 0$, we write $z = x$. Since the inverse transformation is

$$z = \frac{i(1 - w)}{(1 + w)},$$

the quantities $e^{i\phi}$ and x are related by

$$x = i\frac{1 - e^{i\phi}}{1 + e^{i\phi}}.$$

The boundary value $u(x, 0) = \frac{x}{x^2+1}$ along the real axis in the z-plane is transformed into the boundary value $u(e^{i\phi})$ along $|w| = 1$ in the w-plane. Since r and ϕ are related by the above equation, the boundary value $u(e^{i\phi})$ can be expressed as

$$u(e^{i\phi}) = \frac{i\dfrac{1 - e^{i\phi}}{1 + e^{i\phi}}}{1 + i^2\left(\dfrac{1 - e^{i\phi}}{1 + e^{i\phi}}\right)^2} = \frac{i}{4}\frac{1 - e^{2i\phi}}{e^{i\phi}} = \frac{\sin\phi}{2}.$$

The function $u(w)$ that is harmonic inside $|w| < 1$ and satisfying the boundary condition $u(e^{i\phi}) = \frac{1}{2}\sin\phi$ is easily seen to be

$$u(w) = \frac{1}{2}\,\mathrm{Im}\,w.$$

Since the function

$$w = \frac{i - z}{i + z}$$

is analytic in the upper z-plane, its imaginary part

$$u(x, y) = \frac{1}{2}\,\mathrm{Im}\frac{i - z}{i + z} = \frac{x}{x^2 + (y + 1)^2}, \quad z = x + iy,$$

would be harmonic in the upper half-plane. Also, it satisfies the prescribed boundary condition

$$u(x, 0) = \frac{x}{x^2 + 1}.$$

Hence, it is the desired solution.

Remark

(i) The final solution

$$u(z) = \frac{1}{2}\mathrm{Im}\frac{i - z}{i + z}$$

has its singularity at $z = -i$, which is outside the upper half-plane. The true solution can be obtained only when we transfer the domain of the problem from the upper half-plane to the unit disk.

(ii) Suppose the boundary condition along the real axis remains the same, but the domain of the problem is changed from the upper to the lower half-plane. The solution to this new problem can be judiciously deduced

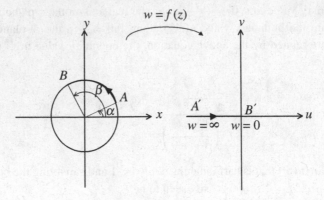

Figure 8.15. The circular arc $e^{i\alpha}$ to $e^{i\beta}$ on the unit circle is mapped onto the negative real axis.

to be

$$u(z) = \frac{1}{2} \operatorname{Im} \frac{z+i}{z-i} = \frac{x}{x^2 + (1-y)^2}.$$

It can be checked easily that the boundary condition

$$u(x, 0) = \frac{x}{x^2 + 1}$$

is satisfied. The singularity of the above solution is now at $z = i$, which is outside the lower half-plane.

Example 8.2.8   Find a bilinear transformation $w = f(z)$ that carries the interior of the unit circle $|z| < 1$ onto the upper half-plane $\operatorname{Im} w > 0$, such that the circular arc $e^{i\alpha}$ to $e^{i\beta}$ ($\alpha < \beta$) in the z-plane is mapped onto the negative real axis in the w-plane.

Solution   As deduced from eq. (8.2.21), the required bilinear transformation takes the form

$$z = F(w) = e^{i\theta} \frac{w - w_0}{w - \overline{w}_0},$$

where θ and w_0 ($\operatorname{Im} w_0 > 0$) are to be determined. To satisfy the requirements that the circular arc $e^{i\alpha}$ to $e^{i\beta}$ is mapped onto the negative real axis and the interior of the unit circle is mapped onto the upper half-plane, we impose two mapping conditions: $z = e^{i\beta}$ is mapped to $w = 0$ and $z = e^{i\alpha}$ is mapped to $w = \infty$ (see Figure 8.15). Note that the left side of the directed circular arc (interior of the unit circle) is mapped onto the left side of the directed line segment along the negative real axis (upper half-plane). Since

$$e^{i\alpha} = F(\infty) = \lim_{w \to \infty} e^{i\theta} \frac{w - w_0}{w - \overline{w_0}} = e^{i\theta},$$

this leads to $\theta = \alpha$. Further, given that

$$e^{i\beta} = F(0) = e^{i\alpha} \frac{w_0}{\overline{w_0}},$$

we obtain Arg $w_0 = \frac{\beta - \alpha}{2}$. We expect w_0 to lie in the upper half-plane since $0 < $ Arg $w_0 < \pi$, given that $0 < \beta - \alpha < 2\pi$. The modulus $|w_0|$ is not fixed, and it may take any real positive value. Suppose we denote $|w_0|$ by r; correspondingly, we write $w_0 = re^{i(\beta - \alpha)/2}$. Putting the results together, the required bilinear transformation becomes

$$z = e^{i\alpha} \frac{w - re^{i(\beta-\alpha)/2}}{w - re^{-i(\beta-\alpha)/2}} \quad \text{or} \quad w = re^{i(\alpha-\beta)/2} \frac{z - e^{i\beta}}{z - e^{i\alpha}},$$

where r is some undetermined real constant. The bilinear transformation leaves one undetermined constant since the problem prescribes only two rather than three mapping pairs.

Example 8.2.9   Find a bilinear transformation that carries the unit circle $|z| = 1$ to a line segment parallel to the imaginary axis, takes the point $z = 9$ to the point $w = 0$, and leaves the circle $|z| = 3$ invariant.

Solution   We are only given one mapping pair: $z = 9$ is mapped to $w = 0$. The symmetry-preserving property can be used to find an additional mapping pair. We know that the circle $|z| = 3$ is mapped onto the circle $|w| = 3$. The symmetric point of $z = 9$ with respect to the circle $|z| = 3$ is $z = 1$, while the symmetric point of $w = 0$ with respect to the circle $|w| = 3$ is $w = \infty$. By virtue of the symmetry-preserving property, the point $z = 1$ is mapped to $w = \infty$. Using the information about the two mapping pairs, the bilinear transformation takes the form

$$w = f(z) = \beta \frac{z - 9}{z - 1},$$

where β is some complex constant to be determined. For a point on the circle $|z| = 3$, we write $z = 3e^{i\theta}$. Since the circle $|z| = 3$ is mapped onto the circle

$|w| = 3$, we have

$$3 = |f(3e^{i\theta})| = |\beta| \left| \frac{3e^{i\theta} - 9}{3e^{i\theta} - 1} \right|$$

$$= |\beta| \, |-3e^{i\theta}| \left| \frac{1 - 3e^{-i\theta}}{1 - 3e^{i\theta}} \right| = 3|\beta|;$$

thus giving $|\beta| = 1$. The image of an arbitrary point $z = e^{i\theta}$ on the unit circle $|z| = 1$ is given by

$$f(e^{i\theta}) = \beta \frac{e^{i\theta} - 9}{e^{i\theta} - 1} = \beta \left(5 + 4i \cot \frac{\theta}{2} \right).$$

We know that this image point always lies on a line segment parallel to the imaginary axis; that is,

$$\text{Re } f(e^{i\theta}) = \text{Re } (\beta(5 + 4i \cot \frac{\theta}{2}))$$

which is a constant, independent of θ. We deduce that β must be real in order to satisfy the above property.

To satisfy both conditions, β has to be real and $|\beta| = 1$, we then have $\beta = 1$ or -1. Hence, the two bilinear transformations

$$w = \frac{z - 9}{z - 1} \quad \text{and} \quad w = -\frac{z - 9}{z - 1}$$

both satisfy all the mapping requirements. They map the unit circle $|z| = 1$ onto the vertical lines $\text{Re } w = 5$ and $\text{Re } w = -5$, respectively.

Circles to circles

We wish to find a bilinear transformation that maps the unit disk $|z| \leq 1$ onto the unit disk $|w| \leq 1$; in particular, the point $z = \alpha$ ($|\alpha| < 1$) is mapped to $w = 0$. By the symmetry-preserving property, the point $z = 1/\overline{\alpha}$ (the symmetric point of $z = \alpha$ with respect to the circle $|z| = 1$) is mapped to the complex infinity $w = \infty$ (the symmetric point of $w = 0$ with respect to the circle $|w| = 1$). The bilinear transformation is seen to assume the form

$$w = f(z) = \beta \frac{z - \alpha}{1 - \overline{\alpha}z},$$

where β is some complex constant to be determined. Let $z = e^{i\theta}$ be a point on the circle $|z| = 1$; the corresponding image point $f(e^{i\theta})$ in the w-plane always lies on the circle $|w| = 1$. This mapping property can be represented by

$$1 = |f(e^{i\theta})| = |\beta| \left| \frac{e^{i\theta} - \alpha}{1 - \overline{\alpha}e^{i\theta}} \right| = |\beta| \, |e^{-i\theta}| \left| \frac{e^{i\theta} - \alpha}{e^{-i\theta} - \overline{\alpha}} \right| = |\beta|;$$

thus giving $|\beta| = 1$. Suppose we write $\beta = e^{i\theta_0}$, then the bilinear transformation can be expressed as

$$w = f(z) = e^{i\theta_0} \frac{z - \alpha}{1 - \bar{\alpha}z},$$

where θ_0 is real and $|\alpha| < 1$.

Example 8.2.10 Find the bilinear transformation $w = f(z)$ that carries $|z| < 1$ to $|w| < 1$, $z = \frac{1}{3}$ to $w = \frac{i}{3}$, and $f'\left(\frac{1}{3}\right) > 0$.

Solution In this problem, we need to construct a sequence of two bilinear transformations; one maps $z = \frac{1}{3}$ to $\zeta = 0$ and the other one maps $\zeta = 0$ to $w = \frac{i}{3}$. First, we try to find the bilinear transformation $\zeta = g(z)$ that carries $|z| < 1$ onto $|\zeta| < 1$, $g\left(\frac{1}{3}\right) = 0$ and $g'\left(\frac{1}{3}\right) > 0$. From eq. (8.2.22), the bilinear transformation takes the form

$$\zeta = g(z) = e^{i\theta_0} \frac{z - \frac{1}{3}}{1 - \frac{z}{3}}, \qquad \theta_0 \text{ is real.}$$

The derivative of $g(z)$ at $z = \frac{1}{3}$ is found to be

$$g'\left(\frac{1}{3}\right) = e^{i\theta_0} \frac{\left(1 - \frac{z}{3}\right) + \frac{1}{3}\left(z - \frac{1}{3}\right)}{\left(1 - \frac{z}{3}\right)^2}\bigg|_{z = \frac{1}{3}} = \frac{9}{8} e^{i\theta_0}.$$

The condition $g'\left(\frac{1}{3}\right) > 0$ implies $e^{i\theta_0} = 1$, so

$$\zeta = g(z) = \frac{3z - 1}{3 - z}.$$

Next, we find another bilinear transformation $\zeta = h(w)$ that carries $|w| < 1$ onto $|\zeta| < 1$, maps $w = \frac{i}{3}$ to $\zeta = 0$ and satisfies $h'\left(\frac{i}{3}\right) > 0$. By judiciously setting $z = \frac{w}{i}$ in $g(z)$, the required bilinear transformation is deduced to be

$$\zeta = h(w) = \frac{3w - i}{3 + iw}.$$

Recall that the inverse of a bilinear transformation remains bilinear and the composition of two bilinear transformations is also bilinear. Since we have

$$\zeta = h(w) = g(z),$$

where both h and g are bilinear, the mapping defined by

$$w = h^{-1}(g(z))$$

remains bilinear, and carries $|z| < 1$ onto $|w| < 1$. Further, we have

$$h^{-1}\left(g\left(\frac{1}{3}\right)\right) = h^{-1}(0) = \frac{i}{3}.$$

Observing that

$$h'(w)\frac{dw}{dz} = g'(z),$$

we then have

$$\frac{d}{dz}h^{-1}(g(z))\Bigg|_{z=\frac{1}{3}} = \frac{g'\left(\frac{1}{3}\right)}{h'\left(\frac{i}{3}\right)} > 0,$$

so all the mapping requirements are satisfied. To express w explicitly in terms of z, we solve for w from the relation

$$\frac{3w - i}{3 + iw} = \frac{3z - 1}{3 - z}$$

and obtain

$$w = \frac{3(i - 1) + (9 - i)z}{(9 + i) - 3(1 + i)z}.$$

This is the required bilinear transformation.

Riemann mapping theorem

We have seen various examples of conformal mappings between the upper half-plane and the unit disk. The natural question is: what are the types of domain that can be mapped conformally onto the unit disk? The Riemann mapping theorem gives the sufficient conditions on the domain that can be mapped conformally onto the unit disk. We state this renowned theorem without proof.

Theorem 8.2.2 (Riemann mapping theorem) *Let $\mathcal{D}$ be any simply connected domain in the complex plane other than the entire plane. There exists a one-to-one analytic function $w = f(z)$ that maps $\mathcal{D}$ in the z-plane onto the interior of the unit disk $|w| < 1$ in the w-plane. The mapping can be chosen to carry an arbitrary point of the domain and a direction through that point onto the center of the unit disc and the direction of the positive real axis, respectively. Under these conditions, the mapping is unique.*

The Riemann mapping theorem guarantees the existence of the mapping function. However, it does not inform us how to derive the required mapping. We can deduce from the theorem that any simply connected domain $\mathcal{D}$ can be

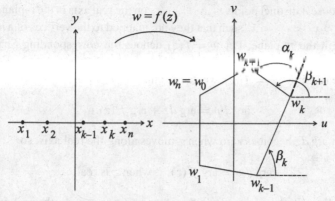

Figure 8.16. The Schwarz–Christoffel transformation carries the upper half-plane onto the interior of an n-sided polygon, where $f(x_k) = w_k$, $k = 1, 2, \ldots, n$. The angle of inclination of the kth side joining the neighboring vertices w_{k-1} and w_k is denoted by β_k. The exterior angle α_k at the vertex w_k is given by $\beta_{k+1} - \beta_k$. For convenience of notation, we also write w_n as w_0 and treat β_{n+1} as β_1.

analytically mapped in a one-to-one manner onto another simply connected domain, provided that neither domain is the whole complex plane.

8.3 Schwarz–Christoffel transformations

We often encounter physical problems in potential flows, electrostatic fields and steady state temperature distributions where the domains of interest are polygons (or open polygons with one or more of the vertices at infinity). The solution of these problems is greatly facilitated if the domain of the problem can be mapped onto the upper half-plane. The first step in the solution procedure is to construct a conformal mapping that carries the upper half-plane onto the polygonal domain (see Figure 8.16). This class of conformal mappings has come to be known as the *Schwarz–Christoffel transformations*.

Consider a closed n-sided polygon with vertices $w_1, w_2, \ldots, w_n$ in the finite w-plane. The angle of inclination of the kth side joining w_{k-1} and w_k is denoted by β_k, $k = 1, 2, \ldots, n$. For convenience of notation, we also write w_n as w_0 so that the first side of the polygon refers to the line segment joining w_n and w_1. The angle of turning of the inclination angles of the two neighboring sides at the vertex w_k is called the exterior angle at w_k. Let α_k denote the exterior angle at w_k; then α_k is given by $\beta_{k+1} - \beta_k$, $k = 1, 2, \ldots, n$. Here, β_{n+1} is taken to be β_1. It is known from elementary geometry that the sum of the exterior angles of a polygon always equals 2π.

We choose n distinct points $x_1, x_2, \ldots, x_n$ on the real axis in the z-plane satisfying $x_1 < x_2 < \cdots < x_n$ such that they are mapped to the vertices of a n-sided polygon in the w-plane. Let $w = f(z)$ denote the corresponding Schwarz–Christoffel transformation to be found, where $f(x_k) = w_k$, $k = 1, 2, \ldots, n$. From the relation $\frac{dw}{dz} = f'(z)$, we deduce that

$$\arg dw = \arg dz + \arg f'(z).$$

Note that $\arg dz$ becomes zero when z moves along the real axis, so

$$\arg dw = \arg f'(z), \quad \text{when } z \text{ is real.}$$

Specifically, when z moves along the segment (x_{k-1}, x_k) on the real axis, the image point w moves along the kth side of the polygon and $\arg dw$ assumes the constant value β_k. However, when z moves across the point x_k along the real axis, $\arg dw$ jumps by an amount α_k to assume a new constant value β_{k+1}. In summary, $\arg dw = \beta_k$ when z assumes real value in (x_{k-1}, x_k), $k = 2, 3, \ldots, n$ and $\arg dw = \beta_1$ when z assumes real value in $(-\infty, x_1)$ or (x_n, ∞).

Now, the problem of finding $\arg f'(z)$ is equivalent to the steady state temperature distribution problem in the upper half-plane with discrete constant values along the real axis [see Problem 3.25]. The solution for $\arg f'(z)$ is given by

$$\arg f'(z) = \frac{\beta_1}{\pi} \arg(z - x_1) + \frac{\beta_2}{\pi} \arg \frac{z - x_2}{z - x_1} + \cdots + \frac{\beta_n}{\pi} \arg \frac{z - x_n}{z - x_{n-1}}$$

$$+ \frac{\beta_{n+1}}{\pi} [\pi - \arg(z - x_n)]$$

$$= \beta_1 - \sum_{k=1}^{n} \frac{\alpha_k}{\pi} \arg(z - x_k), \tag{8.3.1a}$$

where $\beta_{n+1} = \beta_1$ and $\alpha_k = \beta_{k+1} - \beta_k$, $k = 1, 2, \ldots, n$. We then deduce that

$$f'(z) = K e^{i\beta_1} \prod_{k=1}^{n} (z - x_k)^{-\alpha_k/\pi}, \tag{8.3.1b}$$

where K is a real constant to be determined. Integrating with respect to z, the Schwarz–Christoffel transformation formula is found to be

$$f(z) = w_1 + K e^{i\beta_1} \int_{x_1}^{z} \prod_{k=1}^{n} (\zeta - x_k)^{-\alpha_k/\pi} \, d\zeta,$$

where the integration is performed along a simple path joining x_1 and z. To find the real constant K, we set $z = x_2$ and observe that $w_2 = f(x_2)$. Taking the modulus value on both sides of the equation, we then obtain

$$|w_2 - w_1| = K \left| \int_{r_1}^{x_2} \prod_{i=1}^{n} (\zeta - x_k)^{-\alpha_k/\pi} \, d\zeta \right|. \qquad (8.3.2)$$

Remarks

(i) For convenience, x_n is often chosen to be at infinity. Correspondingly, eq. (8.3.1a) has to be modified when the last term $-\frac{\alpha_n}{\pi} \arg(z - x_n)$ disappears; that is,

$$\arg f'(z) = \beta_n - \sum_{k=1}^{n-1} \frac{\alpha_k}{\pi} \arg(z - x_k). \qquad (8.3.3)$$

In this case, the mapping formula reduces to

$$f(z) = w_1 + K e^{i\beta_n} \int_{x_1}^{z} \prod_{k=1}^{n-1} (\zeta - x_k)^{-\alpha_k/\pi} \, d\zeta. \qquad (8.3.4)$$

(ii) Since only the angles of turning at the vertices are involved in the mapping formula, the polygon can be an unbounded polygon with one or more of the vertices at infinity. Since the angles of inclination of the neighboring sides of a vertex (even at infinity) are always known, the corresponding α_k can be found as the difference of the two angles of inclination of the neighboring sides of the polygon.

Example 8.3.1 Find the Schwarz–Christoffel transformation that maps the upper half z-plane to the triangle with vertices w_1, w_2 and w_3 as shown in Figure 8.17, where $w_1 = 0$, $w_2 = 1$, and θ_1 and θ_2 are some given angles. Take $x_1 = 0$, $x_2 = 1$ and $x_3 = \infty$ for ease of computation.

Solution The corresponding mapping pairs of points and the respective turning angles at the vertices are tabulated below:

k	x_k	w_k	α_k
1	0	0	$\pi - \theta_1$
2	1	1	$\pi - \theta_2$
3	∞	w_3	$\theta_1 + \theta_2$

Note that β_3 is the angle of inclination of the side joining w_2 and w_3, and its value equals $\pi - \theta_2$. As a check, the sum of the exterior angles at the vertices

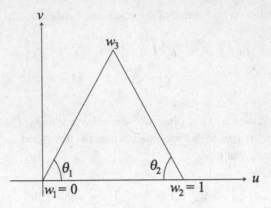

Figure 8.17. The remaining vertex w_3 of the triangle can be determined once w_1, w_2, θ_1 and θ_2 are known.

equals 2π. Using eq. (8.3.4), we obtain

$$f(z) = K e^{i(\pi - \theta_2)} \int_0^z \zeta^{\frac{\theta_1}{\pi} - 1} (\zeta - 1)^{\frac{\theta_2}{\pi} - 1} \, d\zeta,$$

where K is a real constant to be determined. Since $f(1) = 1$, the constant K is given by

$$1 = K \int_0^1 \zeta^{\frac{\theta_1}{\pi} - 1} (1 - \zeta)^{\frac{\theta_2}{\pi} - 1} \, d\zeta,$$

where the factor $e^{i(\pi - \theta_2)}$ has been absorbed into the integral. The above integral is related to the beta function defined by

$$\beta(p, q) = \int_0^1 x^{p-1} (1 - x)^{q-1} \, dx.$$

By setting $p = \frac{\theta_1}{\pi}$ and $q = \frac{\theta_2}{\pi}$, it is seen that

$$K = \frac{1}{\beta \left(\frac{\theta_1}{\pi}, \frac{\theta_2}{\pi} \right)}.$$

Finally, the required Schwarz–Christoffel transformation is found to be

$$f(z) = \frac{1}{\beta \left(\frac{\theta_1}{\pi}, \frac{\theta_2}{\pi} \right)} \int_0^z \zeta^{\frac{\theta_1}{\pi} - 1} (1 - \zeta)^{\frac{\theta_2}{\pi} - 1} \, d\zeta.$$

Example 8.3.2 Find the Schwarz–Christoffel transformation that maps the upper half z-plane Im $z > 0$ onto the semi-infinite strip $-\pi/2 < \text{Re } w < \pi/2$, Im $w > 0$.

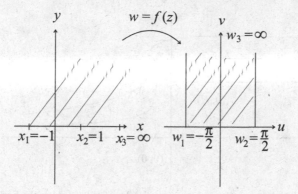

Figure 8.18. The upper half z-plane is mapped onto the semi-infinite strip $-\frac{\pi}{2} <$ Re $w < \frac{\pi}{2}$, Im $w > 0$ (considered as an open triangle with one of the vertices at infinity) in the w-plane.

Solution   The semi-infinite strip can be considered as an unbounded triangle with one of the vertices w_3 at infinity. For convenience, we assign $x_1 = -1$, $x_2 = 1$ and $x_3 = \infty$. The corresponding mapping pairs of points and the respective angles of turning at the vertices are summarized below (also see Figure 8.18):

k	x_k	w_k	α_k
1	-1	$-\frac{\pi}{2}$	$\frac{\pi}{2}$
2	1	$\frac{\pi}{2}$	$\frac{\pi}{2}$
3	∞	∞	π

Note that $\beta_3 = \frac{\pi}{2}$ since the side of the unbounded triangle joining w_2 and w_3 is the vertical line Re $w = \frac{\pi}{2}$. Using the Schwarz–Christoffel formula (8.3.4), we have

$$f(z) = -\frac{\pi}{2} + Ke^{i\frac{\pi}{2}} \int_{-1}^{z} \frac{1}{(\zeta+1)^{1/2}(\zeta-1)^{1/2}} \, d\zeta$$

$$= -\frac{\pi}{2} + K \int_{-1}^{z} \frac{1}{\sqrt{1-\zeta^2}} \, d\zeta$$

$$= K[\sin^{-1} z - \sin^{-1}(-1)] - \frac{\pi}{2}.$$

Applying the condition $f(1) = \frac{\pi}{2}$, K is found to be 1. The required mapping is then given by

$$f(z) = \sin^{-1} z.$$

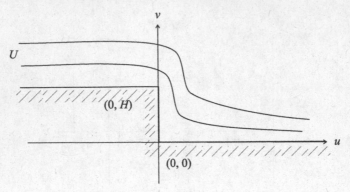

Figure 8.19. The bottom edges of the potential flow field are along the horizontal lines $v = H$ for $u < 0$ and $v = 0$ for $u > 0$. The vertical step of height H is placed along $u = 0$ and $0 < v < H$.

The result agrees with the mapping properties of $w = \sin z$ discussed in Subsection 3.2.1.

Example 8.3.3 Consider the potential flow over a vertical step of height H, the configuration of which is shown in Figure 8.19. The flow upstream far from the step is uniform with constant speed U and parallel to the floor bottom. Find the speed along the bottom edges of the flow field.

Solution First, we construct the Schwarz–Christoffel transformation $w = f(z)$ that maps the upper half z-plane onto the domain of the given step flow (shown in Figure 8.19) in the w-plane. The flow domain can be visualized as an unbounded polygon with true vertices at $(0, 0)$ and $(0, H)$, and a virtual vertex at infinity. We assign $x_1 = -1$, $x_2 = 1$ and $x_3 = \infty$; correspondingly, $w_1 = iH$, $w_2 = 0$ and $w_3 = \infty$. The angles of turning at w_1 and w_2 are $-\frac{\pi}{2}$ and $\frac{\pi}{2}$, respectively, and $\beta_3 = 0$. Using eq. (8.3.1b), we obtain

$$f'(z) = K(z + 1)^{1/2}(z - 1)^{-1/2};$$

and the constant K is determined by [see eq. (8.3.2)]

$$K = \frac{H}{\left| \int_{-1}^{1} (\zeta + 1)^{1/2}(\zeta - 1)^{-1/2} \, d\zeta \right|} = \frac{H}{\left| \left. (z^2 - 1)^{1/2} + \cosh^{-1} z \right|_{-1}^{1} \right|} = \frac{H}{\pi}.$$

The required transformation is then found to be

$$w = f(z) = \frac{H}{\pi} \left[(z^2 - 1)^{1/2} + \cosh^{-1} z \right].$$

Let $F(w)$ and $V(w)$ denote the complex potential and complex velocity of the potential flow field in the w-plane, respectively, where $V(w) = F'(w)$. The complex potential of the corresponding flow field in the z-plane is given by $F(w(z))$, where $w(z)$ is the above Schwarz–Christoffel transformation. By the chain rule of differentiation, the two complex velocities $\frac{dF(w)}{dw}$ and $\frac{dF(z)}{dz}$ are related by

$$\frac{dF(w)}{dw} = \frac{dF(z)}{dz}\frac{dz}{dw}.$$

When $w \to \infty$, $\frac{dF(w)}{dw}$ tends to U; and when $z \to \infty$, $\frac{dw}{dz}$ tends to $\frac{H}{\pi}$. Using the above relation between the complex velocities, we deduce that

$$\lim_{z \to \infty} \frac{dF(z)}{dz} = \frac{H}{\pi}U.$$

Since the flow domain in the z-plane is the whole upper half-plane, the flow velocity is uniform throughout. In other words, the complex velocity $\frac{dF}{dz}$ of the flow field in the z-plane is equal to the constant value $\frac{H}{\pi}U$. Also, we see that

$$\frac{dz}{dw} = \frac{1}{f'(z)} = \frac{\pi}{H}\left(\frac{z-1}{z+1}\right)^{1/2},$$

so the complex velocity of the flow field in the w-plane is given by

$$V(w) = \frac{dF(w)}{dw} = U\left(\frac{z-1}{z+1}\right)^{1/2}.$$

The points along the bottom edges and the vertical step of the flow field in the w-plane are the image points of $z = x$ along the real axis in the z-plane. In terms of x, the speed of the step flow along the bottom edges is given by

$$|V| = U\sqrt{\left|\frac{x-1}{x+1}\right|}.$$

The speed becomes infinite at $x = x_1 = -1$ and zero at $x = x_2 = 1$. These two points correspond to the upper and lower corners of the vertical step, respectively.

Example 8.3.4 Consider the steady state temperature distribution within a thin wall near the corner of a building. When the thickness a of the wall is small compared to its span, the wall can be visualized as having two perpendicular semi-infinite channels (see the configuration shown in Figure 8.20). Let the outer surface be maintained at zero temperature, while the inner surface is kept at constant temperature T_0. Find the steady state temperature distribution inside the wall.

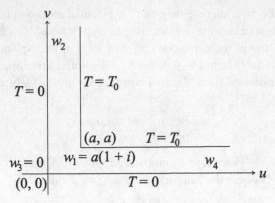

Figure 8.20. The thin wall at the corner of a building is visualized as having semi-infinite channels along both the positive u-axis and v-axis. The domain of interest can be recognized as an unbounded polygon with vertices at $(0, 0)$ and (a, a) and virtual vertices at the two far ends of the channels.

Solution   We recognize the semi-infinite L-shaped thin wall as an unbounded polygon and attempt to construct a Schwarz–Christoffel transformation $w = f(z)$ that carries the upper half z-plane onto the unbounded L-shaped polygon in the w-plane. The two true vertices of the unbounded polygon are $w_1 = a(1 + i)$ and $w_3 = 0$, while the two virtual vertices w_2 and w_4 are at the far ends of the channels (see Figure 8.20). For convenience, we assign $x_1 = -1$, $x_2 = 0$, $x_3 = 1$ and $x_4 = \infty$, and they are mapped to the vertices w_1, w_2, w_3 and w_4, respectively.

The angles of turning at w_1, w_2 and w_3 are $-\frac{\pi}{2}$, π and $\frac{\pi}{2}$, respectively, and $\beta_4 = 0$. By eq. (8.3.4), the corresponding Schwarz–Christoffel transformation is given by

$$w = f(z) = a(1 + i) + K \int_{-1}^{z} (\zeta + 1)^{1/2} \, \zeta^{-1} (\zeta - 1)^{-1/2} \, d\zeta.$$

The appropriate integration variable is

$$\eta = \left(\frac{\zeta + 1}{\zeta - 1} \right)^{1/2} \quad \text{or} \quad \zeta = \frac{\eta^2 + 1}{\eta^2 - 1}$$

so that

$$d\zeta = -\frac{4\eta}{(\eta^2 - 1)^2} \, d\eta.$$

Upon integration, we obtain

$$w = a(1 + i) + K \int_0^\xi \frac{4\eta^2}{1 - \eta^4} \, d\eta$$

$$= a(1 + i) + K \left(\text{Log} \frac{1+\xi}{1-\xi} - 2 \tan^{-1} \xi \right), \quad \xi = \left(\frac{z+1}{z-1} \right)^{1/2},$$

where the principal branches for Log and $\tan^{-1}$ are chosen so that the corresponding values tend to zero as $\xi \to 0$. To determine K, we apply the condition $w_3 = f(1) = 0$. This gives

$$0 = a(1 + i) + K \left[-i\pi - 2 \left(\frac{\pi}{2} \right) \right],$$

and leads to $K = \frac{a}{\pi}$. The required mapping function is then found to be

$$w = f(z) = a(1 + i) + \frac{2a}{\pi} \left(\frac{1}{2} \text{Log} \frac{1+\xi}{1-\xi} - \tan^{-1} \xi \right), \quad \xi = \left(\frac{z+1}{z-1} \right)^{1/2}.$$

In the z-plane, the boundary temperature values along the x-axis are given by

$$T(x, 0) = \begin{cases} T_0 & x < 0 \\ 0 & x > 0 \end{cases}.$$

The steady state temperature function $T(z)$ is seen to be

$$T(z) = \frac{T_0}{\pi} \text{Arg } z, \quad \text{Im } z > 0.$$

Referring back to the w-plane, the temperature function $T(w)$ can be expressed as

$$T(w) = \frac{T_0}{\pi} \text{Arg } f^{-1}(w),$$

where $f^{-1}(w)$ is the inverse of the above Schwarz–Christoffel mapping function.

Example 8.3.5 Suppose two semi-infinite charged rods with electric potential ϕ_1 and ϕ_2 are placed in the configuration shown in Figure 8.21. Find the electric potential of the electrostatic field induced by the two charged rods.

Solution It does require a certain amount of imagination to visualize the whole u-v plane minus the two charged rods as an unbounded polygon. The two true vertices are at $w_1 = ih$ and $w_3 = h$, and the two virtual vertices w_2 and w_4 are at the far bottom left and far top right corners of the w-plane, respectively (see Figure 8.21).

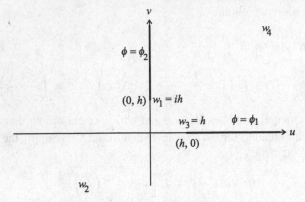

Figure 8.21. The horizontal semi-infinite rod with electric potential ϕ_1 is placed along the positive u-axis with the free end at $(h, 0)$ while the vertical semi-infinite rod with electric potential ϕ_2 is placed along the positive v-axis with the free end at $(0, h)$.

We assign $x_1 = -1$, $x_2 = 0$, $x_3 = 1$ and $x_4 = \infty$ along the real axis in the z-plane, and these points are mapped by a Schwarz–Christoffel transformation $w = f(z)$ to w_1, w_2, w_3 and w_4 in the w-plane, respectively. The angles of turning at the vertices w_1 and w_3 are easily seen to be both equal to $-\pi$. Interestingly, the angle of turning at the virtual vertex w_2 is $\frac{5\pi}{2}$ since the orientation starts northward, then turns one complete revolution plus a final right angle turn to settle in the westward direction. The angle of inclination β_4 of the virtual line segment joining w_3 and w_4 is seen to be zero.

By eq. (8.3.4), the corresponding Schwarz–Christoffel transformation is given by

$$w = f(z) = ih + K \int_{-1}^{z} \frac{(\zeta - 1)(\zeta + 1)}{\zeta^{5/2}} \, d\zeta$$

$$= ih + K \left(2z^{1/2} + \frac{2}{3} z^{-3/2} - \frac{8}{3} i \right).$$

The constant K is determined by $f(1) = h$, and this gives

$$h = ih + K \left(2 + \frac{2}{3} - \frac{8}{3} i \right) \quad \text{or} \quad K = \frac{3}{8} h.$$

The required Schwarz–Christoffel transformation is found to be

$$w = f(z) = \frac{h}{4} z^{1/2} \left(3 + \frac{1}{z^2} \right).$$

In the z-plane, the boundary values for the electric potential $\phi(z)$ along the x-axis are given by

$$\phi(u, 0) = \begin{cases} \phi_2 & x < 0 \\ \phi_1 & x > 0 \end{cases}$$

The solution for $\phi(z)$ is easily found to be

$$\phi(z) = \phi_1 + \frac{\phi_2 - \phi_1}{\pi} \operatorname{Arg} z.$$

Using the relation between z and w as defined by the above Schwarz–Christoffel mapping function, the electric potential $\phi(w)$ at any arbitrary point in the w-plane can be obtained.

8.4 Problems

8.1. For each of the following functions, find the points at which the function is not conformal:
    (a) $w = z^2 + \dfrac{1}{z^2}$; (b) $w = \sin 2z$.

8.2. A mapping that preserves the magnitude of the angle between any two smooth arcs passing through any point in a domain but not the sense is called an *isogonal* mapping in that domain. Suppose $f(z)$ is conformal in a domain $\mathcal{D}$; show that $\overline{f(z)}$ is isogonal in the same domain $\mathcal{D}$.

8.3. For each of the following curves in the z-plane, find the corresponding image curve under the mapping $w = \frac{1}{z}$ in the w-plane:

    (a) a line through the point $z = \alpha \neq 0$;
    (b) the hyperbola $x^2 - y^2 = 1$;
    (c) the parabola $y^2 - 2px = 0$.

8.4. Show that the transformation $w = \frac{1}{z}$ maps the common part of the two disks $|z - 1| < 1$ and $|z + i| < 1$ onto the quarter-plane $\operatorname{Re} w > \frac{1}{2}$ and $\operatorname{Im} w > \frac{1}{2}$.

8.5. Find a transformation that maps the cycloid as defined by

$$x = a(t - \sin t), \, y = a(1 - \cos t), \quad 0 \leq t \leq 2\pi,$$

    onto a line through the origin with slope m.

8.6. Find the curvature of the image contour of the unit circle $|z| = 1$ under the conformal mapping $w = f(z)$.

8.7. Show that the function $w = 2 \operatorname{Log} z - z^2$ maps the upper half z-plane conformally onto the w-plane minus the two lines $v = 2\pi$, $u < -1$ and $v = 0$, $u < -1$.

8.8. Find the area of the closed region in the w-plane which is the image of the closed region $\left\{z : 1 \le |z| \le 2, \ -\frac{\pi}{4} \le \operatorname{Arg} z \le \frac{\pi}{4}\right\}$ in the z-plane under the mapping $w = f(z) = z^2$.

8.9. Find a conformal mapping $w = f(z)$ that maps the region outside the hyperbola

$$\frac{x^2}{\cos^2 \alpha} - \frac{y^2}{\sin^2 \alpha} > 4,$$

where $0 < \alpha < \frac{\pi}{2}$, in the z-plane onto the upper half w-plane.

8.10. Show that the mapping function

$$w = \int_{z_0}^{z} \frac{\sqrt{1 - \zeta^4}}{\zeta^2} \, d\zeta, \quad z_0 \ne 0,$$

maps the disk $|z| < 1$ in the z-plane onto the exterior of a square in the w-plane.

8.11. Find the image in the w-plane of the domain $-\frac{\pi}{2} < x < \frac{\pi}{2}$, $y > 0$ under the mapping $w = (\sin z)^{1/4}$.

8.12. Find the steady state temperature inside the sectoral domain $0 < \operatorname{Arg} z < \frac{\pi}{4}$, where the ray $\operatorname{Arg} z = 0$ is maintained at constant temperature K and the ray $\operatorname{Arg} = \frac{\pi}{4}$ is at constant temperature $\frac{K}{2}$.

8.13. Find the steady state temperature $T(x, y)$ inside the first quadrant where the temperature values along the x-axis satisfy

$$\begin{cases} T(x, 0) = 1, & x > 1 \\ \dfrac{\partial T}{\partial y}(x, 0) = 0, & x < 1 \end{cases},$$

and the temperature values along the y-axis satisfy

$$\begin{cases} T(0, y) = -1, & y > 1 \\ \dfrac{\partial T}{\partial x}(0, y) = 0, & y < 1 \end{cases}.$$

8.14. Find the steady state temperature $T(x, y)$ inside the first quadrant where the temperature along the y-axis is maintained at T_1. The boundary temperature values along the x-axis satisfy

$$\begin{cases} T(x, 0) = T_2, & x > x_0 \\ \dfrac{\partial T}{\partial y}(x, 0) = 0, & x < x_0 \end{cases}.$$

8.15. Find the steady state temperature $T(x, y)$ inside the infinite slab $0 < y < 1$, where the boundary temperature values are given by

$$T(x, 0) = T_0, \ x < 0; \quad T(x, 1) = -T_0, \ x < 0.$$
$$\frac{\partial T}{\partial y}(x, 0) = \frac{\partial T}{\partial y}(x, 1) = 0, \ x > 0.$$

8.16. Find the complex potential of the uniform flow U_∞ parallel to the x-axis that streams past the elliptical cylinder

$$\frac{x^2}{a^2} + \frac{y^2}{b^2} = 1.$$

Hint: Consider the Joukowski mapping $\zeta = \frac{z + \sqrt{z^2 - c^2}}{2}$, $c^2 = a^2 - b^2$.

8.17. Find the complex potential of the flow field with uniform upstream flow $U_\infty > 0$ that streams past an infinite obstacle of parabolic shape defined by

$$y^2 = 2px, \quad p > 0.$$

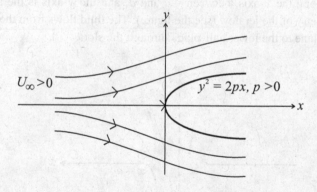

Hint: Let $\mathcal{D}$ denote the domain of the flow field that is external to the parabolic obstacle. The mapping $\zeta = \sqrt{z - \frac{p}{2}}$ takes $\mathcal{D}$ onto the first quadrant above the line segment $\zeta = \sqrt{\frac{p}{2}}i$. The subsequent mapping

$$w = g(\zeta(z)) = \left(\zeta - i\sqrt{\frac{p}{2}}\right)^2 = z - p + \sqrt{p^2 - 2pz}$$

takes $\mathcal{D}$ onto the upper half w-plane. Check that $g'(\infty) = 1$.

8.18. A potential flow field is confined between the parallel lines $y = \pm 1$, with the source and sink of equal strength m placed at $z = -1$ and $z = 1$, respectively. Find the complex potential of the flow field.

8.19. Consider the seepage problem discussed in Subsection 8.1.2. Show that the total amount of seepage from upstream to downstream of the dam is given by

$$\int_L^\infty \frac{\phi_1 - \phi_2}{\pi} \cosh^{-1} \frac{x}{L} \, dx.$$

Hint: The flux across the line segment joining $x = L$ to $x = x_0$ ($x_0 > L$) is given by

$$\int_L^{x_0} \psi(x, 0) \, dx.$$

8.20. This problem investigates the potential flow field of a fluid jet coming through a slot of width $2a$ in a two-dimensional plane. We wish to find the shape of the jet, given that the amount of fluid flowing through the slot per unit time is Q. The coordinate axes are assigned so that the slot lies along the x-axis, between $-a$ and a, and the y-axis is the axis of symmetry of the jet flow (see the figure). The fluid flows from the upper half-plane to the lower half-plane through the slot.

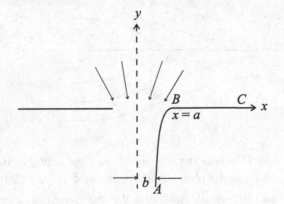

Let AB define the free boundary of the jet flow on the right-hand plane and BC be the wall along the x-axis. The point B corresponds to $x = a$ and the point A is on the free jet boundary far from the slot. The free boundary of the fluid jet is not known in advance but has to be determined as part of the solution procedure. Fortunately, the velocity values along the axis of symmetry, the bounding wall and the free jet boundary can be readily deduced.

(a) Let U_∞ denote the uniform flow speed of the free jet at infinitely far distance from the slot. Explain why

$$U_\infty = \frac{Q}{2b},$$

where $2b$ is the width of the jet at infinity (b will be determined later).

(b) Let u and v denote the x-component and y-component of the fluid velocity, respectively. Show that

   (i) $u = 0$, $-U_\infty \leq v \leq 0$ for $x = 0$ (along the axis of symmetry);

   (ii) $u^2 + v^2 = U_\infty^2$ along the free jet boundary;

   (iii) $-U_\infty \leq u \leq 0$, $v = 0$ for $a \leq x < \infty$, $y = 0$ (along the bounding wall).

(c) Let $w = -u + iv$, and treat w as a mapping. Show that the flow field in the right half z-plane, $z = x + iy$, is mapped onto the quarter circle $C_w = \{w : |w| \leq U_\infty, -\frac{\pi}{2} \leq \operatorname{Arg} w \leq 0\}$ in the lower right quadrant in the w-plane. Check that the free boundary is mapped onto the circular boundary of the quarter circle in the w-plane.

(d) Show that the mapping

$$w = U_\infty(\sqrt{\zeta} - \sqrt{\zeta - 1})$$

carries the upper half ζ-plane onto the quarter circle C_w in the w-plane. Check that the negative real axis is mapped to the vertical segment of C_w, the line segment $(0, 1)$ is mapped onto the circular arc of C_w and the line segment $(1, \infty)$ is mapped onto the horizontal segment of C_w.

   Hint: Consider the Joukowski mapping discussed in Subsection 3.6.1.

(e) By relating the mappings together, we see that the streamline along ABC (the free boundary of the jet plus the bounding wall along the positive x-axis) in the z-plane is mapped onto the positive real axis in the ζ-plane, and the streamline along the y-axis (the axis of symmetry) in the z-plane is mapped onto the negative real axis in the ζ-plane. Without loss of generality, let the stream function value on the streamline along the y-axis assume zero value. Explain why the stream function value on the streamline along ABC assumes the value $\frac{Q}{2}$. Show that the complex potential $f(z) = \phi(x, y) + i\psi(x, y), \; z = x + iy$, is given by

$$f(z) = \frac{Qi}{2} - \frac{Q}{2\pi} \text{Log } \zeta.$$

Hint: Note that when ζ moves along the positive part and negative part of the real axis, Im $f = \psi$ equals $\frac{Q}{2}$ and zero, respectively.

(f) Using the following relations

$$\frac{df(z)}{dz} = u - iv = -w \quad \text{and} \quad w = U_\infty(\sqrt{\zeta} - \sqrt{\zeta - 1}),$$

show that

$$z = \frac{Q}{\pi U_\infty} \left[\sqrt{\zeta} + \sqrt{\zeta - 1} + \tan^{-1} \sqrt{\zeta - 1} - 1 \right] + a,$$

where the branch of the inverse tangent is chosen such that $\tan^{-1} \sqrt{\zeta - 1}$ becomes zero as $\zeta \to 1$.

Hint: The point $z = a$ is mapped to the point $\zeta = 1$.

(g) Verify that the parametric representation of the free boundary of the jet is given by

$$x = \frac{Q}{\pi U_\infty}(\sqrt{t} - 1) + a,$$

$$y = \frac{Q}{\pi U_\infty} \left(\sqrt{1 - t} - \frac{1}{2} \ln \frac{1 + \sqrt{1 + t}}{1 - \sqrt{1 - t}} \right), \quad 0 \le t \le 1.$$

Deduce that the width of the jet at far distance from the slot is given by

$$2b = \frac{2a}{1 + \frac{2}{\pi}}.$$

8.21. Given the straight line whose foot of perpendicular from the origin is $\xi \neq 0$, show that its image under the inversion transformation

$$w = f(z) = \frac{1}{z}$$

is a circle whose center is at $w = \frac{1}{2\xi}$ and which passes through the origin.

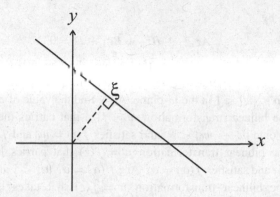

8.22. Show that the composite of two bilinear transformations remains bilinear.

8.23. In each of the following cases, find the corresponding bilinear transformation that maps z_1, z_2 and z_3 to w_1, w_2 and w_3, respectively:

(a) $z_1 = 1$, $z_2 = i$, $z_3 = -1$; $w_1 = 1$, $w_2 = 0$, $w_3 = i$.
(b) $z_1 = \infty$, $z_2 = 0$, $z_3 = 1$; $w_1 = 2$, $w_2 = \infty$, $w_3 = 0$.
(c) $z_1 = -1$, $z_2 = i$, $z_3 = 1 + i$; $w_1 = i$, $w_2 = \infty$, $w_3 = 1$.

8.24. Find the bilinear transformation that has two invariant points 1 and -1, and maps $z = e^{i\pi/3}$ to $w = e^{2i\pi/3}$.

8.25. Find a mapping function that maps the upper half of the unit circle $\operatorname{Im} z > 0$, $|z| < 1$ in the z-plane onto the interior of the unit circle $|w| < 1$ in the w-plane.

8.26. Find a bilinear transformation that maps the circle $|z| < 1$ in the z-plane onto the circle $|w - 1| < 1$ in the w-plane, and takes the points $z = -1$ and $z = 1$ to $w = 2$ and $w = 0$, respectively.

8.27. Find the conditions for the coefficients such that a bilinear mapping of the form $w = \frac{az+b}{cz+d}$ maps the unit circle $|z| = 1$ in the z-plane onto a straight line in the w-plane.

8.28. In each of the following cases, find a transformation that carries

(a) the circle $|z| = 1$ onto the line $\operatorname{Re}(1 + i)w = 0$;
(b) the circle $|z - z_0| = r$ onto the circle $|w| = 1$.

8.29. Given the circle $C : |z - (1 + i)| = 4$, find the inversion point of $z_1 = 2(1 + i)$ with respect to C. Express the equation of the circle in terms of z_1 and its inversion point.

8.30. Show that the necessary and sufficient condition for the two points z_1 and z_2 to be a pair of symmetric points of the circle

$$A z\bar{z} + B\bar{z} + \bar{B}z + D = 0$$

is given by

$$Az_1\bar{z}_2 + B\bar{z}_2 + \overline{B}z_1 + D = 0.$$

8.31. Find a bilinear transformation that carries the region between the two circles $|z - 3| = 9$ and $|z - 8| = 16$ in the z-plane onto the annular region $\rho < |w| < 1$ in the w-plane. Also, find the value of ρ.

8.32. Find the bilinear transformation $w = f(z)$ that carries the upper half z-plane onto $|w - w_0| < R$, and satisfies $f(i) = w_0$ and $f'(i) > 0$.

8.33. Find the bilinear transformation $w = f(z)$ that carries $|z| < 1$ onto $|w| < 1$, and satisfies $f(\alpha) = \alpha$, $\text{Arg} f'(\alpha) = \alpha$, $|\alpha| < 1$ and α is real.

8.34. Find the bilinear transformation $w = f(z)$ that takes $|z| \leq 1$ onto $|w| \leq 1$, where $z = \frac{1}{2}$ is mapped to $w = 0$ and $f'\left(\frac{1}{2}\right) > 0$.

8.35. Find the Schwarz–Christoffel transformation that maps the upper half z-plane conformally onto the interior of the rectangle with vertices $w_1 = -1 + i$, $w_2 = -1$, $w_3 = 1$ and $w_4 = 1 + i$. The preimages of w_1, w_2, w_3 and w_4 along the z-axis are, respectively, $x_1 = -\frac{1}{k}$, $x_2 = -1$, $x_3 = 1$ and $x_4 = \frac{1}{k}$, $0 < k < 1$.

8.36. Find the Schwarz–Christoffel transformation that carries $\text{Im } z > 0$ onto the upper half w-plane minus the semi-infinite strip $\{w : \text{Re } w \geq 0 \text{ and } 0 \leq \text{Im } w \leq H\}$ (see the figure). Here, we choose $x_1 = 0$, $x_2 = 1$, $x_3 = \infty$ and $w_1 = 0$, $w_2 = iH$, $w_3 = \infty$.

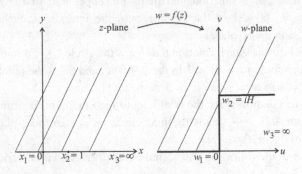

8.37. Consider the domain defined by

$$\{w : \text{Im } w > 0\} \setminus \{w : \text{Re } w = a, \ 0 \leq \text{Im } w \leq h\},$$

which is the upper half w-plane minus a vertical line segment (see the figure). By considering the above domain as a degenerate quadrilateral with vertices $w_1 = a$, $w_2 = a + ih$, $w_3 = a$ and $w_4 = \infty$, find the Schwarz–Christoffel transformation $w = f(z)$ that carries the upper half z-plane onto the degenerate quadrilateral. Here, we choose $x_1 = -1$, $x_2 = 0$, $x_3 = 1$ and $x_4 = \infty$.

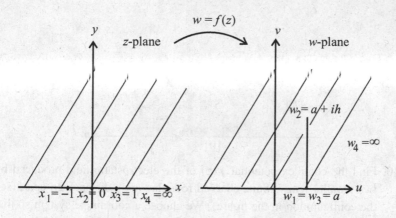

8.38. Find the transformation $w = f(z)$ that maps the infinite strip $-\pi < \operatorname{Im} z < \pi$ in the z-plane onto the polygonal domain in the w-plane bounded within the lines $v = \pm h$, $u < 0$ and $\operatorname{Arg}(w - \pm ih) = \pm \alpha \pi$, $u > 0$, where $0 < \alpha \leq 1$ (see the figure).

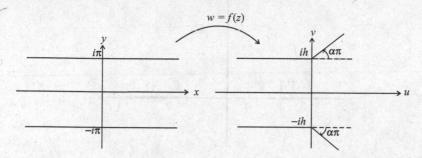

8.39. Find the function $T(w)$, $w = u + iv$, that is harmonic in the semi-infinite strip $\{w : u > 0, 0 < v < H\}$, subject to the following Dirichlet boundary conditions (see the figure):

(i) $T(iv) = K$, $\quad 0 < v < H$;
(ii) $T(u) = T(u + iH) = 0$, $\quad u > 0$.

Hint: Find the Schwarz–Christoffel transformation that carries the upper half z-plane onto the above semi-infinite strip (considered as a degenerate triangle) in the w-plane. For convenience, we choose $x_1 = -1$, $x_2 = 1$, $x_3 = \infty$ and $w_1 = iH$, $w_2 = 0$, $w_3 = \infty$. The solution of the Dirichlet problem in the upper half z-plane is given by

$$\frac{K}{\pi} \operatorname{Arg} \frac{z-1}{z+1}.$$

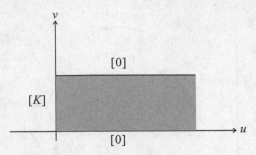

8.40. Find the complex potential $f(z)$ of the electrostatic field produced by two parallel semi-infinite charged rods that are a distance $2a$ apart (see the configuration in the figure). We choose a coordinate system so that the rods lie along the positive and negative x-axis, with end points at $z = a$ and $z = -a$, respectively. The electric potentials of the left rod and right rod are $-V$ and V, respectively.

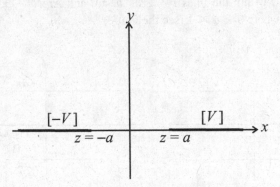

8.41. Find the Schwarz–Christoffel transformation that carries $\operatorname{Im} z > 0$ conformally onto the domain $\{w : -h_1 < \operatorname{Im} w < h_2, \ h_1 > 0 \text{ and } h_2 > 0\} \setminus \{w : \operatorname{Re} w < 0 \text{ and } \operatorname{Im} w = 0\}$ (see the figure).

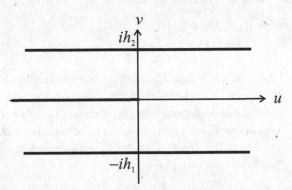

Answers to Problems

Chapter 1

1.1. (a) $-11 - 2i$; (b) i; (c) $2^{(n+2)/2} \cos \frac{n\pi}{4}$;

(d) $-\dfrac{3}{5} - \dfrac{4}{5}i$; (e) $\dfrac{3}{2} - \dfrac{i}{2}$.

1.2. (a) $5\sqrt{2}$; (b) 1; (c) 2.

1.8. $\dfrac{1}{x^2 + y^2} \begin{pmatrix} x & -y \\ y & x \end{pmatrix}$.

1.14. $r = \left[\displaystyle\sum_{j=1}^{n} r_j^2 + 2 \sum_{k=1}^{n} \sum_{\ell=1}^{n} r_k r_\ell \cos(\theta_k - \theta_\ell) \right]^{1/2}$, $k \neq \ell$ in the summation;

$$\theta = \tan^{-1} \left(\sum_{j=1}^{n} r_j \sin\theta_j \bigg/ \sum_{j=1}^{n} r_j \cos\theta_j \right).$$

1.15. When $-\pi < \theta < 0$, the real and imaginary parts of $z^{1/2}$ have opposite signs. In this case

$$z^{1/2} = \pm\sqrt{r} \left(\sqrt{\frac{1 + \cos\theta}{2}} - i\sqrt{\frac{1 - \cos\theta}{2}} \right).$$

The two square roots of $3 - 4i$ are $\pm(2 - i)$.

1.20. Equality holds when (i) $z_1 = z_2 = \cdots = z_n = 0$; (ii) for nonzero z_1, z_k/z_1 are all real positive, valid for $k = 2, \ldots, n$.

1.22. $\text{Im } \dfrac{z_1 - z_3}{z_2 - z_3} = 0$.

1.23. Since $|z_1| = |z_2| = |z_3| = 1$, we have $z_2 = e^{i\alpha} z_1$ and $z_3 = e^{i\beta} z_1$. Together with $z_1 + z_2 + z_3 = 0$, we obtain $e^{i\alpha} + e^{i\beta} + 1 = 0$. One then deduces that $\alpha = -\beta$ and $\cos\alpha = -\frac{1}{2}$.

1.24. As an intermediate step, show that

$$|z - 1| \le \left| z - |z| \right| + \left| |z| - 1 \right|$$

$$= |z| \left| 2 \sin \frac{\theta}{2} \right| + \left| |z| - 1 \right|, \quad \text{where } \theta = \text{Arg } z.$$

1.26. $p = 2 \sin \dfrac{\theta_1 - \theta_2}{2}$, where $\text{Arg } z_1 = \theta_1$ and $\text{Arg } z_2 = \theta_2$.

1.27. $|z|_{max} = |\beta|, \quad |z|_{min} = \sqrt{|\beta|^2 - |\alpha|^2}$.

1.29. $|z|_{max} = \dfrac{\sqrt{a^2 + 4|b|} + a}{2}, \quad |z|_{min} = \dfrac{\sqrt{a^2 + 4|b|} - a}{2}$.

1.30. (a) The region is on or within the two branches of the hyperbola $x^2 - y^2 = 1$.

   (b) The region is a wedge with its vertex at the origin and the angle subtended equal to $\frac{2\pi}{3}$; the axis of symmetry lies on the real axis.

   (c) The region is outside the circle $|z| = \frac{1}{3}$.

   (d) The region lies on the left side of the right branch of the hyperbola $12x^2 - 4y^2 = 3$.

   (e) The region is outside the circle $x^2 + (y - 1)^2 = 2$ and lying in the upper half-plane.

1.31. (a) The family of circles where the ratio of the distances of any point on these circles to z_1 and z_2 is equal to λ.

   (b) The family of circular arcs with end points at z_1 and z_2, including also the straight line segment joining z_1 and z_2.

1.36. $0, \dfrac{n}{\omega - 1}$.

1.39. $b \ne 0, \ d^2 - adb + cb^2 = 0$ or $b = d = 0, \ a^2 - 4c \ge 0$.

1.41. First, establish the relation

$$\frac{1 + \cos(\text{Arg } w_2 - \text{Arg } \Delta w)}{1 + \cos(\text{Arg } w_1 - \text{Arg } \Delta w)}$$

$$= \frac{|\Delta w| + |w_2| - |w_1| \cos(\text{Arg } w_1 - \text{Arg } w_2)}{|\Delta w| + |w_2| \cos(\text{Arg } w_1 - \text{Arg } w_2) - |w_1|},$$

then apply the cosine rule

$$\cos(\text{Arg } w_1 - \text{Arg } w_2) = \frac{|w_1|^2 + |w_2|^2 - |\Delta w|^2}{2|w_1||w_2|}$$

to obtain

$$\frac{1 + \cos(\text{Arg } w_2 - \text{Arg } \Delta w)}{1 + \cos(\text{Arg } w_1 - \text{Arg } \Delta w)} = \frac{|w_1| + |w_2| + |\Delta w|}{|w_1| + |w_2| - |\Delta w|} \frac{|w_1|}{|w_2|}.$$

1.42. (a) no limit point; (b) 0;

(c) $\dfrac{1+i}{\sqrt{2}}$, i, $\dfrac{-1+i}{\sqrt{2}}$, -1, $\dfrac{-1-i}{\sqrt{2}}$, $-i$, $\dfrac{1-i}{\sqrt{2}}$, 1;

(d) same as (c).

1.43. Both are closed sets.

1.44. (a) Interior points: $\{z = x + iy: -3 < y < 0\}$,

exterior points: $\{z = x + iy: y < -3 \text{ or } y > 0\}$,

boundary points: $\{z = x + iy: y = -3 \text{ or } y = 0\}$,

limit points: $\{z = x + iy: -3 \le y \le 0\}$.

(b) Write $z = r(\cos\theta + i\sin\theta)$, $-\pi < \theta \le \pi$;

interior points: $\{z: 0 < \theta < \pi/4 \text{ and } r > 2\}$,

exterior points: $\{z: \pi/4 < \theta < 2\pi\} \cup \{z: r < 2\}$,

boundary points: $\{z: \theta = 0 \text{ and } r \ge 2\} \cup$

$\{z: \theta = \pi/4 \text{ and } r \ge 2\} \cup$

$\{z: 0 < \theta < \pi/4 \text{ and } r = 2\}$,

limit points: $\{z: 0 \le \theta \le \pi/4 \text{ and } r \ge 2\}$.

1.45. (a) closed and unbounded; (b) open and unbounded;

(c) bounded, neither open nor closed.

1.46. $z = 0$ is a limit point.

1.47. (a) $(1/2, 0, 1/2)$; (b) $(0, 1/2, 1/2)$; (c) $(-1/2, 0, 1/2)$;

(d) $(0, -1/2, 1/2)$.

1.48. $z_1 = 1/\overline{z_2}$.

1.51. $x_p(t) = \dfrac{F}{\Delta}\cos(\omega t - \delta)$, where $\delta = \sin^{-1}\dfrac{c\omega}{\Delta}$, $\Delta = \sqrt{(k^2 - \omega^2)^2 + c^2 w^2}$.

1.52. $\omega L = \dfrac{1}{\omega C}$.

Chapter 2

2.1. (a) $u(x, y) = 2x^3 - 6xy^2 - 3x$, $v(x, y) = 6x^2 y - 2y^3 - 3y$;

(b) $u(x, y) = \dfrac{x}{x^2 + y^2}$, $v(x, y) = -\dfrac{y}{x^2 + y^2}$;

(c) $u(x, y) = \dfrac{-x^2 - y^2 + 1}{x^2 + (y - 1)^2}$, $v(x, y) = -\dfrac{2x}{x^2 + (y - 1)^2}$.

2.2. (a) z^2; (b) $\overline{z}^3$; (c) $\dfrac{z}{2z - 1}$.

2.3. $\dfrac{u^2}{\left(r_0 + \frac{1}{r_0}\right)^2} + \dfrac{v^2}{\left(r_0 - \frac{1}{r_0}\right)^2} = 1$;

$x(x^2 + y^2 + 1) = \alpha(x^2 + y^2)$, $y(x^2 + y^2 - 1) = \beta(x^2 + y^2)$.

2.5. $V(z) = \dfrac{ik}{z - \alpha}$, k is the vortex strength.

2.8.   (a) not continuous at $z = 0$;   (b) continuous at $z = 0$.

2.12. Along the line segment joining z_1 and z_2 with period π.

2.14. $\dfrac{x^2}{a^2} + \dfrac{y^2}{b^2} = 1.$

2.15. (a) the whole complex plane;

(b) the whole complex plane except at $z = 0$;

(c) $z = 0$;   (d) the whole complex plane.

2.16. (a) $z = 0$;   (b) $z = 0$;   (c) empty set.

2.17. 0.

2.19. (a) nowhere analytic;   (b) entire, $f'(z) = 2z - 3$;

(c) analytic everywhere except at $z = 0$, $f'(z) = -\dfrac{1+i}{z^2}$.

2.20. (b) $2(1 + i)$; $f'(z)$ exists only along the line $x - y = 1$.

2.22. No.

2.23. $f'(z) = \begin{cases} 2z & \text{for } |z| < 1 \\ 1 & \text{for } |z| > 1 \end{cases}$.

f is not differentiable on the circle $|z| = 1$. Define $D = \{z : |z| \neq 1\}$, which is an open set since its complement $D^c = \{z : |z| = 1\}$ is closed. Since every point inside D is an interior point and f is differentiable throughout D, f is analytic in D.

2.25. (a) no;   (b) yes, the harmonic conjugate is also constant.

2.30. (a) $iz^3 + i$;   (b) $i\left(\dfrac{1}{z} - 1\right)$;   (c) $(1 - i)z^3 + ic$, c is any real number.

2.31. (a) $\dfrac{3}{2}x^2y^2 - \dfrac{x^4 + y^4}{4} = \beta$;   (b) $2e^{-x}\cos y + 2xy = \beta$;

(c) $(r^2 - 1)\sin\theta = \beta r$.

2.32. (a) $a = 2$, $b = -1$, $c = -1$, $d = 2$;   (b) $k = 1$;   (c) $\ell = 1$.

2.33. $v(x, y)$ is not a harmonic conjugate of $u(x, y)$.

2.34. u^2 is not harmonic.

2.35. No.

2.36. Yes.

2.38. $v = \dfrac{1}{2}\ln\dfrac{(x + a)^2 + y^2}{(x - a)^2 + y^2} + c$, c is a real constant,

$\theta = \tan^{-1}\dfrac{2ay}{x^2 + y^2 - a^2}$.

2.39. $\dfrac{\partial^2 u}{\partial r^2} + \dfrac{1}{r}\dfrac{\partial u}{\partial r} + \dfrac{1}{r^2}\dfrac{\partial^2 u}{\partial \theta^2} = 0.$

2.40. (a) $u(x, y) = (y - 2)xe^y$;   (b) $u(r, \theta) = \dfrac{r^4\cos\theta}{15}$.

2.44. $T(r, \theta) = k\ln r, r \neq 0$, k is real. The flux lines are $y = \beta x$, β is constant.

Chapter 3

3.4. (a) $\ln 2 + \left(\dfrac{1}{2} + 2k\right)\pi i$, k is any integer,

(b) $\left(\dfrac{1}{2} + 2k\right)\pi + 3i$, k is any integer.

3.6. (a) $\text{Im}(\sin z) = 0$ if $\text{Re } z = (2k+1)\pi/2$ or $\text{Im } z = 0$, k is any integer,

$\text{Re}(\sin z) = 0$ if $\text{Re } z = k\pi$, k is any integer;

(b) $\text{Im}(\tan z) = 0$ if $\text{Im } z = 0$,

$\text{Re}(\tan z) = 0$ if $\text{Re } z = \frac{k\pi}{2}$, k is any integer;

(c) $\text{Im}(\coth z) = 0$ if $\text{Im } z = \frac{k\pi}{2}$, k is any integer,

$\text{Re}(\coth z) = 0$ if $\text{Re } z = 0$.

3.7. (a) $|\tan z| = \dfrac{\sqrt{\sin^2 2x + \sinh^2 2y}}{\cos 2x + \cosh 2y}$,

(b) $|\tanh z| = \dfrac{\sqrt{\sinh^2 2x + \sin^2 2y}}{\cosh 2x + \cos 2y}$.

3.10. $2i$.

3.15. (a) $\dfrac{\pi}{4} + 2k\pi - i\ln(\sqrt{2} \pm 1)$; (b) $\dfrac{3\pi}{4} + 2k\pi - i\ln\dfrac{3 \pm \sqrt{7}}{2}$;

(c) $\dfrac{\pi}{4} + 2k\pi - i\ln\dfrac{\sqrt{3}-1}{\sqrt{2}}$ and $-\dfrac{3\pi}{4} + 2k\pi - i\ln\dfrac{\sqrt{3}+1}{\sqrt{2}}$;

(d) $2k\pi i$; (e) $-\ln 2 + (2k+1)\pi i$;

(f) $\left(2k + \dfrac{1}{2}\right)\pi i$ and $-\ln 3 + \left(2k - \dfrac{1}{2}\right)\pi i$;

(g) $k\pi(1 \pm i)$; (h) $k\pi(1+i)$ and $\dfrac{(2k+1)\pi}{1+i}$;

(i) $\dfrac{(4k+1)\pi}{2(1+2i)}$ and $\dfrac{(4k-1)\pi}{2(1-2i)}$.

In all cases, k is any integer.

3.16. (a) limit does not exist;

(b) (i) unbounded as $y \to \pm\infty$; (ii) oscillatory;

(c) $|\sin z| \le \sqrt{1 + \sinh^2 \alpha}$.

3.17. (a) $\dfrac{1}{2}\ln 13 - \tan^{-1}\dfrac{3}{2}i$;

(b) $\dfrac{1}{2}\ln 13 + \left[(2k+1)\pi - \tan^{-1}\dfrac{3}{2}\right]i$, k is any integer;

(c) $\cos 2\cosh 1 - i\sin 2\sinh 1$; (d) $\dfrac{\sinh 4 - i\sin 2}{\cosh 4 - \cos 2}$; (e) $\dfrac{40 + 9i}{41}$;

(f) $\dfrac{\pi}{2} + 2k\pi - i\ln(\sqrt{2}+1)$ and $-\dfrac{\pi}{2} + 2k\pi - i\ln(\sqrt{2}-1)$, k is any integer;

(g) $2k\pi + i\ln(\sqrt{2} + 1)$ and $(2k + 1)\pi - i\ln(\sqrt{2} - 1)$, k is any integer;

(h) not defined.

3.18. (a) analytic everywhere except at 0, $\dfrac{\pi}{2} + k\pi$, k is any integer;

(b) analytic everywhere except at $-\dfrac{\pi}{4} + k\pi$, k is any integer;

(c) analytic everywhere except at $-\dfrac{\ln 3}{2} + i\dfrac{2k + 1}{2}\pi$, k is any positive integer.

3.21. The image is the whole complex plane except the origin. The inverse function is $\dfrac{1}{2\pi}\mathrm{Log}\, z$.

3.23. $\mathrm{Log}\, z$ is multi-valued.

3.24. The heat source and heat sink are at z_1 and z_2, respectively. The temperature value is zero along the perpendicular bisector of the line segment joining z_1 and z_2.

3.25. $T(z) = \dfrac{U_1}{\pi}\mathrm{Arg}(z - x_1) + \dfrac{U_2}{\pi}\mathrm{Arg}\,\dfrac{z - x_2}{z - x_1} + \cdots$

$\qquad + \dfrac{U_n}{\pi}\mathrm{Arg}\,\dfrac{z - x_n}{z - x_{n-1}} + \dfrac{U_{n+1}}{\pi}[\pi - \mathrm{Arg}(z - x_n)].$

3.26. $T(z) = \dfrac{U}{\pi}\tan^{-1}\dfrac{2\,\mathrm{Im}(\sin z)}{|\sin^2 z| - 1}$

or $T(x, y) = \dfrac{2U}{\pi}\tan^{-1}\dfrac{\cos x}{\sinh y}$, $z = x + iy$.

3.27. $-\sqrt[3]{2}$.

3.28. Both $z = 0$ and $z = 1$ are branch points of order 3; $z = \infty$ is not a branch point. Choose the line segment joining the two branch points as the branch cut. The Riemann surface consists of four sheets.

3.30. Both $z = i$ and $z = -i$ are branch points, while $z = \infty$ is not a branch point. The function is not defined at $z = \pm i$, while $\tan^{-1}\infty = \frac{\pi}{2} + k\pi$, k is any integer. The branch cut of the Riemann surface is the line segment joining $z = -i$ and $z = i$.

3.31. $\dfrac{1}{1 - z^2}$.

3.32. (b) $\ln 3 + i\pi$; (c) No, the starting point of the semi-infinite branch cut can be chosen to be any point that lies on the branch cut joining $z = -1$ and $z = 1$.

3.33. No. $[z(z + 1)]^{1/2}$ is a double-valued function while $z^{1/2}(z + 1)^{1/2}$ is the product of two double-valued functions.

3.37. One branch of the hyperbola $\dfrac{x^2}{\cos^2\theta_0} - \dfrac{y^2}{\sin^2\theta_0} = 1$.

Chapter 4

4.2. (a) $e + \dfrac{1}{e}$; (b) $\dfrac{-5 + 7i}{3}$.

4.3. (a) 1; (b) $1 + i$

4.4. The length of the line segment joining any two points is always less than or equal to the arc length of any curve joining the same two points.

4.6. $6\pi(\pi + \ln 3)$.

4.7. $2\pi e$.

4.8. (a) $7\pi/2$; (b) $\pi \cosh 1$; (c) $2/e$.

4.11. (a) 0; (b) $6\pi i$; (c) $\dfrac{\pi i}{\sqrt{2}}$; (d) $-\pi e^{-i}$; (e) $-\dfrac{\pi i}{\sqrt{2}}$; (f) 0;

(g) 0; (h) 0; (i) πi; (j) 8.

4.12. $\dfrac{2\sqrt{2}}{3} i$.

4.18. $2\pi i e^{i\pi/3}$, $2\pi i e^{-i\pi/3}$, 0.

4.20. (a) $4\pi i$; (b) 0. $g(2)$ does not exist.

4.26. No; for example, for $n \neq 1$, $\displaystyle\oint_{|z|=r<1} \dfrac{1}{z^n}\, dz = 0$ but $\dfrac{1}{z^n}$ is not analytic at $z = 0$.

4.30. $\sqrt{13}$ at $z = \pm i$.

4.31. $\left(\dfrac{\pi}{2}, 1\right)$

4.36. $\phi(x, y) = \dfrac{K_1}{2\pi} \ln\left|\dfrac{z - \alpha}{z - \beta}\right| + \dfrac{K_2}{2\pi} \operatorname{Arg} \dfrac{z - \alpha}{z - \beta}$,

$\psi(x, y) = -\dfrac{K_2}{2\pi} \ln\left|\dfrac{z - \alpha}{z - \beta}\right| + \dfrac{K_1}{2\pi} \operatorname{Arg} \dfrac{z - \alpha}{z - \beta}$;

the streamlines are families of logarithmic spirals.

4.40. $f(z) = U_\infty \sqrt{z^2 + 1}$; the streamlines are $y^2 = c^2 + \dfrac{c^2}{x^2 + c^2}$, c is any constant.

4.41. $u(x, y) = \dfrac{K}{2\pi} \displaystyle\sum_{j=1}^{n} \left[S_j \ln \dfrac{r_{j+1}}{r_j} - C_j(\theta_{j+1} - \theta_j) \right]$,

$v(x, y) = -\dfrac{K}{2\pi} \displaystyle\sum_{j=1}^{n} \left[C_j \ln \dfrac{r_{j+1}}{r_j} - S_j(\theta_{j+1} - \theta_j) \right]$,

where $r_j^2 = (\xi_j - x)^2 + (\eta_j - y)^2$, $\theta_j = \tan^{-1} \dfrac{\eta_j - y}{\xi_j - x}$, $j = 1, 2, \ldots, n$,

$C_j = \dfrac{[(\eta_{j+1} - \eta_j)(\xi_j - x) - (\xi_{j+1} - \xi_j)(\eta_j - y)](\xi_{j+1} - \xi_j)}{(\xi_{j+1} - \xi_j)^2 + (\eta_{j+1} - \eta_j)^2}$,

$S_j = \dfrac{[(\eta_{j+1} - \eta_j)(\xi_j - x) - (\xi_{j+1} - \xi_j)(\eta_j - y)](\eta_{j+1} - \eta_j)}{(\xi_{j+1} - \xi_j)^2 + (\eta_{j+1} - \eta_j)^2}$.

4.43. The greatest speed occurs at $(a, \frac{\pi}{2})$ and $(a, -\frac{\pi}{2})$. The equipotential lines are $c(x^2 + y^2) = x(x^2 + y^2 + a^2)$, c is any constant.

4.45. $\Omega(z) = -\dfrac{\rho}{2\pi\epsilon}[\text{Log}(z - z_1) - \text{Log}(z - \overline{z_1})]$; the equipotential lines form the family of coaxial circles defined by $\left| \dfrac{z - z_1}{z - \overline{z_1}} \right| = k$, k is any constant.

4.46. $\phi(r) = \dfrac{\phi_b - \phi_a}{\ln \frac{b}{a}} \ln r + \dfrac{\phi_a \ln b - \phi_b \ln a}{\ln \frac{b}{a}}$.

4.48. $g_h = G\rho \, \text{Re} \displaystyle\oint_{|\zeta|=1} \dfrac{r(\overline{\omega}_0 \zeta + r)}{\zeta(\omega_0 + r\zeta)} \, d\zeta$, $\quad \omega_0 = \xi_0 + i\zeta_0$,

$\qquad = \dfrac{2G\rho\pi r^2 \zeta_0}{\xi_0^2 + \zeta_0^2}$, $\quad |\omega_0| > r$.

Chapter 5

5.4. (a) absolute convergence; (b) divergence;
(c) absolute convergence.

5.5. (a) when $|z| < 1$, the series converges to -1;
(b) when $z = 1$, the series converges to 0;
(c) when $z = -1$, the series diverges;
(d) when $z = e^{i\theta}$, $\theta \neq 0$, the series diverges;
(e) when $|z| > 1$, the series diverges.

5.8. (a) R; (b) $R/2$; (c) when $R = 0$, the new radius of convergence becomes infinite; when $R > 0$, the new radius of convergence can be any value; (d) R^k.

5.9. Divergence on $|z| = 1$.

5.11. (a) ∞, (b) 1, (c) 1/4, (d) e.

5.14. (a) $z - z^2 + \dfrac{z^3}{3}$, (b) $\ln 2 + \dfrac{z}{2} + \dfrac{z^2}{8} - \dfrac{z^4}{192}$.

5.15. (a) $1 - 2z + z^2 + z^4 - 2z^5 + z^6 + z^8 - 2z^9 + z^{10} + \cdots$,

(b) $z - \dfrac{7}{6}z^3 + \dfrac{47}{40}z^5 - \cdots + \cdots$, (c) $z^3 + \dfrac{z^9}{3!} + \dfrac{z^{15}}{15!} + \cdots$,

(d) $\dfrac{z^2}{4} - \dfrac{z^4}{96} + \dfrac{z^6}{4320} - \cdots$.

5.16. $E_0 = 1$, $E_2 = -1$, $E_4 = 5$, $E_6 = -61$; circle of convergence is $|z| < \dfrac{\pi}{2}$.

5.17. $|z| = 1$.

5.18. $c_n = \dfrac{1}{\sqrt{5}} \left[\left(\dfrac{1 + \sqrt{5}}{2} \right)^{n+1} - \left(\dfrac{1 - \sqrt{5}}{2} \right)^{n+1} \right]$, $n \geq 0$, $R = \dfrac{\sqrt{5} - 1}{2}$.

5.19. $\sin^{-1} z = z + \frac{z^3}{6} + \frac{3z^5}{40} + \cdots$, valid for $|z| < 1$.

5.20. $c_0 = \dfrac{a_0}{b_0}$, $c_1 = \dfrac{b_0 a_1 - b_1 a_0}{b_0^2}$, $c_2 = \dfrac{b_0^2 a_2 - b_0 b_1 a_1 + (b_1^2 - b_0 b_2)a_0}{b_0^3}$;

$$b_0 c_n + b_1 c_{n-1} + b_2 c_{n-2} + b_3 c_{n-3} = 0 \quad n = 3, 4.$$

5.22. The complex extension of the function, $f(z) = \dfrac{1}{1+z^2}$, has singularities at $z = \pm i$.

5.24. $\dfrac{1}{1-z} = 1 + z + z^2 + \cdots$, valid for $|z| < 1$; the series diverges at $z = -1$ but the sum function $\dfrac{1}{1-z}$ is analytic at $z = -1$.

5.28. (a) $0 < |z| < 1$, $(1 + z + z^2 + \cdots) - \dfrac{1}{2}\left(1 + \dfrac{z}{2} + \dfrac{z^2}{4} + \cdots\right)$;

$$1 < |z| < 2, \quad -\dfrac{1}{z}\left(1 + \dfrac{1}{z} + \dfrac{1}{z^2} + \cdots\right) - \dfrac{1}{2}\left(1 + \dfrac{z}{2} + \dfrac{z^2}{4} + \cdots\right);$$

$$2 < |z| < \infty, \quad \dfrac{1}{z}\left(1 + \dfrac{2}{z} + \dfrac{4}{z^2} + \cdots\right) - \dfrac{1}{z}\left(1 + \dfrac{1}{z} + \dfrac{1}{z^2} + \cdots\right);$$

(b) $0 < |z - i| < 1$, $\displaystyle\sum_{n=1}^{\infty}(-1)^{n-1}\dfrac{n(z-i)^{n-2}}{i^{n+1}}$;

$$1 < |z - i| < \infty, \quad \sum_{n=0}^{\infty}(-1)^n\dfrac{(n+1)i^n}{(z-i)^{n+3}}.$$

5.29. (a) (i) $a_n = \dfrac{1}{2^{n+1}(i-2)}$, $n \geq 0$; $b_n = \dfrac{i^{n-1}}{i-2}$, $n \geq 1$;

(ii) $a_n = 0$, $n \geq 0$; $b_n = \dfrac{i^{n-1} - 2^{n-1}}{i-2}$, $n \geq 1$;

(iii) $a_n = \dfrac{-1}{(2-i)^{n+2}}$, $n \geq 0$; $b_1 = \dfrac{1}{i-2}$, $b_n = 0$ for $n \geq 2$;

value of the integral $= \dfrac{2\pi i}{i-2}$.

(b) $a_n = \left(\dfrac{-1}{2}\right)^{n+2}$, $n \geq 0$; $b_1 = -\dfrac{1}{2}$, $b_n = (-1)^n$ for $n \geq 2$;

(c) $a_n = b_n = \displaystyle\sum_{k=0}^{\infty}\dfrac{1}{k!(n+k)!}$, $n \geq 1$; $a_0 = \displaystyle\sum_{k=0}^{\infty}\dfrac{1}{(k!)^2}$.

5.30. $\sin 1\left[1 - \dfrac{1}{2!}\dfrac{1}{(z-1)^2} + \cdots + (-1)^n\dfrac{1}{(2n)!}\dfrac{1}{(z-1)^{2n}} + \cdots\right] + \cos 1$

$$\times\left[\dfrac{1}{z-1} - \dfrac{1}{3!}\dfrac{1}{(z-1)^3} + \cdots + (-1)^n\dfrac{1}{(2n+1)!}\dfrac{1}{(z-1)^{2n+1}} + \cdots\right].$$

5.32. $2\pi i$.

5.34. The two power series expansions are valid within their respective circles of convergence which are non-overlapping. Therefore, it is meaningless to add these two series together.

Chapter 6

6.2. (a) removable singularity; (b) removable singularity;
      (c) essential singularity.

6.3. (a) When $n = m$, $z = 0$ can be a removable singularity, for example,
          $f_1(z) = -f_2(z)$; it can be a pole of order k, $k \le m$, for example,
          $f_1(z) = -f_2(z) + \frac{1}{z^k}$.
      (b) It is a pole of order $m + n$.
      (c) When $n > m$, $z = 0$ is a pole of order $n - m$; when $n \le m$, it is a
          zero of order $m - n$.

6.4. It becomes a removable singularity if $f_1 = -f_2$, and a pole if $f_1 = -f_2 + (z - z_0)^{-k}$, k is a positive integer.

6.5. (a) The function is entire;
      (b) $z = 2k\pi i$ is a simple pole, k is any nonzero integer;
      (c) $z = 0$ is an essential singularity;
      (d) $z = \frac{1}{k\pi}$ is an essential singularity, k is any nonzero integer.

6.6. The classification of an isolated singularity and the computation of its
      residue should be done with reference to the Laurent series expansion of
      the function in a deleted neighborhood of the singularity.

6.7. (a) $\mathrm{Res}\left(\dfrac{z^2 - 1}{z^3(z^2 + 1)}, i\right) = -1$, $\mathrm{Res}\left(\dfrac{z^2 - 1}{z^3(z^2 + 1)}, -i\right) = -1$,

      $\mathrm{Res}\left(\dfrac{z^2 - 1}{z^3(z^2 + 1)}, 0\right) = 2$;

      (b) $\mathrm{Res}\left(\dfrac{\tan z}{1 - e^z}, n\pi + \dfrac{\pi}{2}\right) = \dfrac{1}{e^{n\pi + \frac{\pi}{2}} - 1}$, n is any integer;

      (c) $\mathrm{Res}\left(\dfrac{e^{i\alpha z}}{z^4 + \beta^4}, \dfrac{\beta(1 \pm i)}{\sqrt{2}}\right) = \mp \dfrac{\sqrt{2}}{8\beta^3}(1 \pm i)e^{\alpha\beta(i \mp 1)/\sqrt{2}}$;

      (d) $\mathrm{Res}\left(\dfrac{1}{z \sin z}, n\pi\right) = (-1)^n \dfrac{1}{n\pi}$, n is any non-zero integer,

      $\mathrm{Res}\left(\dfrac{1}{z \sin z}, 0\right) = 0$;

      (e) $\mathrm{Res}\left(\dfrac{e^{\frac{1}{z}}}{z}, 0\right) = 1$.

6.8. (a) n, (b) $-n$; (a) $ng(\alpha)$, (b) $-ng(\alpha)$.

6.9. (a) essential singularity; (b) pole of order 6.

6.10. Write $f(z) = \displaystyle\sum_{n=0}^{\infty} f_n(z - z_0)^n$ and $g(z) = \displaystyle\sum_{n=2}^{\infty} g_n(z - z_0)^n$,

      $$\mathrm{Res}\left(\dfrac{f(z)}{g(z)}, z_0\right) = \dfrac{f_1 g_2 - f_0 g_3}{g_2^2}.$$

6.11. $z = 0, \pm 1, \pm 2, \cdots$, represent poles of order one of $f(z) = \pi \cot \pi z$; $\mathrm{Res}\,(\pi \cot \pi z, n) = 1$.

6.12. The function has a pole of order 2 at $z = 0$ and a pole of order 3 at

$$\mathbf{[}\; \text{res}\,\mathbf{R}_{\mathrm{es}}(f, 0) = \frac{\partial}{\pi^4},\; \mathbf{R}_{\mathrm{es}}(f, \pi) = \frac{11^2}{2\pi^4}\,\mathbf{]}.$$

6.13. 8.

6.14. (a) -1; (b) $\dfrac{(2n)!}{(n-1)!(n+1)!}$.

6.16. $2\pi i \displaystyle\sum_{k=1}^{n} \dfrac{f(a_k)}{(a_k - a_1)\cdots(a_k - a_{k-1})(a_k - a_{k+1})\cdots(a_k - a_n)}$.

6.17. (a) $\dfrac{1}{z-1} - \dfrac{3}{z-2} + \dfrac{2}{z-3}$;

(b) $-\dfrac{1}{5}\left(\dfrac{z_2}{z - z_1} + \dfrac{z_5}{z - z_2} + \dfrac{z_3}{z - z_3} + \dfrac{z_1}{z - z_4} + \dfrac{z_4}{z - z_5} \right)$,

where $z_k = e^{k\pi i}$, $k = 1, 3, 5, 7, 9$;

(c) $\dfrac{1}{n}\left(\dfrac{z_1}{z - z_1} + \cdots + \dfrac{z_n}{z - z_n} \right)$, $z_k = e^{2k\pi i}$, $k = 1, 2, \ldots, n$.

6.19. (a) $10\pi i$; (b) $-\frac{6}{5}\pi i$; (c) 0; (d) $\frac{e}{3}\pi i$; (e) $2\pi i$;

(f) $\dfrac{2\pi i}{(n-1)!} \sin\!\left(i + (n-1)\dfrac{\pi}{2} \right)$, n is any positive integer.

6.20. (a) $\dfrac{2\pi a}{(a^2 - b^2)^{3/2}}$; (b) $\dfrac{(2a + b)\pi}{[a(a + b)]^{3/2}}$;

(c) $\dfrac{2\pi}{1 - a^2}$ if $|a| < 1$, $\dfrac{2\pi}{a^2 - 1}$ if $|a| > 1$, 0 (principal value) if $|a| = 1$ but $a \neq \pm 1$. When $a = \pm 1$, the principal value of the integral does not exist.

6.21. (a) $-\dfrac{\pi}{27}$; (b) $\dfrac{\pi}{4a}$; (c) $\dfrac{1 \cdot 3 \cdot 5 \cdots (2n - 3)}{2 \cdot 4 \cdot 6 \cdots (2n - 2)}\dfrac{\pi}{2}$ if $n > 1$ and $\dfrac{\pi}{2}$ if $n = 1$;

(d) $\dfrac{\pi}{ab(a + b)}$; (e) $\dfrac{\pi}{\sqrt{2}}$; (f) $\dfrac{\pi}{n \sin \frac{\pi}{n}}$; (g) $\dfrac{\pi}{\sqrt{2}} \sin \dfrac{\pi}{8}$.

6.23. (a) $\pi e^{-|\omega|}$, ω is real; (b) $\sqrt{\pi}e^{-\omega^2/4}$; (c) $\dfrac{2 \sin \beta\omega}{\omega}$.

6.27. (a) $\dfrac{\pi}{3e^3}(\cos 1 - \cos 3)$; (b) $\dfrac{\pi}{2e^4}(2\cos 2 + \sin 2)$; (c) $\dfrac{\pi e^{-ab}}{2b}$;

(d) πe^{-ab}; (e) $(1 - e^{-a})\dfrac{\pi}{2}$.

6.29. $-\dfrac{\pi^2}{8\sqrt{2}}$.

6.30. $\frac{\pi}{4} \ln 2$.

6.33. $I = \dfrac{\pi}{2}$ when $p = \pm 1$.

6.39. (a) $-i\pi$;   (b) $\pi i(e^{2mi} - e^{mi})$.

6.42. $\dfrac{3\pi}{4}$.

Chapter 7

7.13. $u(x, y) = \dfrac{K}{2\pi}\left[\displaystyle\int_{-1}^{1} \ln((x-t)^2 + y^2)\, dt - 2\ln(x^2 + y^2)\right]$.

7.14. $u(r, \theta) = \dfrac{1}{2\pi}\displaystyle\int_0^{\pi} [P(R, r, \phi - \theta) + P(R, r, \phi + \theta)] f(\phi)\, d\phi$.

7.16. (a) $2s\dfrac{dY(s)}{ds} + (s^2 + \lambda + 2)Y(s) = sy_0 + y_0'$;

$\quad$ (b) $s(s-1)\dfrac{dY(s)}{ds} + (s - \lambda - 1)Y(s) = 0$.

7.18. (a) $\tan^{-1}\dfrac{1}{s}$;   (b) $\text{Log}\dfrac{1}{\sqrt{s^2 + 1}}$.

7.19. (a) $\dfrac{\tan^{-1} s}{s} - \dfrac{\pi}{2s}$;   (b) $\dfrac{\text{Log}\sqrt{s^2 + 1}}{s}$.

7.22. (a) $\dfrac{1 - e^t}{t}$;   (b) $\dfrac{2}{\pi}\displaystyle\int_0^a \dfrac{\cos ty}{\sqrt{a^2 - y^2}}\, dy$.

7.23. $y(t) = \dfrac{1}{\pi}\displaystyle\int_0^t \dfrac{\phi'(\tau)}{(t - \tau)^{1/2}}\, d\tau$.

7.26. $T(x, t) = 2 + e^{-\pi^2 t}\sin \pi x$.

7.27. $T(x, t) = \displaystyle\int_0^{\infty} \dfrac{e^{-(x-\xi)^2/4a^2 t} + e^{-(x+\xi)^2/4a^2 t}}{\sqrt{4\pi a^2 t}} f(\xi)\, d\xi$.

7.28. $T(x, t) = e^{-x\sqrt{\frac{\omega}{2a^2}}} \sin\left(\omega t - x\sqrt{\dfrac{\omega}{2a^2}}\right)$.

7.29. (a) $T(x, t) = \dfrac{q_0}{k}\displaystyle\int_x^{\infty} \text{erfc}\left(\dfrac{\xi}{2\sqrt{t}}\right) d\xi$

$\quad\quad = \dfrac{q_0}{k}\left[2\sqrt{\dfrac{t}{\pi}} e^{-x^2/4t} - x\,\text{erfc}\left(\dfrac{x}{2\sqrt{t}}\right)\right]$;

$\quad$ (b) $T(0, t) = \dfrac{q_0}{k\sqrt{\pi}}\displaystyle\int_0^t \dfrac{\sin \omega(t - \tau)}{\sqrt{\tau}}\, d\tau$,

$\quad\quad T(x, t) = \dfrac{x}{\sqrt{4\pi}}\displaystyle\int_0^t T(0, \tau)\dfrac{1}{(t - \tau)^{3/2}} e^{-x^2/4(t - \tau)}\, d\tau$.

7.30. $T(x, t) = T_1\left(1 - \dfrac{x}{L}\right) + \dfrac{x}{L}T_2$

$$+ \frac{2}{\pi} \sum_{n=1}^{\infty} \frac{1}{n}(T_2 \cos n\pi - T_1)e^{-n^2\pi^2a^2t/L^2} \sin \frac{n\pi x}{L}$$

7.32. $u(x, t) = t \sin x.$

7.33. $u(x, t) = \dfrac{1}{\pi^2}(1 - \cos \pi t) \sin \pi x.$

7.35. $u(x, t) = \begin{cases} t & \text{if } t < \frac{x^2}{2} \\ \frac{x^2}{2} & \text{if } t \geq \frac{x^2}{2} \end{cases}.$

Chapter 8

8.1. (a) $z = \pm 1, \pm i$;   (b) $z = n\pi, \left(n + \frac{1}{2}\right)\pi$, n is any integer.

8.3. (a) a circle that passes through $w = 0$ and $w = \frac{1}{2}$;

    (b) $(u^2 + v^2)^2 = u^2 - v^2$;   (c) $(1 - 2pu)v^2 = 2pu^3.$

8.5. One possible choice of the mapping function is

$$z = x + iy = a\left(\frac{w}{1 + im} + i - ie^{-i\frac{w}{1+im}}\right).$$

8.6. $\dfrac{\text{Re}\left(1 + \frac{zf''(z)}{f'(z)}\right)}{|zf'(z)|}.$

8.8. $\dfrac{15}{2}\pi.$

8.9. $w = \left(e^{-i\alpha}\dfrac{z + \sqrt{z^2 - 4}}{2}\right)^{\pi/(\pi - 2\alpha)}.$

8.11. An infinite wedge: $0 < \text{Arg } w < \dfrac{\pi}{4}.$

8.12. $T(z) = K - \dfrac{K}{2\pi}\text{Arg } z^4, 0 < \text{Arg } z < \dfrac{\pi}{4}.$

8.13. $T(z) = \dfrac{2}{\pi}\text{Re}(\sin^{-1} z^2).$

8.14. $T(x, y) = \dfrac{T_1 + T_2}{2} + \dfrac{T_2 - T_1}{\pi}\text{Re}\left(\sin^{-1}\left(\dfrac{2z^2}{x_0^2} - 1\right)\right)$, $z = x + iy.$

8.15. $T(z) = -\dfrac{2T_0}{\pi}\text{Re}(\sin^{-1}(-e^{-\pi z})).$

8.16. $f(z) = U_\infty\left[\zeta + \dfrac{(a + b)^2}{4\zeta}\right],$

    where $\zeta = \dfrac{z + \sqrt{z^2 - c^2}}{2}$ and $c^2 = a^2 - b^2.$

8.17. $f(z) = U_\infty(z - p + \sqrt{p^2 - 2pz}).$

8.18. $f(z) = m \ln\left(\dfrac{e^{\pi(z+1)} - 1}{e^{\pi(z-1)} - 1}\right).$

8.23. (a) $w = \dfrac{(1+i)(z-i)}{2z}$; (b) $w = \dfrac{2(z-1)}{z}$; (c) $w = \dfrac{iz+3}{(2+i)(z-i)}.$

8.24. $w = \dfrac{1 - 2z}{e^{i\pi/3}z - 1}.$

8.25. $w = \dfrac{z^2 + 2iz + 1}{z^2 - 2iz + 1}.$

8.26. $w = 3\dfrac{z-1}{z-2}.$

8.27. $|d| = |c|.$

8.28. (a) $w = (1-i)\dfrac{z-1}{z+1}$; (b) $w = \dfrac{z - z_0}{r}.$

8.29. Inversion point is $5(1+i)$; $\left|\dfrac{z - 2(1+i)}{z - 5(1+i)}\right| = 2\sqrt{2}.$

8.31. $w = e^{i\theta}\dfrac{2z}{z+24}$, θ is real; $\rho = \dfrac{2}{3}.$

8.32. $w = w_0 + Ri\dfrac{z-i}{z+i}.$

8.33. $\dfrac{w - \alpha}{1 - \overline{\alpha}w} = e^{i\alpha}\dfrac{z - \alpha}{1 - \overline{\alpha}z}.$

8.34. $w = \dfrac{2z - 1}{2 - z}.$

8.35. $w = \dfrac{\displaystyle\int_0^z \dfrac{1}{\sqrt{(1 - z^2)(1 - k^2 z^2)}}\, dz}{\displaystyle\int_1^{\frac{1}{k}} \dfrac{1}{\sqrt{(z^2 - 1)(1 - k^2 z^2)}}\, dz}.$

8.36. $w = i\dfrac{2H}{\pi}[\sin^{-1}\sqrt{z} + \sqrt{z(1 - z)}].$

8.37. $w = a + h\sqrt{z^2 - 1}.$

8.38. $w = \dfrac{h}{\pi}\left[\pi i + \int_{\pi i}^z (e^{\zeta} + 1)^{\alpha}\, d\zeta\right];$

in particular, when $\alpha = 1$, $w = \dfrac{h}{\pi}(e^z + z + 1).$

8.39. The required Schwarz–Christoffel transformation is

$$w = iH\left(\frac{1}{2} - \frac{1}{\pi}\sin^{-1}z\right),$$

or, equivalently,

$$z = \cosh\frac{\pi}{H}w.$$

$$T(w) = \frac{K}{\pi}\text{Arg}\,\frac{\cosh\frac{\pi}{H}w - 1}{\cosh\frac{\pi}{H}w + 1} = \frac{K}{\pi}\text{Arg}\left(\tanh^2\frac{\pi}{2H}w\right).$$

8.40. $f(z) = \dfrac{2V}{\pi}\text{Log}(z + \sqrt{z^2 - a^2}).$

8.41. $w = \dfrac{h_1}{\pi}\left[\text{Log}(1 - z) + \dfrac{h_2}{\pi}\text{Log}\left(1 + \dfrac{h_1}{h_2}z\right)\right].$

Index

Printed in the United States
By Bookmasters